D0856717

Protein Biotechnology

Biological Methods

Protein Biotechnology

Isolation, Characterization, and Stabilization

Edited by

Felix Franks • *Pafra Ltd, Cambridge, England*

Humana Press, Totowa, New Jersey

TP
248.65
P76
P72
1993

© 1993 The Humana Press Inc.
999 Riverview Drive, Suite 208
Totowa, New Jersey 07512

All rights reserved.

No part of this book may be reproduced, stored in a retrieval system, or transmitted in any form or by any means, electronic, mechanical, photocopying, microfilming, recording, or otherwise without written permission from the Publisher.

Printed in the United States of America. 10 9 8 7 6 5 4 3 2 1

Library of Congress Cataloging-in-Publication Data

Main entry under title:

Protein biotechnology : isolation, characterization, and stabilization
/ edited by Felix Franks.
p. cm. — (Biological methods)
Includes index.
ISBN 0-89603-230-2
1. Proteins—Biotechnology. I. Franks, Felix. II. Series.
[DNLM: 1. Proteins—analysis. 2. Proteins—physiology.
3. Biotechnology—methods. QU 55 P9653 1993]
TP246.65.P76P72 1993
660'.63--—dc20
DNLM/DLC
for Library of Congress   93-3448
                                                           CIP

OLSON LIBRARY
NORTHERN MICHIGAN UNIVERSITY
MARQUETTE, MICHIGAN 49855

Preface

Proteins are the servants of life. They occur in all component parts of living organisms and are staggering in their functional variety, despite their chemical similarity. Even the simplest single-cell organism contains a thousand different proteins, fulfilling a wide range of life-supporting roles. Additions to the total number of known proteins are being made on an increasing scale through the discovery of mutant strains or their production by genetic manipulation.

The total international protein literature could fill a medium-sized building and is growing at an ever-increasing rate. The reader might be forgiven for asking whether yet another book on proteins, their properties, and functions can serve a useful purpose. An explanation of the origin of this book may serve as justification. The authors form the tutorial team for an intensive postexperience course on protein characterization organized by the Center for Professional Advancement, East Brunswick, New Jersey, an educational foundation. The course was first mounted in Amsterdam in 1982 and has since been repeated several times, in both Amsterdam and the US, with participants from North America and most European countries.

In a predecessor to this book, emphasis was placed on the role of protein isolation in the food industry, because at the time this reflected the interests of most of the participants at the course. Today, isolated proteins for food use are extracted from yeasts, fungal sources, legumes, oilseeds, cereals, and leaves. Large-scale isolation methods have been developed and are operated on the basis of cost and efficiency. Any improvement in the product is measured against the cost of achieving such an improvement, and this cost features largely in the marketing equation along with other factors, such as the perceived (by the consumer) benefit of such an improvement. This latter attitude is of considerable importance in protein isolation technology.

In recent years interest in proteins has shifted away from food components, to their role in fighting disease and their uses as analytical tools and biocatalysts. These major changes are reflected in the present book. It is a pleasure, however, to acknowledge the indirect contribution to this volume by two of the authors of the previous book. Some of the subject matter relating to food proteins and earlier discussed by Peter J. Lillford and Chester Myers has been incorporated, particularly in Chapters 2 and 5. We are also grateful to S. Fischer, a former tutor in our protein characterization course, for permitting us to use some material from his course notes in Chapter 2.

Probably the most important advances in recent years have been the developments in recombinant DNA technology that have led to massive commercial interest in the production (or creation) of highly specific proteins, such as interferon, hormones, and blood factors, and to the current boom in biotechnology. Here again the emphasis is on profitability (and a quick return on capital), so that allowable costs incurred in extraction and recovery must be determined by the value of the product in the marketplace.

Market information on protein materials is closely guarded and hard to come by and, in any case, soon becomes out of date. We have attempted to include such economic aspects as were available at the time of writing. The most comprehensive data refer to the production and marketing of enzymes.

Although much of the subject matter contained in this book can be found in other publications, its combination in one volume has probably never been attempted before. The viewpoint taken here is that methods developed for the isolation, purification, processability, analysis, stabilization, and characterization of proteins rely on our present knowledge of their structures, stabilities, and performance under a variety of physical and chemical conditions. These methods are described and related to economic, clinical, and technological criteria. In other words, we are here concerned with the methodologies of relevance to the "protein business." Protein-converting is a young, but rapidly growing, industry; some of the quantitative estimates here represented are therefore based on extrapolations of presently available data, which may turn out to be wide off the mark, but this is surely the general nature of industrial and economic forecasts.

In any discussion of the technology of protein isolation and processing from a business perspective, the value of the final product needs to be considered, since this will probably determine the method chosen for the extraction and purification of the protein. Another factor that needs to be considered is the scale of the operation.

An attempt is made to provide a unified approach to the problems of protein characterization, stability, analysis, isolation, and processing. Differences are drawn between in vivo and in vitro methods of characterization and between analytical procedures that are appropriate to the quantities of material to be handled. Certain basic physical and chemical principles exist, whatever the scale of the operation, and these principles are summarized in the opening chapters.

Finally, a word of warning: Historically, much of our knowledge regarding protein structure, stability, and performance has been derived from in vitro studies on isolated proteins in dilute solution, acting on isolated substrates. An increasing number of such proteins is becoming available in bottled form, obtainable from laboratory suppliers, a prime example being lysozyme, on which much of the pioneering work was done. Whatever useful information can be obtained about lysozyme from model studies in aqueous solutions of the enzyme, it must be remembered that, in vivo, a lysozyme molecule is most unlikely ever to "see" another lysozyme molecule. It is questionable, therefore, whether a study of the protein–protein interactions that are responsible for the crystallization of globular molecules has any relevance to physiological situations. On the other hand, such interactions are of great importance in isolation and purification technology.

We emphasize, therefore, that despite some common physico-chemical principles, clear distinctions must be drawn between physiological and technological function and performance and between results obtained from in vivo and in vitro characterization and analysis.

Felix Franks
Cambridge, England
January 1993

Contents

Contributors

Johannes F. M. G. Aerts • *University of Amsterdam, Amsterdam, the Netherlands*

James L. Dwyer • *Ventec Inc., Marlborough, MA*

Linda A. Fothergill-Gilmore • *University of Edinburgh, Edinburgh, Scotland*

Felix Franks • *Pafra Ltd., Cambridge, England*

André W. Schram • *Zaadunie BV, Enkhuizen, the Netherlands*

Gert Van Duijn • *Center for Phytotechnology, Leiden, the Netherlands*

One

Proteins

Description and Classification

Felix Franks

1. Chemical Constitution

Proteins are linear condensation polymers of amino acids and are formed by the reaction

$$H_2N . CHR_1 . COOH + H_2N . CHR_2 . COOH \rightarrow$$
$$H_2N . CHR_1 . CONH . CHR_2 . COOH + H_2O$$

The amino acids (some 20) that occur in proteins differ only in the nature of the residue R. Residues are linked by identical trans-peptide bonds:

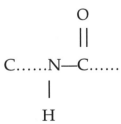

Proteins thus have identical backbone structures and resemble synthetic homopolymers. They are distinguished only by the distribution of the variable side chains along the main chain. It is interesting to compare this structure with the covalent structures of the other biopolymers.

From: *Protein Biotechnology* • F. Franks, ed. © 1993 The Humana Press Inc.

	Proteins	Carbohydrates	Nucleotides
Monomer variety	20+	Usually 1–3	4
Linkage	indentical	variable	identical
Chain branching	never	common	never

Proteins derive their chemical variety from the amino acid composition and the distribution of the amino acid residues along the peptide chain. Differences between carbohydrates, on the other hand, arise mainly from different types of linkages between sugar residues and from chain branching. Thus, cellulose and amylose are chemically identical, being homopolymers of glucose, but they are physically very distinct. The former is a $\beta1$, 4-1inked polyglucose, and the latter has $\alpha1$, 4 glucosidic linkages. Nucleic acids, like proteins, have identical chemical linkages without chain branching. They can, however, exist as cyclic polymers, e.g., plasmids. In any one nucleic acid variety, the number of residues is limited to four bases (two purines and two pyrimidines), variety being achieved by the sequence distribution.

Since the amino acids that make up proteins contain two or more polar or ionogenic functional groups, such as $-NH$, $>CO$, $-OH$, $-COOH$, and $-SH$, they are chemically reactive and subject to chemical attack, for instance, by oxidizing and reducing agents. This sensitivity renders proteins somewhat unstable in vitro.

2. Types of Classification

The relationships between the chemical composition, the structure, the stability, and the function of proteins is still somewhat obscure and is the subject of much current study and speculation. As has often been the case in other scientific disciplines, taxonomy and nomenclature take the place of understanding. Proteins are thus classified according to their amino acid composition, their sequences and three-dimensional structures, and their functions. Attempts are made to discover common themes that relate the chemical composition to the structure and, hence, to the biological function.

2.1. Amino Acid Classification (Common Amino Acids)

This type of classification is somewhat arbitrary, but in practice three groups of amino acids are distinguished, as shown in Fig. 1. Hydrophobic amino acids contain substantial apolar side chains, e.g., three or four carbon alkyl or benzyl radicals. Methionine is often included in this group, but neither alanine nor glycine is generally considered a hydrophobic amino acid. Acidic and basic amino acids have side chains that contain ionogenic groups, i.e., terminal–COOH or NH_2 groups that give the protein its polyelectrolyte character. Whereas five basic residues occur, there are only two acidic amino acids (asp and glu). The third group contains all those residues that do not fit into either of the above two classes. It comprises the polar residues, with –OH or –NH groups capable of hydrogen bonding or reacting with sugars; the two amino acids, proline and hydroxyproline, which differ from amino acids in that they contain the –NH group as part of a cyclic structure; and cysteine, glycine, and alanine. Cysteine is unique: Its –SH groups are responsible for the only covalent crosslinking (-S-S-) found in proteins.

2.1.1. Rare Amino Acids

Apart from the common amino acids shown in Fig. 1, other residues are also sometimes found in proteins. They include 5-hydroxylysine and some other lys and glu derivatives (mainly phosphorylated); desmosine, which provides the crosslinks in elastin, is a condensation product in which four lysine residues condense to form a pyridine ring.

2.1.2. Nonprotein Amino Acids

Many amino acids that do not occur in proteins are found naturally in the free state; they are common in plants, fungi, and microorganisms. They are often involved as intermediates in amino acid synthesis or metabolism. *See* Table 1 for examples.

2.2. Conjugated Proteins

Most proteins are conjugated, usually covalently, to other chemical groups, organic or inorganic, referred to as prosthetic groups. They are shown in Table 2. The proportion of nonamino acid material can vary from a fraction of a percent to 80%. The

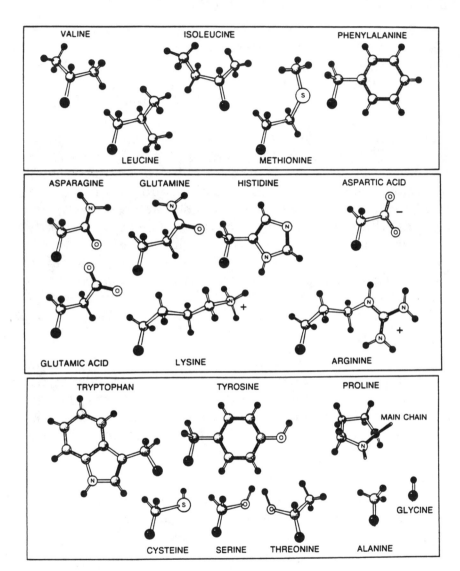

Fig. 1. Three groups of amino acids; reproduced with permission from A. C. T. North, in *Characterization of Protein Confirmation and Function* (1979) (F. Franks, ed.) Symposium Press, London.

Table 1
Nonprotein Amino Acids

Amino acid	Function
α-Aminobutyric acid	Accumulates under stress conditions
β-Alanine	Coenzyme A synthesis
β-Cyanoalanine	Neurotoxin
Ornithine $H_2N(CH_2)_3CH(NH_2)COOH$	Arginine biosynthesis
Homoserine $HO(CH_2)_2CH(NH_2)COOH$	Threonine, isoleucine, and methionine biosynthesis
O-Acetylserine	Carbon source for the production of sulfur amino acids
Albizzine $NH_2CONHCH_2CH(NH_2)COOH$	Nitrogen storage in legumes
Mimosine	Plant toxin

$$O = \overset{=}{\underset{=}{\bigcirc}} NCH_2CH(NH_2)COOH$$

OH
Strombine

	Accumulates in snail hemolymph under stress conditions
Glycine betaine $(CH_3)_3 . N^+ . CH_2 . CO . O^-$	Stress (desiccation) protectant in plants

occurrence of some of the rarer metals is especially interesting. Thus, molybdenum is associated with the process of nitrogen fixation through the reduction of nitrates by bacteria. The same enzymes are also able to catalyze the reduction of acetylene to ethylene. Presumably, the molybdenum (4 atoms/nitrogenase molecule) functions in the chemical reaction by virtue of its many possible valence states.

Table 2
Compositions of Some Conjugated Proteins

Class	Prosthetic group	Approx % of wt
Nucleoprotein systems		
Ribosomes	RNA	50–60
Tobacco mosaic virus	RNA	5
Lipoproteins		
Plasma β_1-lipoproteins	Phospholipid, cholesterol, neutral lipid	79
Glycoproteins		
γ-Globulin	Hexosamine, galactose, mannose, sialic acid	2
Plasma orosomucoid	Galactose, mannose, N-acetylgalactosamine, N-acetylneuraminic acid	40
Seed lectin (potato)	β-Arabinose, galactose,	50
(soya bean)	N-acetylgalactose	6
Fish antifreeze protein	N-acetylgalactosamine, galactose	50
Immunoglobulins	N-acetylglucosamine, mannose, fucose, and so on	15
Collagen (tendon)	Galactose, glucose	0.5
Phosphoproteins		
Casein (milk)	Phosphate esterified to serine residues	4
Hemoproteins		
Hemoglobin	Iron protoporphyrin	4
Cytochrome c	Iron protoporphyrin	4
Flavoproteins		
Succinate dehydrogenase	Flavin nucleotide	2
D-Amino acid oxidase	Flavin nucleotide	2
Metalloproteins		
Ferritin	$Fe(OH)_3$	23
Tyrosine oxidase	Cu	0.2
Alcohol dehydrogenase	Zn	0.3
Pyruvate carboxylase	Mn	0.02
Nitrate reductase	Mo	0.2
Glutathione	Se	3

Table 3
Carbohydrate–Peptide Linkages in Glycoproteins

Linkage	Anomeric type	Occurrence	Stability to acid	Stability to alkali
N-Glycosydic N-Acetylglucosaminyl- asparagine	β	Widespread	+	+
O-Glycosidic N-Acetylgalactosaminyl- serine or threonine	α	Mucins, blood group, substances, fetuin, antifreeze glyco- proteins	+	–
Xylosyl-serine	β	Proteoglycans	+	–
Galactosyl-serine	α	Plant cell walls, earthworm cuticle	+	–
Galactosyl-5-hydroxylysine	β	Collagens	++	++
L-Arabinofuranosyl-O- hydroxyproline	β	Plant cell walls	–	

Most proteins in their in vivo state are believed to exist as glycoproteins, e.g., lysozyme contains 3% sugar, which is probably hydrolyzed off during isolation and purification. The function of such short sugar chains is still subject to debate. Sugar residues are added to the wholly or partially preformed peptide by the action of enzymes (glycotransferases). This leads to a certain degree of irregularity in the resulting glycoproteins. Glycosylation occurs in animal cells, in plants, and some yeasts, but not in microorganisms. Two types of peptide–sugar linkages can be distinguished and are summarized in Table 3: N-linkages (asn) and O-linkages (thr, ser, hypro, hylys).

3. Amino Acid Profiles

One of the most important properties of a protein is its amino acid profile. Analytical methods for the determination of the amino acid composition are discussed in Chapter 3. Amino acid profiles of some representative proteins are collected in Table 4, with small

Table 4
Amino Acid Profiles of Representative Classes of Proteins

	Lysozyme	Chymotrypsinogen	Carbonic anhydrase	Alcohol dehydrogenase	Pepsin	Cytochrome c	β-Lactoglobulin	Serum albumin	Ferredoxin	Myoglobin (human)	TMV protein	α-Casein	Soya bean trypsin inhibitor	Histone H2A	Ubiquitin	Myosin	Silk fibroin	Collagen	Elastin	Wool keratin
Ala	12	22	19	28	18	6	30	3	9	12	14	9	8	18	2	78	334	107	58	46
Val	6	23	17	39	21	3	18	45	7	7	14	11	14	7	4	42	31	29	118	40
Leu	8	19	20	25	28	6	44	58	8	17	12	17	15	16	8	79	7	28	56	86
Ile	6	10	10	24	27	6	17	9	4	8	9	11	14	6	7	42	8	15	26	—
Pro	2	9	17	20	15	4	17	31	4	5	8	17	10	5	3	22	6	131	136	83
Phe	3	6	11	18	14	4	9	33	2	7	8	8	9	1	2	27	20	15	29	22
Trp	6	8	6	2	6	1	3	1	1	1	3	2	2	0	0	4	0	0	—	9
Met	2	2	2	9	5	2	8	6	0	3	0	5	2	0	1	22	—	5	0	5
Gly	12	23	16	38	38	12	7	15	6	15	6	9	16	11	4	39	581	363	376	87
Asx	21	23	31	25	44	8	32	46	13	11	18	15	26	7	6	85	21	47	4	54
Glx	5	15	22	29	27	12	48	80	13	21	16	39	18	10	111	155	15	77	22	96
Lys	6	14	18	30	1	19	29	58	4	20	2	14	10	13	7	85	5	31	3	19

Arg	11	4	7	12	2	2	6	25	1	2	11	6	9	12	2	41	6	49	6	60
Ser	10	28	13	26	44	0	14	22	7	7	16	16	11	4	3	41	154	32	9	95
Thr	7	23	14	24	28	10	16	27	8	4	16	5	7	55	6	41	13	19	10	54
Cys	8	10	1	14	4	2	6	20	5	1	1	0	4	0	0	4	—	0	—	—
Tyr	3	4	8	4	18	4	8	18	4	2	4	10	4	3	1	18	71	5	8	36
His	1	2	11	7	1	3	4	16	1	9	0	5	2	4	0	15	2	5	0	7
Total	129	245	260	374	513	104	316	513	97	153	158	199	181	129	74	840	1274	598	861	799
Percent nonpolar	46	49	45	44	50	44	50	55	36	49	46	40	50	50	26	29	78	72	93	48
No. positively charged	17	20	36	49	4	24	39	99	6	31	13	32	20	36	5	141	13	85	9	86
No. negatively charged	26	33	27	38	71	20	80	126	20	32	34	25	46	8	122	240	36	124	26	150
% positively charged	13	8	14	13	1	23	13	20	6	20	8	16	11	28	7	18	1	10	2	11
% negatively charged	20	13	11	10	21	19	25	24	20	21	21	13	25	6	16	29	3	13	3	19

globular proteins on the left and structural proteins on the right. Apart from the classification in Fig. 1, amino acids fall into two groups: common, e.g., glycine, alanine, leucine, and rare, e.g., histidine and tryptophan.

Few patterns can be discerned in the profiles of the globular proteins, which appear to be rather uniform in composition. A common feature is the proportion of apolar residues, which is in the region of 50%. This seems to be a requirement for a stable globular structure. The proportions of acidic and basic residues depend on the environment of the protein. Thus, pepsin is exceptionally acidic, whereas nucleohistones (which bind to the phosphate groups of nucleotides) are strongly basic.

Fibrous proteins display a chemical and structural rhythm, being composed of relatively few amino acid residues that occur in high proportions, e.g., glycine and proline. Most proteins contain few cysteine residues, most of them being involved in disulfide links. Among structural proteins cysteine is even rarer.

4. Classification by Size

No sharp dividing line can be drawn between what is normally referred to as a small peptide and a protein. Small peptides are rapidly gaining in importance; some examples are shown in Table 5. They display many types of biological activities, extending even to dipeptides, such as the artificial sweetener aspartame. It has also been reported that a dipeptide that is removed from the blood during hemodialysis is implicated in promoting schizophrenia.

From a commercial point of view, small peptides have distinct advantages over proteins: In principle, they can be synthesized from readily available starting materials, although the synthesis of pure chiral compounds by organic chemistry methods can present difficulties. They can also be produced by enzymatic methods, and there is currently much debate as to the most effective methods for their production. As regards the stability during and after processing, small peptides are believed to have advantages over large proteins, which are easily denatured during their isolation and purification. However, recent experience shows that even peptides can be subject to inactivation through aggregation, conformational changes, or by chemical means.

Table 5
Structures of Small Peptides and Their Biological Activities

Peptide	Sequence	Main biological activities
Thyrotropin releasing factor (TRF)	Glu-His-Pro-NH$_2$	Thyrotropin and prolactin release
Gastrin (14-17)-NH$_2$	-Trp-Met-Asp-Phe-NH$_2$	Stimulant of acid secretion
Cholecystokinin (CCK) (24-33)NH$_2$	-Asp-Arg-Asp-Tyr-Met-Gly-Trp-Met-Asp-Phe-NH$_2$	Stimulant of pancreatic enzyme secretion and gallbladder contraction
Enkephalin	Tyr-Gly-Gly-Phe-Leu(Met)	Analgesic
Oxytocin	Cys-Tyr-Ile-Gln-Asn-Cys-Pro-Leu-Gly-NH$_2$	Uterus contraction and milk ejection
Vasopressin Lys(Arg)-Gly-NH$_2$	Cys-Tyr-Phe-Gln-Asn-Cys-Pro-	Vasoconstrictor (constricts blood vessels) and antidiuretic
Angiotensin II	Asp-Arg-Val-Tyr-Ile-His-Pro-Phe	Vasoconstrictor
Luteinising hormone releasing hormone (LHRH)	Glu-His-Trp-Ser-Tyr-Gly-Leu-Arg-Pro-Gly-NH$_2$	Luteinising hormone (LH) and follicle stimulating hormone (FSH) release
Substance P	Arg-Pro-Lys-Pro-Gln-Gln-Phe-Phe-Gly-Leu-Met-NH$_2$ gastrointestinal systems	Effects on central nervous system, cardiovascular and
Somatostatin	Ala-Gly-Cys-Lys-Asn-Phe-Phe-Trp-Lys-Thr-Phe-Thr-Ser-Cys	Inhibits insulin, glucagon, growth hormone, and acid secretion
Bradykinin	Arg-Pro-Pro-Gly-Phe-Ser-Pro-Phe-Arg	Affects smooth muscle, lowers arterial blood pressure, increases vascular permeability
Bombasin	Glu-Gln-Arg-Leu-Gly-Asn-Gln-Trp-Ala-Val-Gly-His-Leu-Met-NH$_2$	Smooth muscle contraction, stimulation of gastric and pancreatic secretion, induction of hyperglycemia, various behavioral effects

Table 6
Molecular Weights and Subunit Structures of Proteins

	Mol wt	No. of residues	No. of subunits
Insulin (bovine)	5,733	51	2
Ribonuclease (bovine pancreas)	12,640	124	1
Lysozyme (egg white)	13,930	129	1
Myoglobin (horse heart)	16,890	153	1
Chymotrypsin (bovine pancreas)	22,600	241	3
Hemoglobin (human)	64,500	574	4
Serum albumin (human)	68,500	~550	1
Hexokinase (yeast)	96,000	~800	4
Hemerythrin	107,000	~890	8
Tryptophan synthetase (*E. coli*)	117,000	~975	4
Soybean lectin	120,000	1,000	4
γ-Globulin (horse)	149,900	~1,250	4
Cholesterol esterase	400,000	~3,300	6
Glycogen phosphorylase (rabbit muscle)	495,000	~4,100	4
Myosin	620,000	~5,160	3
Glutamate dehydrogenase (bovine liver)	1,000,000	~8,300	~40
Fatty acid synthetase (yeast)	2,300,000	~20,000	~21
Tobacco mosaic virus	40,000,000	~336,500	2,130

In proteins, the limiting number of residues per peptide chain is approx 1000. Any further increase in size is achieved by multi-subunit aggregation, usually by noncovalent interactions, but some subunits are held together by permanent or labile -S-S- links (immunoglobulins, fibrinogen). Subunit compositions of some proteins are shown in Table 6.

5. Classification by Function

Distinction must be made between in vivo biological function and in vitro technological function (often referred to as functionality).

5.1. Versatility and Limitations

Even a simple organism, *E. coli*, contains 1000 different proteins. It is believed that the total number of proteins is of the order of 10^{10}–10^{12}. This is only a small fraction of the total possible number of polypeptides, as illustrated by two examples: For a peptide containing one of each of 20 *different* amino acid residues, there are $20! = 2 \times 10^{18}$ possible amino acid sequences. For a protein of molecular weight 34,000 containing equal numbers of 12 different amino acid residues, there are 10^{300} sequence isomers. Since the advent of protein "engineering," the way has been cleared for the possible production of a vast number of new peptides (either mutants or completely new structures).

5.2. Biological Function

Table 7 provides a small selection of representative functions, of which, enzymes have received the most intensive study. They are the in vivo catalysts of biochemistry and act by perturbing the stability of the substrate molecule(s), weakening covalent bonds.

Storage proteins are chemically very stable, with a low turnover rate. They are usually concentrated in so-called protein bodies under conditions of low moisture. Transport proteins can be localized, as in membranes, where they regulate the flow of material in and out of the cell, such as ions, molecules or electrons, or they can themselves be transported, supplying nutrients (e.g., oxygen) to distant parts of the organism and removing waste products.

Transducer proteins interconvert different types of energy. Thus, actomyosin and flagellin convert chemical into mechanical energy, and opsin converts electromagnetic radiation into electrical impulses as part of the photoreceptor system of the eye. Structural proteins are load bearing systems, usually characterized by a high tensile strength.

Defense proteins recognize and inactivate foreign microbes and materials (antigens), rendering them harmless. They are also implicated in the repair of injury (blood clotting factors), the removal of damaged proteins (ubiquitin), and the response to physiological stress (heat shock proteins). Toxins act in the opposite sense, by chemical attack or by blocking mechanisms. They are usually small, flexible molecules.

Table 7
Classification of Proteins by Biological Function

Type and example	Occurrence or function
Enzymes	
Ribonuclease	Hydrolyzes RNA
Cytochrome *c*	Transfers electrons
Trypsin	Hydrolyzes some peptides
Storage proteins	
Ovalbumin	Egg-white protein
Casein	Milk protein
Ferritin	Iron storage in spleen
Gliadin	Seed protein of wheat
Zein	Seed protein of corn
Transport proteins	
Hemoglobin	Transports O_2 in blood of vertebrates
Hemocyanin	Transports O_2 in blood of some invertebrates
Myoglobin	Transports O_2 in muscle
Serum albumin	Transports fatty acids in blood
β-Lipoprotein	Transports lipids in blood
Iron-binding globulin	Transports iron in blood
Ceruloplasmin	Transports copper in blood
Transducer proteins	
Myosin	Stationary filaments in myofibril
Actin	Moving filaments in myofibril
Dynein	Cilia and flagella
Opsin	Photoreceptor
Defense proteins	
Antibodies	Form complexes with foreign proteins
Complement	Complexes with some antigen–antibody systems
Fibrinogen	Precursor of fibrin in blood clotting
Thrombin	Component of clotting mechanism
Interferon	Antiviral agent
Lectins	Seeds; agglutinating agents
Toxins	
Clostridium botulinum toxin	Causes bacterial food poisoning
Diphtheria toxin	Bacterial toxin
Snake venoms	Enzymes that hydrolyze phosphoglycerides
Ricin	Toxic protein of castor bean
Hormones	
Insulin	Regulates glucose metabolism

Table 7 *(continued)*

Type and example	Occurrence or function
Adrenocorticotrophic hormone (ACTH)	Regulates corticosteroid synthesis
Growth hormone	Stimulates growth of bones
Structure proteins	
Viral-coat protein	Sheath around chromosome
Glycoproteins	Cell coats and walls
α-Keratin	Skin, feathers, nails, hoofs
Sclerotin	Exoskeletons of insects
Fibroin	Silk of cocoons, spider webs
Collagen	Fibrous connective tissues (tendon, bone cartilage)
Elastin	Elastic connective tisue (ligaments)
Mucoproteins	Mucous secretions, synovial fluid
Chromosomal proteins	
Histones	Eukaryotes
Protamines	Fish sperm, some mammals
Nonhistone proteins	For example, enzymes for DNA replication
HMG proteins (structural?)	Thymus, kidney, liver
Ubiquitin	Removal of damaged proteins (?)

5.2.1. Functional Diversity Within Classes

Each group in Table 7 can be further subdivided according to function. Enzymes are the most intensively studied class of proteins; common bioreactions that are catalyzed by enzymes are shown in Table 8. Hydrolases are the simplest in function; they are usually composed of single peptide chains of low molecular weight, and are easy to immobilize. Complex chemical reactions, especially those involving transfer of chemical groups (synthetases, transferases, kinases), require multifunctional enzymes, usually oligomers composed of several different subunits, each fulfilling a specific function.

Each group in Tables 7 and 8 can once again be functionally subdivided; this is illustrated for proteolytic enzymes in Table 8 and for the families of growth factors in Table 9.

Table 8A
Classification of Enzymes and Subclassification of Proteolytic Enzymes

Class	Group affected	Example
Oxidoreductases	CHOH	Lactate dehydrogenase
	CH—CH	Acyl CoA dehydrogenase
	CH—NH$_2$	Amino acid oxidase
Transferases	Methyl	Guanidinoacetate methyltransferase
	Hydroxymethyl	Serine hydroxymethyl-transferase
	Acyl	Choline acetyltransferase
	Amino	Transaminase
Hydrolases	Carboxylic ester	Esterase, lipase
	Phosphoric monoester	Phosphatase
	Phosphoric diester	Ribonuclease
	Glycosidic bond	Amylase
	Peptide	*see below*
Lyases	C—C	Aldolase
	C—O	Fumarate hydratase
Isomerases	Aldose-ketose	Phosphoglucoisomerase Ligases
Forms C—O bonds	Amino acid activating	enzyme
	Forms C—N bonds	Glutamine synthetase
	Forms C—C bonds	Acetyl CoA carboxylase

Table 8B
Proteolytic Enzymes

Source	Enzyme	Specificity
Mammalian	Trypsin	R_1 = arg or lys
	Chymotrypsin	R_1 = phe or tyr (preferred)
	Pepsin	R_2 = phe, tyr, or trp
	Prolidase	R_1 or R_2 = glycine (preferred)
	Carboxypeptidase A	R_2 = Any amino acid except arg, lys, and pro
	Carboxypeptidase B	R_2 = lys or arg
	Iminopeptidase	Liberates proline
	Leucine amino peptidase	R_1 = Any amino acid except pro and gly
	Aminopeptidase M	Nonspecific[b]
Plant	Papain	Nonspecific[b]
	Ficin	Nonspecific[b]
	Carboxypeptidase C	Unknown
Bacterial	Subtilisin[a]	Nonspecific[b]
	Nagarase[a]	Nonspecific[b]
	Pronase[a]	Nonspecific[b]

[a]Commercial commonly used names.
[b]Most proteolytic enzymes are inactive toward the imide link of proline and attack glycine peptides very slowly.

Table 9
Growth Factor Families

Family	Members
Epidermal growth factor (EGF)	EGF, Transforming growth factor (TGF-α) Vaccinia growth factor (VGF) Shope fibroma growth factor (SFGF) Myxoma growth factor (MGF) Amphiregulin (AR)
Platelet-derived growth factor (PDGF)	PDGF-AA PDGF-AB PDGF-BB
Transforming growth factor-β (TGF-β)	TGF- β1 TGF- β2 TGF- β3 Inhibins Activins Mullerian inhibiting substance (MIS) Bone morphogenic proteins (BMPs)
Fibroblast growth factor (FGF)	Acidic FGF Basic FGF <u>hst</u>-gene product <u>int</u>-gene product
Insulin-like growth factor (IGF)	IGF-I IGF-II Insulin Relaxin

6. Technological Function

In the case of technological function, the protein is regarded as a chemical material with desirable technological attributes, to be exploited in fabricated products, such as foods, textiles, chemicals, or pharmaceutical preparations. Some of the specific characteristics exploited in food processing are summarized in Table 10

Table 10
General Classes of Attributes of Proteins Important in Food Applications

General property	Special attribute	Examples
Organoleptic	Color, flavor, odor, texture, mouth-feel, smoothness, grittiness, turbidity	Beverages, creams, pastes
Hydration	Solubility, dispersibility, wettability, water sorption, swelling, thickening, gelling, water-holding capacity, syneresis, viscosity, dough formation	Dough, meat products, desserts, frozen/thawed products
Surface	Emulsification, foaming, aeration, whipping, protein/lipid film formation, lipid binding, stabilization	Processed meat, coffee whitener, whipped desserts
Structural, textural, rheological	Elasticity, grittiness, cohesion, chewiness, viscosity, adhesion, network cross-binding, aggregation, stickiness, gelation, dough formation, texturizability, fiber formation, extrudability, elasticity	Beverages, toppings, textured and frozen products
Other	Compatibility with additives, enzymatic activity/inertness, modification properties	

(nutritional attributes are not included). The exploitation of the technological attributes requires the destruction and subsequent modification of the natural protein structure (and function). Processing stages can include some or all of the following: extraction, isolation, fractionation and purification, chemical modification, and reassembly into a new structure. Usually the end product will still be susceptible to microbial attack and should be suitably protected.

Bibliography

1. Werner E. H. et al. (1986) Immunomodulating peptides. *Experientia* **42,** 521, in German.
2. Schlesinger M. J. (1986) Heat shock proteins: the search for functions. *J. Cell Biol.* **103,** 321.
3. Burdon R. H. (1986) Heat shock and heat shock proteins. *Biochem. J.* **240,** 313.

4. Pegg M. et al. (1986) Confirmation that catalase is a glycoprotein. *Biochem. Internat.* **12**, 831.
5. Schülke N. and Schmid F. X. (1988) Effect of glycosylation on the mechanism of renaturation of invertase from yeast. *JBC* **263**, 8827, 8832.
6. Grafl R. et al. (1987) Mechanism of folding of RNase is independent of the presence of covalently linked carbohydrate. *JBC* **262**, 10625.
7. Alagona G. et al. (1986) Simple model for the effect of a glu 165 → asp mutation on the rate of catalysis in triose phosphate isomerase. *J. Mol. Biol.* **191**, 23.
8. Franco R. et al. (1986) A computer program for enzyme kinetics. *Biochem. J.* **238**, 855.
9. Potter M. (1986) Myeloma proteins. *Experientia* **42**, 967.
10. Hol W. J. G. (1986) Protein crystallography and computer graphics toward rational drug design. *Angew. Chem.* **25**, 767.
11. Pfleiderer G. (1986) Isoenzymes. *Naturwiss.* **73**, 643.
12. "Protein Structure, Folding and Design." (1988) UCLA Symp. 30 chapters reprinted from *J. Cell Biochem.* Liss, New York.
13. "Enzymes and isoenzymes in pathenogenesis and diagnostics." (1988) *Adv. Clinical Enzymol.* Karger, Basle.
14. Dubey A. K. et al. (1987) Sources, production, and purification of restriction enzymes. *Process Biochem.* **22**, 25.
15. Salton M. R. J. (1987) Bacterial membrane proteins. *Microbiol. Sci.* **4**, 100.
16. D'Souza V. T. and Bender M. L. (1987) Miniature organic models of enzymes. *Acc. Chem. Res.* **20**, 146.
17. Knowles J. R. (1987) Tinkering with enzymes: what are we learning? *Science* **236**, 1252.

Two

In Vitro Characterization

Economics and Technology

Felix Franks

1. Economic Considerations: Operating Scales and Product Values

The science and technology of proteins in the industrial scene need to be introduced with some history and current commercial positioning of the "protein business." Functional protein powders for food use were virtually unknown prior to 1940 except for dried "natural" materials (milk, egg). The war led to the development of a chemically hydrolyzed milk powder, marketed as a partial egg white replacer. Today, isolated food proteins are extracted from yeasts, fungal sources, legumes, oil seeds, cereals and leaves, and recovered from offal (meat, fish). As regards proteins destined for therapeutic, diagnostic, or research uses, the range of raw materials is much wider, including human sources (e.g., placenta) and, increasingly, materials derived from recombinant DNA technology.

Studies of the biochemistry of disease, coupled with advances in recombinant DNA technology have led to massive commercial interest in the isolation of highly specific proteins, such as interferons, interleukins, hormones, and blood factors —a biotechnology boom with a constantly changing range of values and pre-

From: *Protein Biotechnology* • F. Franks, ed. © 1993 The Humana Press Inc.

dictions. Cancer chemotherapeutic agents and antiinfective agents were (correctly) predicted several years ago to grow to a business worth $ 2.5–4 billion by 1990.

In this context the question must be asked: How reliable are predictions? In 1980 the predictions for 1985 were that 16 enzymes would be produced industrially (mainly proteases and amylases) with an annual value of $500M. Actually 22 enzymes were produced with a total value of $400M. Market percentage shares were Novo 25–40, Gist Brocades 15–20, Miles 5–10, Haegen (DK) 5, Sanofi 5, Finnish Sugar 5, others 15. The picture changes when only diagnostic enzymes are considered. In 1987 their total turnover was $80M; three companies (Boehringer Mannheim, Toyo Jozo, and Toyobo) commanded 65% of the world market.

Some of the more important human therapeutic products are now being addressed by recombinant DNA technology:

insulin	diabetes
growth hormone	small stature
serum albumin	surgery, shock, burns
Factor VIII	hemophilia
urokinase, tPA	heart attack, stroke, pulmonary embolism
interferons	cancer, viral infection
lymphokines	cancer, autoimmune diseases, infectious diseases

In principle, the use of human material for therapeutic purposes is preferable, because it need not be as rigorously purified as material from other sources, which might give rise to immune responses. On the other hand, human raw material might carry with it viral contaminants that can give rise to serious problems. This was the case with blood products (Factor VIII), containing HIV and/or hepatitis virus contamination, which were distributed to hemophiliacs in the UK during the 1970s.

1.1. Market Sectors

The total market in protein products (other than foods and animal feeds) can be divided into ethical and generic drugs, veterinary products, research biochemicals, ethical diagnostic products, and processing aids for production-scale biotransformations.

The total world generic drug market stood at $22 billion in 1990, of which the US, Japan, and Europe had 25, 50, and 12% shares, respectively. Within Europe, France has by far the largest market share (28%). The antiinfective drug market ($4 billion worldwide) is segmented in a similar manner and is projected to rise to $8.3 billion by 1992. Of this sum, immunologicals are estimated to amount to $530 million.

Veterinary products include biologicals, antibiotic and therapeutic drugs, feed additives, and other miscellaneous products. The total market is $8 billion, with an annual growth rate of 3%. The geographical segmentation is North America 32%, Western Europe 23%, Eastern Europe/USSR 15%, Central and South America 9%, Japan 9%, other 12%.

Research biochemicals fall into four groups:

1. *Molecular Biology*
   Restriction enzymes
   Modifying enzymes
   Reverse transcriptases
   Cloning vectors
   Linkers, adaptors
   Sequencing primers
   Probes
   Translation/transcription kits
   DNA synthesis reagents

   World market: $220 million

2. *Immunology*
   Polyclonal antibodies
   Purified antibodies
   Monoclonal antibodies
   Conjugated antibodies
   Protein-A material
   Purified serum
   Proteins

   World market: $120 million

3. *Cell Biology*
   Antibiotics
   Lectins
   Tissue dissociation enzymes
   Growth hormones
   Culture media
   Synthetic media
   Fetal calf serum

   World market: $160 million

4. *Biochemicals*
   Enzymes
   Coenzymes, nucleotides
   Carbohydrates
   Lipids
   Toxins
   Synthesis reagents
   Buffer reagents

   World market: $230 million

The main enzymes used in diagnostic applications are classified according to their fields of use:

1. *Clinical Chemistry.* Alcohol dehydrogenase, asparaginase, creatine kinase, β-fructosidase, β-glucosidase, galactose dehydrogenase, glucose-6-phosphate dehydrogenase, glutamate dehydrogenase, hexokinase, isocitrate dehydrogenase, lactate dehydrogenase, malate dehydrogenase, and pyruvate kinase. The world market value is $70 million, of which the US has a 50% share. Individual world market values of the most used clinical diagnostic enzymes are (in $M): pyruvate kinase 2, lactate dehydrogenase 2, glycerol-3-phosphate dehydrogenase 5, glutamate dehydrogenase 5.5, uricase 8, and cholesterol oxidase 12.

2. *Immunoassays.* Alkaline phosphatase, horseradish peroxidase, luciferase, and β-galactosidase, with a world market value of $25 million.

3. *New Enzymes.* Cholesterol oxidase/esterase, GCPDX-glucose amylase, urease, glycerol-3-phosphate dehydrogenase/oxidase, amylase, elastase, phospho- and lipoprotein lipases, and urokinase.

The commercial value of an enzyme is expressed in terms of its activity in Dollars per unit, the International Unit (IU) being the biological function it can perform per unit weight and in unit time. Unit price comparisons for important immunoassay enzymes are given below:

Enzyme	*Surface Price*	*Activity Price (per million units)*
Peroxidase	40–170/g	$1–4K
Alkaline phosphatase	4.5–14/mg	$4–10K
Glucose oxidase	$8–20/g	$50–80

The IU activity/unit weight of a commercial product is a measure of its purity and the yield associated with the isolation and purification process employed. Thus, it is usually found that an enzyme that is marketed in concentrated ammonium sulfate has a higher activity per gram of protein than has the same enzyme in a lyophilized form, the inference being that lyophilization is accompanied by a partial inactivation.

Table 1
The Scale of Operations and the Current Annual Demand
for Some Purified Proteins

Scale	Applications	Annual demand
Fine chemicals	medical/clinical research enzymes	g–kg
Intermediate	Industrial enzymes, plasma proteins, cosmetics, pharmaceuticals	kg–ton
Large	Industrial enzymes food specialties	>10 ton
Commodities	Food/feed (nutritional)	>10,000 ton

The scale of operations and the current annual demand for some purified proteins are summarized in Table 1.

1.2. Industrial Enzymes

Although the development of new, more stable, more active enzymes opens interesting scientific vistas, the commercial profitability of such activities is not necessarily assured *(1)*. There is a consensus that the producers of engineered enzymes should aim at increasing stability, possibly in mixed organic-aqueous media. New developments include bacterial nitrile hydratases that convert acrylonitrile into acrylamide, new enzymes for the production of antibiotics by nonfermentation processes, and "expandases," which are able to expand rings and can be used in the preparation of new antibiotics from synthetic tripeptides. Enzyme membranes are also coming into more frequent use, for the production of single-isomer drugs. Eighty-eight percent of leading drugs are still marketed as racemic mixtures, causing regulatory authorities to ask questions about "excess chemical baggage," to the point that tests on all isomers in a composition may soon become mandatory. On the ecological front, enzymes may be introduced into the Kraft-pulp papermaking process to reduce or eliminate emission of dioxins and other chlorinated compounds.

On the commercial side, the increase in enzyme sales between 1975 and 1990 in inflation-adjusted terms has only been 20%, hardly

OLSON LIBRARY
NORTHERN MICHIGAN UNIVERSITY
MARQUETTE, MICHIGAN 49855

impressive. According to authoritative sources, the expansion of the enzyme business has been held back by 13 factors: pricing, profitability, creation of new markets, patents, time and funding requirements for R&D, the decision-making process, impact of low profits, facilities and personnel, problem identification, capital and technology requirements, technical service requirements, locked-in customers, and lack of efficient mechanisms for technology transfer. Over and above, enzymes, as used in industrial processes, add little, sometimes less than 1%, to the cost of the final product produced and sold.

A recent (1991) study of the US enzyme market projects industrial enzyme sales over 6% annually to 1994, to $400 M and to $640 M by 2000. The market leaders will be enzymes other than proteases, carbohydrases, and lipases. Detergent enzyme sales will increase steadily, especially with the introduction of lipases into detergents, because of their "environment-friendliness," as compared to phosphates. Other markets that will figure in the growth of enzyme usage include cellulases in cutting down bleaching in paper mills, animal feeds, and ethanol production.

To develop a commercial enzyme system may require 5–10 years of effort, with an annual expenditure of $1–3M. A vicious circle operates: because of low profits, R&D expenditures by the companies involved in industrial enzyme development are not very large, by any standards. This then results in a lack of development and marketing of new enzyme-based products and processes. Another facet of the problem is the general lack of industry-oriented R&D personnel. Neither the educational system nor the training process in industry develops technologists with the necessary experience and interest in the commercial aspects of enzymology.

1.3. Process Economics

Market demand/product value define allowable extraction and recovery costs. Source materials are more restricted for "therapeutics"and "biochemicals" than for "commodities" (surfactants, thickeners, gels), not least because of government regulatory restrictions. For instance, oil seeds are a large source of proteins because oil recovery pays for the crop: Meal left after oil extraction may be used as animal feed, fertilizer, or fuel—usually at little extra cost. Methods of extraction and scales of operation depend on the value of the purified product and its end use. Consideration of the sensitivity of proteins to processing conditions and the use of computers will give improvements in functional quality and a "cost/unit functional property" evaluation. This can be misleading if a high percentage of pure product has inevitably to be wasted by the user. Taking the example of alkaline phosphatase, the minimum quantity of lyophilized enzyme commercially available is 0.4 mg, equivalent to approx 500 IU. Since one assay only requires 0.1–1.0 IU, the sample, when dissolved, would be sufficient for 500–5000 assays. Most of the material would thus be discarded, because, once diluted, the enzyme rapidly deteriorates and cannot be stored for later use (2). A more realistic value estimate would be cost/unit of *useful* property.

Unlike laboratory processing, large scale methods of protein isolation are often primarily concerned with cost and efficiency. In handling biological materials there is constant "negotiation" between the engineer and the chemist. The former is primarily concerned with efficient heat and mass transfer—because of its efficient utilization of energy, whereas the latter is concerned with maintenance of protein activity. It is usually not easy to reconcile these divergent criteria. Furthermore, the transient and nonuniform conditions in production-scale plants have not been mathematically modeled, so that most progress is still achieved by a combination of past experience and empiricism.

One concept on which both engineer and chemist can agree is that each step in the process must be optimized for yield, and the number of steps must be cut to a minimum. For an isolation and purification process consisting of n steps, each one of which can be carried out with a 90% efficiency, the final percentage yield of product is given by

$$\text{Yield (\%)} = 9^n \times 10^{(2-n)} = 59\% \text{ for } n = 5 \text{ and } 53\% \text{ for } n = 6$$

For biochemical and therapeutic uses purity is the overriding consideration. In the production of therapeutic proteins fractionation, purification, and stabilization (lyophilization) generally account for 50% of the total production cost. However, since such products command high premiums in the marketplace, produc-

tion hardly figures in the cost equation. Taking restriction endonucleases as a typical example of high value products that have to
be presented in a highly purified state, the crude enzyme preparation would normally be subjected to 4–6 purification stages. A typical procedure is given below for *Bst*31, ex *B. stearothermophilus (3)*:

> Crude preparation
> Fractional ammonium sulfate ppt. (30–60%)
> Precipitate-TEM buffer
> dialysis
> Enzyme solution
> Elution with TEM buffer and dialysis against PED buffer
> DEAE-Biogel A
>
> Phosphocellulose
> Salt gradient elution in PED buffer
> Pure enzyme

Where products are obtained by recombinant methods, the
estimated relative costs of cloning research are (1 team year ty =
$0.2 – 0.5M) *(4)*:

> Make a gene library (series of clones
>   with large DNA fragments that
>   make up the required sequence)   0.1 ty
>
> Make a probe (DNA sequence
>   from the desired gene)   0.2 ty
>
> Probe the library (find desired gene
>   and splice into plasmid)   0.5–0.8 ty
>
> Make the gene express (plasmids
>   replicate AND produce
>   the desired protein)   0.4 ty
>
> Isolate (and renature) the protein   1.0 ty

A generalized isolation and purification scheme for rDNA
proteins is given in Fig. 1.

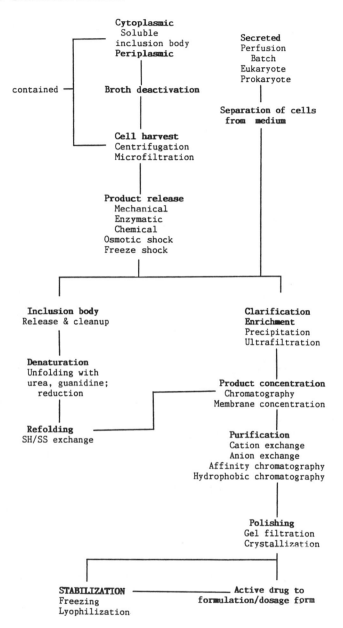

Fig. 1. Generalized isolation and purification process flows for rDNA proteins.

2. Technological Considerations for Isolation Processes

Microbial fermentation has developed into a huge business and supplies many healthcare products: peptide antibiotics, antifungal and anticancer drugs (*see* Table 2). At the present time, chemical synthesis is preferred for small molecules where there are no problems with chirality, but enzymatic methods may well take over in the future.

2.1. Choice of Source Material

The quality of fractions for medical, clinical, or research use is defined in terms of their added value in accordance with regulatory demands. In principle, all separation methods can be upscaled to preparative levels, but raw materials containing high percentages of the desired protein are required. As shown in Fig. 2, the selling price of the final product exhibits a linear correlation with the concentration of the active substance in the starting material, except for genetically engineered products that command higher market prices (5).

Plant breeding and genetic engineering can be considered as means of "source pretreatment." *E. coli* requires cell disruption, whereas *B. subtilis* secretes proteins. Cultured mammalian cells also secrete, they also process the product, e.g., glycosylation. In the isolation of food proteins, antinutritional factors have to be removed: phytic acid, protease inhibitors, lectins, and gossypol, a pigment in cottonseed that reacts with lysine to form toxic cyclopropenoid fatty acids. Table 3 summarizes some of the technical factors to be considered in the choice of source material.

2.2. Cell Disruption and Extraction

Except for liquid feedstocks or the recovery of extracellular enzymes from fermentation broths, the first step in protein isolation requires disruption of the structure in which the proteins are compartmentalized. Unfortunately, most disruption processes release not only the protein but other, undesirable fractions that, at best, contaminate, and at worst crossreact with the required product. On a large scale, the selective efficiency of disruption may domi-

Table 2
Peptide Antibiotics Produced on an Industrial Scale

	Use or molecular mechanism/targets/action
Actinomycins	DNA intercalating, inhibition of DNA/RNa synthesis (most potent antitumor drug)
Albomycins	Membrane transport, Fe(III) complex
Bacitracin	Cell wall synthesis (lipid carrier)
Bestatin	Aminopeptidase inhibitor (immunoenhancing)
Bicyclomycin	Affects biosynthesis of lipoproteins
Blasticidin S	Protein synthesis (agricultural use)
Bleomycins	Inhibit DNA/RNA protein synthesis
Capreomycins	Inhibit protein synthesis
Cycloserine	Cell wall synthesis (D-ala analog)
Cyclosporins	Immunosupressors
Distamycins	Inhibit T2 multiplication, R-factor transfer
Duaxomycins	Purine synthesis, glutamine antagonist
Gramicidin	Topical se
Hadacidin	Nucleotide biosynthesis (antileukemic)
Iturin A	Dermatomycose treatment
Negamycin	Translational termination, misreading
Neocarcinostatin	DNA degradation/synthesis
Netropsin	DNA complex, inhibits DNA/RNA protein synthesis
Nisin	Antiprotozoal, antimalarial
Ostreogrycin A/B	Inhibits cell wall/protein/DNA synthesis
Piperazinedione	Treatment of human lymphomas
Polymyxins	Membrane active (urinary tract infections)
Saramycetin	Antifungal
Thiostrepton	Inhibits protein synthesis (veterinary use)
Tuberactinomycins	Inhibits protein synthesis
Tyrocidines	Topical (hemolysis of erythrocytes)
U-42126	Glutamic antagonist (antileukemic)
Valinomycin	K^+ carrier, uncouples oxidative phosphorylation
Viomycin	Inhibits protein synthesis

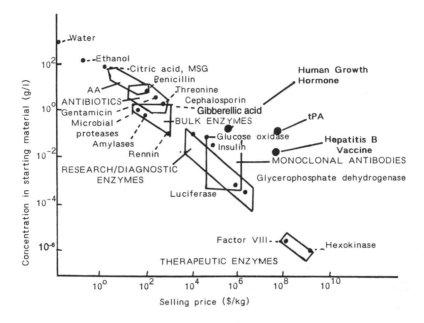

Fig. 2. Concentration in starting materials (pre-purification) of various substances and their selling price (1988) in the purified state. After J. Van Brunt; *Bio/Technology* **6**, 479 (1988).

nate the whole process efficiency. Ironically, since the emphasis on yield is less important in laboratory practice, little information exists about large-scale operations. Figure 3 shows methods that have been used; the choice obviously relates to the mechanical properties of the source material.

3. Solubility Considerations

Solubility is usually a function of pH, ionic strength, temperature, denaturants, lyotropism, excluded volume ("exclusion polymers"), and dielectric permittivity. Some of these various factors are treated in more detail elsewhere in this volume. Conditions for extraction must be chosen that avoid irreversible denaturation. On the other hand, reversible denaturation, e.g., by urea, is frequently used as an initial means of solubilization, if the protein in the source material occurs in an insoluble form, such as inclusion bodies.

Table 3
Technological Considerations in Choice of Starting Material

	E. coli	Yeast	Mammalian
Cell numbers	+	+	−
Cost of media	+	+−	
Expression of protein	+	+	+
Secretion of protein	−	+/−	+
Folding (renaturation)	−	−/+	+
Glycosylation	−	−	+

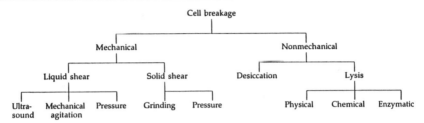

Fig. 3. Classification of cell disruption methods (from ref. 2).

3.1. Precipitation Methods

The manipulation of salt concentration provides for a means of separation. It is referred to as salting-in and salting-out, according to the effect of the particular salt on the solubility of a given protein. In the literature such effects are often, but mistakenly, ascribed to the ionic strength. Salting-out is generally performed with ammonium sulfate, according to the empirical relationship

$$\log S = b - Kc$$

where S is the solubility of the protein, c is the salt concentration (depends on pH and temperature), K is a constant that depends on the salt (it is referred to as the salting-out constant), and β is dependent on the nature of the protein, temperature, and pH but independent of the nature of the salt. As seen in Fig. 4, the relationship does not hold for dilute solutions, deviations typically appear below an ionic strength of 0.15.

In practice, the salt may be added as a solid or as a saturated solution to give a "% saturation" in the protein dispersion. On any

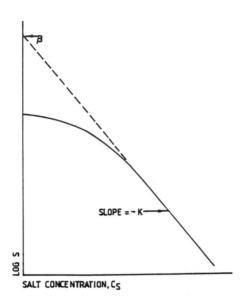

Fig. 4. Typical salting-out curve for soluble proteins.

scale of operation, colloidal precipitation is not instantaneous and the mechanical procedure of salt addition modifies the apparent yield; *see* Fig. 5.

Isoelectric precipitation might be thought to be a useful technique for protein recovery, because at pH = pI protein solubility goes through a minimum. Although this is true for some proteins (globulins), most enzymes are albumins (i.e., they are soluble at low ionic strength) so that isoelectric precipitation can only be used with simultaneous temperature adjustment, otherwise losses of activity would limit the usefulness of the method. It must also be borne in mind that very high acid concentrations are used, typically 10M and that the acid anion influences the degree of protein denaturation, because it acts in a salting-in/out manner. Thus, best recoveries are achieved with sulfuric acid (the SO_4^{2+} ion is strongly salting-out), whereas trichloracetic acid, while precipitating proteins, also tends to denature them in the process.

Thermal coagulation, although very effective and not requiring further dilution of an already dilute solution, is not much favored because of its general denaturing effect. It is however used

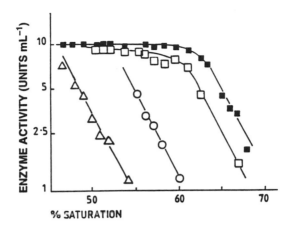

Fig. 5. The effect of contacting procedure and time on the salting out of fumarase by ammonium sulfate. △, solid, batch. 12 h contact: ○, saturated solution. batch. 12 h contact; □, saturated solution, continuous, 12 h contact: ■, saturated solution, continuous 1.6 min contact. Initial protein concentration was 40 mg/mL at pH 5.9 (reproduced from ref. 5).

where denaturation does not matter or may even be desirable, e.g., for some food and nutritional protein uses.

Protein solubility is affected by the addition of water-miscible organic compounds, in a manner similar to salting-out (6). An ingenious example of its use, in combination with pH manipulation, is the famous Cohn method for plasma fractionation (7). It is not generally realized that blood is the largest and richest available source of proteins. The outline of the Cohn method and the fractional recoveries are given in Table 4 and Fig. 6. The transfer of this laboratory exercise to a large-scale method provides an excellent case study for would-be biochemical engineers. The particular problems initially encountered were:

1. The temperature change produced by the heat of mixing of alcohol and water affects the relative solubilities of the component proteins.

2. Flow metering and pH control are general problems that are also encountered with isoelectric precipitation.

Table 4

Distribution of Plasma Proteins into Fractions by Various Methods Estimated by Electrophoretic Analysis and a Nitrogen Factor of 6.25 for All Proteins

Fraction	Method	Albumin	α-Globulin	Cholesterol	β-Globulin	γ-Globulin	Fibrinogen
Plasma		33.2	8.4	1.6	7.8	6.6	4.3
I	1	0.3	0	0	0.2	0.7	3.0
I	2	1.0	0.3	—	0.4	0.2	2.4
I	5	0.2	0.2	0.02	0.8	0.5	2.6
I	6	0.3	0.3	0.01	0.6	0.3	2.3
II	1	1.5	0	0.2	1.3	3.7	0.3
III	1	1.5	0.1	0.3	2.6	2.6	0.2
II + III	2	0.6	0.9	—	5.9	4.7	1.4
II + III	5	0.7	1.8	1.1	6.2	6.0	1.6
II + III	6	0.6	0.9	1.3	6.7	5.7	1.6
IV	1	5.6	4.2	0.7	3.0	0.4	0.2
IV	2	4.9	4.9	—	3.7	0.6	0
IV	5	1.0	5.4	0.4	3.1	0.2	0
IV-1	6	—	3.9	0.4	3.1	0.2	0
IV-4	6	0.9	2.7	0.04	2.2	0	0
V	1	26.0	0.3	0	0.4	0	0
V	2	27.0	0.6	—	0.3	0	0
V	5	29.0	0.6	<0.01	0	0	0
V	6	28.4	1.2	<0.01	0.3	0	0
VI	1	2.2	0.1	0.02	0	0	0
VI	2	0.5	0.1	—	0	0	0
VI	5	0.3	0.3	—	0	0	0
VI	6	0.7	0.2	—	0	0	0

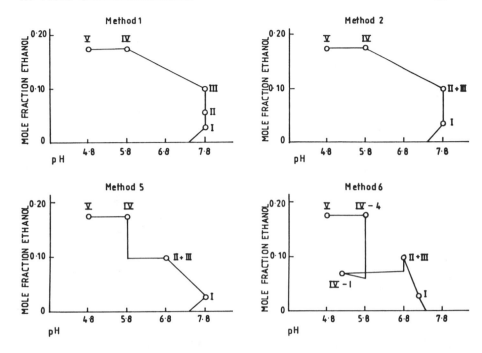

Fig. 6. Ethanol concentration and pH for separation of blood plasma fractions by various methods (from ref. 7).

3. Good mixing, without excessive turbulence that would cause foaming and denaturation.

4. Clean separation to allow efficient dewatering and purification of fractions.

The process was optimized for albumin recovery. Despite these process difficulties, the advent of improved separation methods (e.g., chromatography) and the increasing importance and value of minor components, this original process is still practiced, although for the recovery of minor components (blood coagulation factors), cryoprecipitation is now the favorite initial separation step.

Exclusion polymer precipitation consists of the addition of high molecular weight neutral polymers (dextrans, poyethylene glycol, polyvinyl pyrrolidone, polyethylene imine, and so on) to protein dispersions, resulting in the separation of a protein-rich phase by mutual exclusion (8). For human albumin the linear dependence of log S on polyethylene glycol concentration suggests that the mechanism resembles that of traditional salting-out (*see* Fig. 4). Indeed, coacervation (phase separation of lyophilic colloids by salts or nonelectrolytes), salting-out, and organic solvent precipitation are frequently considered within the same scientific framework, i.e., a competition with the protein for water. We consider this not only an oversimplification, but an incorrect interpretation of the observed effects.

In plant scale separation processes up to 90% yields can be obtained in a single extraction step, with contaminating proteins being removed together with cell debris. Partition coefficients of the order of 3–20 have been measured. In most cases PEG/salt systems are used *(9)*. A specific example is the purification of hydroxyisocaproate dehydrogenase from *Lactobacillus confusus*. The cells are disrupted by wet milling; 20% biomass (specific activity 0.13 IU/mg) is extracted with 18% PEG 6000 and 7% potassium phosphate (pH 7). A purification factor of 16 can be achieved, giving a product of specific activity 2.1 IU/mg. The second extraction is performed on the top phase with 10% potassium phosphate, the bottom phase consisting of $0.3M$ NaCl (pH 6). The yield of enzyme is 80% with a purification factor of 24 and the final product has a specific activity of 3.1 IU/mg.

The following purification factors have been achieved (with number of single-stage extraction steps in parentheses): fumarase 22 *(2)*, aspartase 18 *(3)*, lactate dehydrogenase 1.9 *(2)*, pullulanase 6.3 *(4)*, penicillin acylase 10 *(2)*. Yields varied between 70–90%.

Sometimes superior separations can be achieved by the addition of a high molecular weight PEG to the second extraction mixture. Thus, lactate dehydrogenase can be completely separated from 2–hydroxyisocaproate dehydrogenase through the use of PEG 10,000 for the second extraction stage.

Instead of polymer/salt systems, some processes (e.g., formate dehydrogenase extraction) employ two incompatible polymer phases, usually PEG and dextran. This type of separation is preferable, because polymers are more easily removed from the final protein product than salts. However, the comparatively high cost of dextrans (at least 15 times the cost of PEG) has prevented polymer/polymer systems from replacing polymer/salt extraction systems; *see* Fig. 7 *(10)*.

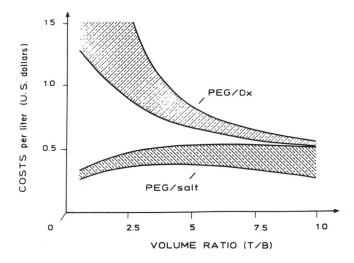

Fig. 7. Cost (1984) of phase systems used for protein purifications as a function of the volume ratio of the phases (top/bottom). Reproduced, with permission, from ref. 10.

3.2. Rate of Precipitation

In the laboratory and batch processing the time scales are not particularly significant, but in continuous, large-scale operations, *rates* of precipitation become dominant. The theoretical and technological treatments of rates of precipitation are based on theories of colloidal flocculation; both nucleation and growth rates of aggregates need to be considered (11,12). Under conditions leading to self-association, molecules collide and aggregate by random motion or thermal agitation. Growth continues to a size at which fluid motion becomes important in promoting further collision or aggregate disruption. This first growth step is termed "perikinetic" and applies to size ranges of up to 0.1–10 μm, respectively, for high- and low-shear conditions. For large particles, fluid motion promotes collisions and hence the aggregates grow (orthokinetic aggregation), i.e., the total number of particles decreases, initially very rapidly. A point is reached at whch further collisions promote the breakup of particles. Simple theories relate the maximum particle diameter in a stirred vessel to the impeller speed, the power dissipation or the mean velocity gradient (13,14).

4. Solid/Liquid Separation

Solid/liquid separation is necessary for the clarification of extracts and the recovery of precipitates. Usually, several passes through a single device or multiple-separation processes are used. They rely mainly on filtration and centrifugation, depending mainly on the scale of the operation.

Centrifugation is sometimes used in combination with filtration, with the centrifugal field acting as the pressure (15). Protein precipitates are essentially soft solids, often gel-like, so that an increased centrifugal pressure can cause blinding of the filters, by extrusion of the precipitate into the filter pores. By comparison, the modification of the precipitate structure by aging (16), ultrasonic conditioning (17), or even freezing (18), tends to have a greater effect on efficiency than higher pressure. In general, the precipitate for efficient dewatering by filtration has an open structure, strong enough to withstand the applied pressure and allowing for efficient drainage of the occluded liquid. This is rarely achievable for globular proteins.

The commercial advantages are low operating costs, lack of contamination (by filter aids), and the opportunities for continuous operation. The disadvantages are that most large-scale equipment was designed for the separation of small amounts of inorganic solids from dilute solutions. This is exactly contrary to the requirements for protein purification, where high solids contents of a colloidal nature are to be separated from saturated solutions. This has led to a reinvestigation of centrifuge efficiency. Frequently, separators are used in series: a low-speed device is followed by a high-speed "polisher" (12).

5. Liquid/Liquid Separation

5.1. Ultrafiltration

Ultrafiltration is a process in which a porous membrane, or filter, separates components on the basis of size and shape (15). The development of controlled pore sizes in the 1–100 μm range allows for the separation of molecular colloids "dissolved" in an aqueous solution. Despite its undoubted advantages, ultrafiltration on a large scale has its own problems (19).

Large-scale processing relies mainly on two types of membrane design: tubular and hollow fibers. Tubular fibers are robust, allowing for high liquid flow rates and simple clean-in-place operation, whereas hollow fibers offer greater flexibility of pore sizes. Membrane characteristics are usually described in terms of their retention characteristics, quoted as the molecular weight of a globular protein that is retained (>90%) by the membrane. Solute retention (R) under optimized conditions is operationally defined as

$$R = 1 - (C_p / C_f)$$

where C_p and C_f are the permeate and feed concentrations, respectively. Under plant conditions other factors also affect transmembrane flux and membrane selectivity. For example, high permeate flux rates can seldom be maintained as the solute concentration increases, because of concentration polarization, i.e., the solute accumulates at the surface and thus reduces the water flow through the membrane. For macromolecules this effect can become extreme since gels may form at the membrane surface, thus producing a secondary membrane that will then determine the selectivity.

In practice, the large-scale fractionation of proteins by ultrafiltration appears unlikely, though concentration, desalting, and recovery are widely practiced under conditions of very turbulent flow (in tube membranes) or back-flushing (hollow fibers) to minimize gel polarization.

5.2. Chromatography

The essential principles of chromatography are discussed elsewhere in this book, as are also details of its many applications in the field of protein separation and purification. The range and subtlety of separation are enormous and still growing. It shows great promise for upscaling, the main problems being: 1. Zone spreading, which can be minimized by the geometry of columns and packing; and 2. Physical stability of the particle bed (most solid phases are relatively soft). Decreasing the particle size improves resolution but increases the pressure drop and the compaction of the bed. Recent developments show promise in creating mechanically stronger systems.

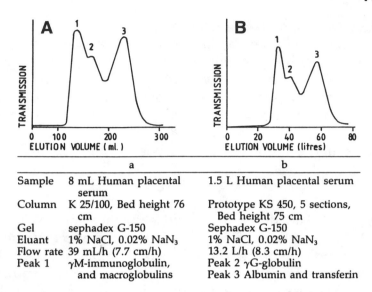

a	b
Sample	8 mL Human placental serum	1.5 L Human placental serum
Column	K 25/100, Bed height 76 cm	Prototype KS 450, 5 sections, Bed height 75 cm
Gel	sephadex G-150	Sephadex G-150
Eluant	1% NaCl, 0.02% NaN$_3$	1% NaCl, 0.02% NaN$_3$
Flow rate	39 mL/h (7.7 cm/h)	13.2 L/h (8.3 cm/h)
Peak 1	γM-immunoglobulin, and macroglobulins	Peak 2 γG-globulin
		Peak 3 Albumin and transferin

Fig. 8. Separation of proteins from human serum in (a) a typical laboratory scale separation, and (b) a separation using a five-section stack. Nearly 200 times more material was treated in (b), but the time was the same.

Successful separations on the 1g–1 kg scale are already possible with reasonable efficiency and high selectivity (*see* Fig. 8). For high value proteins, required in relatively small quantities, chromatography will probably continue to be the preferred separation method, with exciting development prospects.

6. Concentration and Drying

For laboratory and fine chemicals use, proteins may be marketed as wet precipitates, at subzero temperatures in concentrated glycerol solution or in the frozen state. Preferably, further concentration and drying is required, especially to achieve reasonable shelf lives of therapeutic products. Various drying techniques have been employed, but because of the low temperatures that have to be used, the efficiency is low. In energy terms, ultrafiltration has emerged as the most efficient method for concentration and is frequently employed prior to the final drying stage. The operational stabilization of proteins by drying is treated in detail elsewhere in this volume.

Table 5
Impurities and Contaminants
that Might Be Present in rDNA-Derived Proteins

Extraneous contaminants
  Endotoxin
  Host cell protein
  Protein impurities (cell culture media)
  DNA
  Microbes/viruses, and so on
  Media and buffer components
  Leachable matrix components from chromatographic columns

Protein-derived contamination/degradation products
  Proteolytic fragments
  Oxidized (met,cys), deamidated forms (asn, gln)
  Denatured/aggregated forms

7. In Vitro Characterization and Quality Control

The characterization of biotechnology products for therapeutic use, especially in combination with recombinant DNA techniques, has led to the development of guidelines produced by the national and international regulatory authorities to satisfy the requirements of quality and safety of these products (for references *see* (20, 21). These recommendations are sound in principle but in practice there are no detailed guides at all. The diverse (both qualitatively and quantitatively) impurities and contaminants that may often have to be shown to be absent from a purified product are collected in Table 5.

If the final product is a protein, a variety of analytical, biochemical and chemical techniques can be applied, all of which may have to be taken into account to prove the identity, purity, and potency of the substance. A summary is given in Table 6 of methods in common use and their diagnostic features (22).

The literature devoted to in vitro analytical methods describes techniques to determine size, charge, isoelectric point, compositional hydrophobicity, conformation, state of aggregation, and primary and secondary structure. The methods, however, do not readily fulfill the requirements for the quantitative determination of these characteristics.

Table 6
Methods for Protein Analysis and Their Diagnostic Value

Method	Property
Electrophoresis (PAGE)	size, charge
Isoelectric focusing (IEF)	isoelectric point
Size exclusion chromatography (SEC)	size, aggregation status
Ion exchange chromatography (IEC)	charge
Hydrophobic interaction chromatography (HIC)	"hydrophobicity"
Affinity chromatography	identity
Reversed phase chromatography (RPC)	"hydrophobicity," charge
Peptide mapping	correct and complete translation
Amino acid analysis	correct composition
Amino acid sequence	primary structure
IR spectroscopy	secondary structure
UV spectroscopy	typical spectrum
Fluorescence spectroscopy	structure, Trp content
NMR spectroscopy	tertiary structure (domains)
Light scattering	aggregation status
Circular dichroism	secondary structure
Biological activity	activity, potency
Protein determination	concentration, specific activity
Immunochemical methods	cross reaction, identity

At present there are no original data that can be used as guidelines. It is up to the manufacturer to decide which methods should be applied. Owing to the fact that different proteins may display individualities when analyzed by one or several of these methods, every technique has to be evaluated in order to achieve a full characterization of a given product.

The typical characteristics and the required reliability of the techniques can be assessed by the validation of the appropriate methods. Validation could be defined as: "The documented evidence that the system/method under consideration really does what it is supposed to do."

7.1. Polyacrylamide Gel Electrophoresis (PAGE)

SDS-PAGE determines the size, identity (*see* Western blot), and purity of proteins. There are limitations, such as variations in the staining/destaining procedure, difference in the staining intensity of individual proteins, and the precision of quantitative measurements by densitometry. An extensive validation with reference standards is recommended.

7.2. Isoelectric Focusing

This technique can reflect heterogeneity in the charge of proteins such as deamidation of asn and gln, or variations in glycosylation. One has to evaluate carefully whether an apparent heterogeneity is a result of actual differences in structure or is an artifact that may have arisen during isolation, e.g., from denaturation or precipitation.

7.3. Liquid Chromatography

7.3.1. Size Exclusion Chromatography (SEC)

This method allows the estimation of the molecular weight (in terms of the stokes radius) under native conditions and the determination of the aggregation status. The identity of the eluted fractions and the yields have to be confirmed by alternative methods (activity, PAGE, and so on).

7.3.2. Ion Exchange Chromatography (IEC), Hydrophobic Interaction Chromatography (HIC)

The separation by these techniques reflects the surface charge/hydrophobicity. Especially IEC can be a sensitive measure of purity of proteins. As before, the combination with other methods can confirm identity and yields.

7.3.3. Reversed Phase Chromatography (RPC)

RP-HPLC can be used for the very sensitive detection of degradation products or contaminants, especially when monitored at 220 nm. The selectivity for many proteins is not very high, they elute in a narrow composition range of the organic phase. The evaluation of selectivity and detection limits is necessary (with the aid of other techniques).

7.4. Peptide Mapping

Peptide mapping provides information about correct translation of the protein. This typical fingerprint, however, is only reproducible if the quality of the proteinases used is guaranteed. The isolation and confirmation of the C-terminal peptide can also prove the complete translation of the protein.

7.5. Amino Acid Analysis

The amino acid composition is a characteristic pattern for proteins, but it cannot show minor contaminants or variations. The main information that can be obtained is to provide the exact amounts of amino acids that could have been chemically modified during processing, such as met, trp, and lys.

7.6. Amino Acid Sequencing

The N-terminal sequence is important for the identification of the product and, in some cases, the correct processing (signal peptides). Single contaminating proteins may be identified. The C-terminal sequence can be determined with carboxypeptidases or by the identification of the C-terminal peptide from a fingerprint.

7.7. Immunological Methods

7.7.1. Western Blot

In combination with SDS-PAGE the identity of proteins and their degradation products can be determined. The method is not quantitative and the antibodies used have to be evaluated as to their specificities.

7.7.2. Enzyme-Linked Immunoassay (ELISA)

An ELISA for the product can detect the active and the inactive/degraded protein. Major contaminants are also determined with a separate ELISA, specific for the contaminating protein.

7.8. Spectroscopic Methods

7.8.1. Circular Dichroism (CD)

CD spectra reflect the secondary structure of proteins (α-helical content). A reasonable result concerning the identity and intact structure can only be obtained by comparing the spectrum with a reference sample (23).

7.8.2. Infrared Spectrum (IR)

Frequencies of the typical amide I and II bands monitor the degree and nature of secondary structure.

7.8.3. Ultraviolet Spectrum (UV)

UV spectra are similar for all proteins. The analysis can be useful for the detection of nonprotein contaminants and as a simple method for the determination of protein concentration. Trp, tyr, and phe contents can be obtained by the use of derivative and/or difference spectroscopy.

7.8.4. Fluorescence Spectrum

This spectrum can be typical for a protein when compared with a reference sample. In the context of quantitative analysis it provides a simple and accurate method for the determination of trp (24). Fluorescence spectra are extremely sensitive and can be used to detect very low protein concentrations.

7.8.5. NMR Spectrum

NMR spectra provide the most powerful (and expensive) technique for detailed structural and dynamic investigations on small proteins or protein fragments. A large variety of methods, all based on NMR, can be brought to bear on many problems relating to structural detail, denaturation, binding, and so on.

7.9. Light Scattering

In addition to SEC, light scattering can provide quantitative data on the aggregation status. Other types of radiation scattering (X-rays, neutrons) provide information on supermolecular structures, e.g., chromatin, lipoproteins.

7.10. Summary

The individual behavior of proteins in analytical systems does not allow the recommendation of only a limited set of methods for a quantitative analysis of purity, identity, or structure. The evaluation and validation of particular methods, including data processing systems for integration, is necessary.

Once a variety of methods have been evaluated, it may be possible to reduce the number of different techniques for a routine batch control, depending on more comprehensive results of the in-depth investigation.

Bibliography

1. Anon. (1990), *D-J-M-Enzyme Report* **9(5)**, 1–8; **9(6)**, 4-8.
2. Hatley R. H. M., Franks F., and Mathias S. F. (1987), *Process Biochem.* **22(6)**, 169–172.
3. Dubey A. K., Mukhopadhyay S. N., Bisaria V. S. and Ghose T. K. (1987), *Process Biochem.* **22**, 25–34.
4. Ano. *Genetic News*, Jan. 1983.
5. Van Brunt J. (1988), *Bio/Technology* **4**, 479–483.
6. Franks F. and Eagland D. (1975), CRC *Crit Rev. Biochem.* **3**, 165–219.
7. Cohn E. J., Strong L. E., Hughes W. L., Mulford D. J., Ashworth J. N., Melin M. and Taylor H. L. (1946), *J. Am. Chem. Soc.* **68**, 459–475.
8. Walter H., Brooks D. E., and Fisher D., eds. (1985), *Partitioning in Aqueous Two-Phase Systems, Academic Press,* Orlando FL.
9. Hustedt H., Kroner K. H., Schütte H., and Kula M. -R. (1983), *Enzyme Technol. Rotenburg Ferment. Symp.*, 3rd, Sept. 1982 Vol. 1, pp. 135-145, Verlag Chemie.
10. Kroner K. H., Hustedt H., and Kula M. -R. (1984), *Process Biochem.* **19**, 170–179.
11. Ives K. J. ed. (1978), *The Scientific Basis of Floccculation,* Sijthoff and Noordhoff, Alphen an den Rijn, The Netherlands.
12. Bell D. J., Hoare M., and Dunnill P. (1983), *Adv. Biochem. Eng. Biotech.* **26**, 2–72.
13. Tamb N. and Hozum H. (1979), *Water Res.* **13**, 421–427.
14. Tomi D. J. and Bagster D. F. (1978), *Trans. Inst. Chem. Eng.* **56**, 9–18.
15. Lillford P. J. (1988), in *Characterization of Proteins* (Franks F., ed.), Humana Press, Clifton, NJ, p. 427.
16. Hoare M. (1982), *Trans. I. Chem. Eng.* **60**, 79–87.
17. Jewett W. R. (1974), US Patent No. 3826740.
18. Boyer R. A., Crupl J., and Atkinson W. T. (1945), US Patent No. 2377853.
19. Beaton N. C. and Steadly H. (1982), in *Recent Development in Separation Science vol. VII* (Li N. N., ed.), CRC, New York.
20. Duncan M. E., Charlesworth F. A. and Griffin J. P., (1987), *Trends in Biotechnology* **5**, 325–328.
21. Bodgansky F. M. (1987), *Pharmaceuticals Today* **11(9)**, 72–74.
22. Garnick R. L., Rass M. J., and de Mec C. P. (1988) in: *Encyclopedia of Pharmaceutical Technology,* Vol. (Swarbrick J. and Boylan J. C. eds.), Dekker, Inc. New York.
23. Utsumi H., Yamazaki S., Hosoi K., Kimura S., Hanada K., Shimazu T., and Shimizu H. (1987), *J. Biochem.* **101**, 1199–1208.
24. Myers C. (1988), in *Characterization of Proteins* (Franks F., ed.), Humana Press, Clifton, NJ, p. 491.

Three

Analytical Chromatography of Amino Acids, Peptides, and Proteins

James L. Dwyer

1. Introduction

A broad range of separatory modes and media are utilized to identify and characterize the polypeptides and amino acids. The analytical chromatography of polypeptides also encompasses various preparative modes, as chromatography is frequently used to isolate materials for other analytical procedures. High molecular weight proteins are separated both in denatured and active configuration states. In this chapter an arbitrary distinction is made between the low molecular weight peptides (<20,000 Da) and the high molecular weight proteins. Both the kinetic and chemical behavior of the polypeptides within the chromatographic environment varies markedly as molecular weight is increased. The chapter is organized according to the principal chromatography modes, with recognition of special considerations that encompass the differences in physical/chemical behavior between small peptides and protein macromolecules.

2. Basis of Separation

Chromatographic separation is based on the flow of solutes plus solvent through a packed bed, or column. The chromatographer refers to the solvent system as the *mobile phase*. The packing material is called the *stationary phase*. As a solute molecule migrates through the column, it undergoes multiple interactions with the

From: *Protein Biotechnology* • F. Franks, ed. © 1993 The Humana Press Inc.

stationary phase. Similar interactions simultaneously between solvent-stationary phase and solvent-solute. The net result is a differential retardation of the various solute species, based on the nature and strength of these interactions. The high resolving power of chromatography arises not so much from energy differences of these interactions as from their multiplicity. An individual solute molecule can undergo literally thousands of interactions with the stationary phase as it moves through the column. Very slight interaction differences between two similar solute species can thus result in complete separation.

Resolution of species by chromatography thus depends on the triad of interactions (mobile phase/solute/stationary phase) for each solute.

One of the reasons that chromatography is such a powerful analytical and process tool is its versatility. There are a number of chromatographic modes available, and by selecting judicious combinations of these modes, one can frequently separate species of great similarity. Listed below are the major modes used in the separation of peptides.

1. *GPC (Gel Permeation, or Size Exclusion chromatography).* The mechanism of separation based on diffusive accessibility of stationary phase pore volume to solute species. Large molecules are excluded from the pores of the packing, thus passing rapidly through the column in the interstitial volume. Small molecules can enter packing pores, and thus exit the column later.

2. *Hydrophobic RP-HPLC (reverse phase chromatography).* The hydrophobic surface of these packings interacts with the hydrophobic regions of the solute molecule. Separation of solutes is based on differences in the hydrophobicity of the molecules. There is an interaction of nonpolar region of protein with nonpolar *stationary phase*. The selected *mobile phase* has effect of displacing this interaction.

3. *Ion Exchange.* The chromatography medium possesses ionizable sites on its surface. These sites can be either anionic or cationic. Retention is based on the formation of ion pairs between the solute and the packing. This in turn is conditioned by the pH and ionic strength of the mobile phase.

4. *Affinity Chromatography.* Affinity chromatography refers to adsorption/elution of a solute with a specific ligand that is affixed to the packing surface. This can be a general class process (triazine dye ligand with nucleotide binding sites), or highly specific (a monoclonal antibody ligand with its corresponding antigen).

5. *Normal Phase Chromatography.* Normal phase chromatography utilizes a highly polar stationary phase and nonpolar (nonaqueous) solvents. Separation is based on differences in polar regions of solute molecules.

6. *Chiral.* Chiral chromatography exploits the energy differences and differential interaction of stereoisomeric molecular sites with ligands. Such ligands can either be immobilized on the packing surface or carried in the mobile phase.

In many situations it is unlikely that one will develop a complete, unambiguous isolation with a single chromatographic procedure. It is frequently possible to combine two (or more) chromatographic modes into a protocol that achieves homogeneity of isolation of a peptide species from a complex mixture. This is sometimes called "multidimensional chromatography." Say, for example, that the combination of anion exchange and hydrophobic chromatography is used. Neither method itself can achieve isolation of the desired solute. Because the methods are based on different separation principles, sets of solutes that coelute by one method can be well separated by the other. In general, the use of combinations of fundamentally different separation modes provides much more selectivity than combinations of similar modes.

3. Media

A chromatographic column is simply a cylindrical container filled with the chromatographic media. The end plates of the cylinder are porous, to permit flow of the mobile phase; and sufficiently rigid to withstand the mechanical loads imposed by fluid pressure within the column during operation. Design of these porous end members should be such that the column can maintain a uniform velocity profile of fluid flow through the stationary phase. Ideally one achieves "plug flow" (analogous to toothpaste as it is extruded from a tube), with no channeling of flow at the wall/ stationary phase interface.

Although the packing particles are in direct contact with one another, they nonetheless must form an open porous bed, with flow channels in the interstitial space between the particles. The liquid (mobile phase) in the bed is thus divided into two components: the *pore volume* and the *interstitial volume*. The pore volume is that liquid filled space within the solid matrix of the particle, and the interstitial volume is the liquid in the spaces between particles. Together these two volumes comprise the *void volume* of the column. This, of course, is the total volume of the column *(column volume),* less the volume taken up by the solid volume of the packing.

Particle size has a strong effect on packing resolving power. The lesser intraparticle and interparticle distances of small packings afford more effective mass transport, with a greater degree of solute/stationary phase interaction. Resolution of a separation can be increased with smaller particle sizes, albeit with higher column pressure drops.

Pore size and total porosity of rigid packings give rise to conflicting performance factors. Large molecules are more readily accommodated into the internal pore structure of large pore packings. Balanced against this, the larger diameter pores provide less total surface area, hence lower loading capacity. A large pore bead is mechanically weaker, and able to withstand less pressure drop.

The highest resolving power is achieved with nonporous packings, but such materials have considerably lower surface area and loading capacity than the porous beads.

The resolving power of a column is a function both of particle size and packing regularity. Irregular intraparticle flow paths and channeling at the column inner walls will produce more band spreading. Indeed the column packing techniques can have a marked effect on achievable resolving power. Bed homogeneity and resolving power is favored by monodisperse or uniform packing particle size. The resolving power is also favored by spherical particles. Both of these factors (for a given particle size) reduce the interstitial dimensions and promote more uniform hydraulic flow profiles through the bed.

3.1. Gels

Historically, the first chromatography media employed were various mineral granules and powders. As the resolving power of chromatography became recognized, researchers developed a broad array of synthesized media possessing tailored properties. Present day chromatographic media can be broadly divided into two classes; *gels,* or deformable media and rigid media. Within the rigid media it is useful to distinguish between inorganic materials of formation, such as silica, and polymeric media.

Porath and Flodin pioneered the synthesis of the first hydrogel packing materials. Their original work was with finely chopped water-swollen gels, which could be packed into open tubes, but they went on to develop spherical dextran beads. Dextran is a water soluble biopolymer. By incorporating crosslinking reagents into an aqueous dextran solution, and dispersing this solution in a nonmiscible organic phase it was possible to form spherical beads thart can be dried and reconstituted. The effective pore size of these structures is controlled by crosslink density between the polymer chains.

The original application of these materials was separation of various mol size solutes (*see* chromatography modes below). These highly swollen beads (void volume ~95%) can be derivitized with a variety of functional chemical groups. This allowed the development of separation modes such as ion exchange.

Presently, gel media are available that have various polymer matrices, including agarose, agarose-dextran, and polyacrylamides. Most of these materials have excellent biocompatibility with peptides and proteins.

A fundamental limitation of the gels is their soft deformable nature. Columns packed with these materials must operate at very low pressure drops, lest the beads be crushed and the packing bed collapse. Because of this limitation, flow velocity of a gel column is very low, and bead size must be comparatively large. This large particle size fundamentally limits the achievable resolution of the gel column.

3.2. Inorganic Media

By contrast, the inorganic media are completely rigid. A chromatographic column utilizing a silica-based packing can readily withstand pressure drops of several hundred bars without damage or deformation. This property led to the development of High Performance Liquid Chromatography (HPLC). Packing particle size in HPLC is on the order of a few microns. The small particle size permits much higher resolution than can be achieved with gels.

Most of the rigid inorganic packings are based on silica. Silicic acid gels are first formed. When such gels are fired the result is a highly porous structure, with very large internal surface area. The silica is crushed and classified. Alternately, one can produce spheroidal silica particles by dispersion of silicic acid gels in a nonmiscible phase. Such materials can be used directly in normal phase chromatography. By reacting silica particles with various organofunctional silanes, the surfaces of the particle can be modified to yield a variety of packings for the various LC modes discussed in this chapter.

Other inorganic media useful for chromatography include various glasses, alumina, and hydroxyapatite.

Ideally the derivitization of a packing should completely cover the base matrix surface, presenting the specific, desired chemistry to a solute molecule. In practice this is difficult to achieve, and there is some degree of undesirable nonspecific interaction of a solute with the highly polar silica matrix. Silica also possesses the Achilles' heel of a limited pH stability. Recently there has become available a family of modified silica packings that address this issue. These packings are formed by contacting the silica particles with a water soluble polymer such as dextran, which will adsorb onto the silica. The dextran can be permanently affixed by crosslinking. Alternately, one can prederivitize the surface of the silica with functional groups that can be covalently bonded to the dextran. A similar approach is to generate functional groups on the silica, and then to polymerize hydrogels (such as hydroxy polyacrylates) at these sites. With either the coating or the surface polymerization, the end result is a thin skin of hydrogel that completely masks the silica structure from biomolecule solutes. These hydrogel coatings are similarly capable of derivitization with a variety of functional chemistries.

3.3. Rigid Polymers

We are presently seeing the emergence of another type of rigid packing, based on nonswelling polymer structures. Typically, these rigid polymer beads are formed as spheroidal, macroreticular structures possessing considerable porosity and high surface areas. They are available with various surface chemistry modifications, to provide the major separation modes to the user. Such materials possess both the desired compatibility with biosolutes, and the desired physical properties of the silica HPLC packings. Base polymers include polystyrenes, methacrylates, and azlactams. Such rigid polymer packings will doubtless assume increasing importance in protein chromatography.

4. Chromatography Theory

The objective of a chromatographic procedure is the resolution of a mixture of components into the individual pure molecular species that can be identified, quantified, and/or collected for further study or use. Conceptually it is useful to understand the physical process of a chromatographic separation, and to be familiar with the measures of chromatographic performance.

Consider the idealized chromatographic column, a cylindrical packed bed. A constant rate of mobile phase flow is applied to the column. A sample of mixed solutes is next introduced onto the head of the column, ideally as a thin disk of liquid distributed across the head of the column.

As liquid flows through the column the solutes interact with the stationary phase and mobile phase. The result of this is differential retardation of the solute species. As individual solute species emerge from the column (as discrete bands or disks), they pass through a detector. The detector permits quantitation. The only qualitative identification information provided is the elution time of a solute, which must be evaluated against elution time of known standards.

The resolution of a chromatography system is characterized by three variables: efficiency, capacity factor, and selectivity. Efficiency is a physical/mechanical concept, while capacity factor and selectivity relate to chemical interactions.

4.1. Efficiency

Consider the nonidealized real world situation. The flow profile through a cross section of the column will not be completely uniform, and there may be some preferred flow or channeling at the packing/column wall interface. Inlet flow to the column emerges abruptly from a narrow bore inlet tube to the column cross section, and a similar situation holds at the discharge transition from the column. Despite best efforts there is bound to be some level of axial fluid mixing as mobile phase flows through injectors, detectors, and interconnecting plumbing. The result of this is that it is impossible to completely maintain plug flow of a solute through the chromatographic system. There will be some dispersion or mixing of solute along the axis of mobile phase flow.

Further, the flow through the interstitial volume of the packing is convective. Solute (and mobile phase) molecules will diffuse by Brownian motion into and out of the pores of the stationary phase as they travel through the bed. If there is no chemical/physical interaction of solute and stationary phase, there will be no retardation. The diffusive process is stochastic, however, and the inevitable result is an additional broadening of the solute band.

Next, consider the case of solute interaction with the stationary phase. When a solute molecule diffuses into the pore of a packing particle, it will spend a finite additional time within the packing before it is released. The result is a retardation, relative to mobile phase molecules having no stationary phase interaction. The interacting solute molecules will have a lower migration velocity, and will consequently appear to be spread out more as they emerge from the column and pass through the detector.

With proper attention to plumbing, dispersion can be minimized in the external elements, and the column will be the principal source of dispersion. The measure of this (lack of dispersion) is known as *efficiency*. As a first approximation, one can consider a solute peak to be Gaussian as it moves through a column.

Figure 1 depicts the detector response of a nonretained peak as it emerges from the column. If there were no dispersion it would look like a square wave, rather than the bell-shape shown. Using concepts borrowed from mass transport processes, such as distillation, one can characterize dispersion in terms of the *height of an*

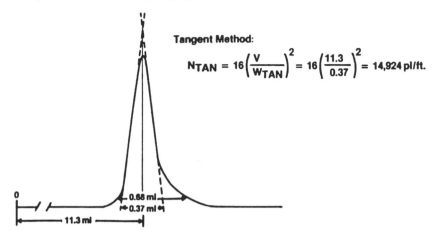

$$\text{Tangent Method:}$$

$$N_{TAN} = 16\left(\frac{V}{W_{TAN}}\right)^2 = 16\left(\frac{11.3}{0.37}\right)^2 = 14{,}924\,pl/ft.$$

Fig. 1. Calculation of plate height. Reprinted with the permission of MILLIPORE Corp.

equivalent theoretical plate (HETP). There are several ways used to arrive at this measure, and the numerical result is dependent on the method used. This example shows a calculation based on the Tangent method.

The elution volume of the peak is measured (injection to peak maximum). Tangents are constructed at the inflection points of the curve (approx one-half the height of the peak), and the width between the two tangent lines where they intersect the baseline is determined.

The ability of a column to resolve components of a sample depends directly on the efficiency of the column, and every effort is made to maximize the plate count. The single most important factor in achieving high efficiency is particle size of the packing. The smaller the particle size, the greater the number of solute/stationary phase interactions that can be achieved. Diffusion times into and out of the packing pores is minimized, resulting in less axial dispersion. As a rule of thumb, one can achieve a plate height of approx 2 particle diameters in a well-packed bed. HPLC chromatography derives its high resolving power from this fundamental fact. There is a price to be paid for high efficiency and that is a large pressure difference across the column. The pressure drop gener-

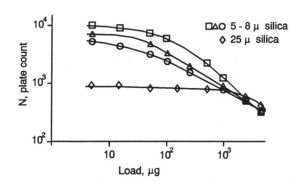

Fig. 2. Plate count as a function of solute load. Reprinted with the permission of MILLIPORE Corp.

ated by a given flow through a porous bed can be shown to be inversely related to the fourth power of particle size. As a consequence, the HPLC column operates at differential pressures of several thousand psi, and requires high strength packing materials that will not crush under the applied force loads. Packing particle diameters of 3–5 µ are commonly employed in high resolution analytical columns.

Although particle size is the chief determinant of efficiency, particle morphology is also a determinant. A monodisperse size distribution will tend to maximize the efficiency/pressure drop tradeoff. Spherical particles will generally yield somewhat higher efficiencies, in part because they will more readily pack in a column.

With any column, the achieved efficiencccy is a function of sample load. As load is increased, the equilibrium relationship of solute/mobile-phase/stationary-phase is perturbed. Solute molecules compete for available sites on the packing, with inevitable spreading of solute bands. Figure 2 shows the relationship of plate count as a function of particle size and solute loading for low molecular weight solutes. In general, HPLC column efficiency will start to degrade when mass loading is above 100 µgm/gm packing.

Still higher efficiencies can be achieved with pellicular packings. These are spherical, nonporous particles with an appropriate surface chemistry. This means that solute molecules contact sites on the particle in a direct, convective mode, without the additional dispersion generated by pore diffusion. For that matter, very high efficiencies can be obtained using narrow bore (<100 μ) capillary tubes as the stationary phase. The problem here is the limited surface available. HPLC silica based packings commonly have total internal surface areas of 50–100 meters^2/g. The much lower surface areas of pellicular columns or capillaries means that they have correspondingly lower sample capacities.

Bed geometry is also a way to influence efficiency. It is apparent that increasing bed length will result in a linear increase in plate count. For a given mobile phase flow, pressure drop across the column and residence time also increase linearly. Decreasing the particle size (as described above) generates a stronger influence on both efficiency and pressure. The practical result is that analytical HPLC columns are typically 10–25 cm in length, and utilize 3–10 μ packings. The practical length of gel columns is 3×–4× greater.

The other aspects of controlling efficiency are:

1. Column packing technique;
2. Column inlet/outlet flow distribution;
3. Downstream fluidics;
4. Mobile phase velocity; and
5. Sample load.

Essentially, these elements are concerned primarily with maintaining the separation efficiency inherent in the medium itself. The mechanical specifics of column design and fluidics control are dealt with in the literature of chromatography. Briefly, column packing methods strive to achieve a uniform bed of particles with a minimum interstitial volume, and an absence of voids. A well packed column will not exhibit significant channeling at the packing/column–cylinder interface. Rigid packing materials can be dry packed, or wet slurry packed; the latter being the mode of choice. Typically, the slurry is introduced on column at very high pressure (15,000–20,000 psi) and velocity. This energetic process results in a bed of maximum density. Gels and deformable media must be slurry packed, but this must perforce be done at low pressure. Care must be taken to degas the packing slurry liquid, lest air bubbles form voids in the gel column. When a bed has been formed, it is judiciously compressed, by using the inlet distribution plate as a piston. Maximum applied compression must not exceed the crush strength of the gel beads.

In both gel columns and rigid packing columns, precautions must be taken to avoid collapse of the formed bed structure. Any voiding of a column bed will virtually eradicate plate count. Any column will have a pressure maximum. Above that pressure, the forces imposed on the particles can crush them. Hydraulic surges on a column should similarly be avoided. With gel media columns, care must be taken to avoid gas bubble formation in the column in any operations subsequent to packing.

Plumbing upstream and downstream of a column should have minimum practical volume, and maintain streamline flow profiles through tubing, valves, detectors, and so on. Tubing diameter for HPLC analytical systems is typically ca. 0.1 mm. Volumes of elements such as detectors are kept as small as possible. Abrupt 90° flow direction changes are to be avoided if possible. The object of all these efforts is to avoid postcolumn fluid mixing, which destroys the spatial resolution of separated sample components.

The plate count of any column will be influenced by both sample load and mobile phase flow rate. At very low concentrations of applied sample, there is no effect on efficiency. When the load is increased beyond a critical concentration, the plate count starts to decline. Detector output will also show a change in peak shape. The eluted peaks become broader, but they also become asymmetric. Typically the leading edge of a peak will sharpen, whereas the trailing side of the peak will tail out. The high concentrations of solute molecules in these bands are responsible for these artifacts. Solute molecules are competing for a limited number of stationary phase sites, and the band of solute must expand to accommodate the solute molecule population.

Flow rate effects on efficiency are a function of the kinetics of mol diffusion. At very low mobile phase velocities, Brownian motion of solute molecules in the interstitial phase will have the effect of band spreading. As flow is increased, this effect will diminish (the solute molecules will have less time on-column to diffuse axially).

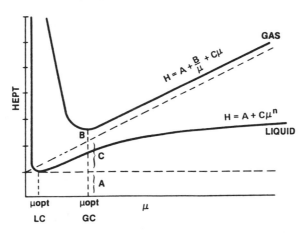

Fig. 3. Plot of HETP against fluid velocity. Reprinted with the permission of MILLIPORE Corp.

Opposing this is the effect of intrapore diffusion. Once a solute molecule enters a stationary phase pore, it ceases migrating down-column until the molecule reemerges into the interstitial fluid. The result of this is band spreading, and the higher the mobile phase flow rate, the greater the degree of spreading. Figure 3 depicts these effects. It is known as a Van Deempter plot, a technique originally developed to characterize gas chromatography columns. The inverse and direct flow rate dependence of these two diffusive effects generates a minimum plate height (or maximum efficiency) at a determined flow rate.

In liquid chromatography systems the mean free path of mol solute diffusion is an order of magnitude smaller than it would be in a gaseous system. As a result the axial dispersion component is much smaller in the liquid chromatography system. At practical liquid mobile phase flow rates one will not measure an increase in plate height with decreasing flow.

As would be expected, larger molecules will experience longer intrapore diffusion times. This results in a greater flow rate dependence for high molecular weight solutes. Figure 4 is a Van Deempter plot for various molecular weight species. Note the greater slopes of the high molecular weight species.

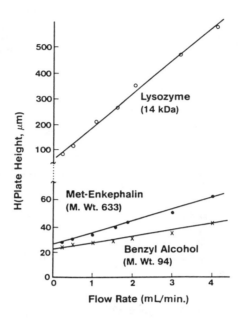

Fig. 4. HETP vs flow for various mol wt solutes. This figure is reprinted with permission of the authors and Eaton Publishing, from an article by Dong, M. W., Gant, J. R., and Larsen, B. R., entitled "Advances in fast reversed-phase chromatography of proteins," which appeared in *BioChromatography* **4**: 19–34, 1989.

4.2. Capacity Factor

The capacity factor is an expression of the partition coefficient of a solute between the stationary phase and the mobile phase. A solute molecule that is strongly bound by the stationary phase will spend more time adsorbed to that phase, than it will in association with mobile phase molecules. As a consequence, that solute will travel more slowly through the column than mobile phase molecules. Consider as an example the interaction of a peptide possessing some degree of hydrophobicity and a reverse phase packing. The peptide molecule, as it encounters stationary phase sites, adsorbs for a finite time. Molecular kinetics cause the molecule to adsorb, but the net effect is that the peptide travels more slowly down the column than the mobile phase molecules.

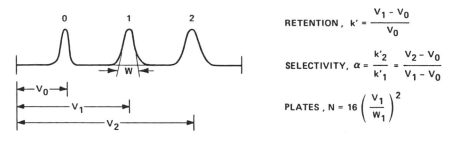

Fig. 5. Calculation of column capacity factor and selectivity. Reprinted with the permission of MILLIPORE Corp.

One can alter the capacity factor for a separation either by selection of the packing chemistry (increase or decrease the hydrophobicity of a reverse phase packing), or by selection of a mobile phase that will more (or less) readily displace the solute from the stationary phase.

A mixture of solute molecules will manifest a somewhat different capacity for each species. This leads to different migration velocities on column, hence separation. In general, increasing the capacity factor of the system will have a similar effect on all solute components. By increasing the capacity factor, elution time for all components is proportionally increased, and resolution of poorly separated components may be achieved.

Although the capacity factor is based on chemical equilibrium considerations, it is determined by measuring elution volumes of a solute.

Figure 5 depicts the determination of capacity factor (k'). In general, one would like a k' of 3–6 for a separation. Too low a capacity factor, and little separation can be expected, whereas too high a value leads to unreasonably long separation times and broad dilute peaks that elute. (*Note:* The width of high k' peaks does not imply axial dispersion. It is rather owing to the fact that the solute band is emerging from the column at a considerably slower velocity than the mobile phase. The detector output presentation is time-based and not spatially based, and so shows this as a very broad peak. The calculation of efficiency compensates for this in that it is based on the ratio of peak width to elution volume.)

In the section on efficiency it was noted that increasing solute load increases plate height. The k' of a solute is also effected by the sample load onto the column. As load is increased, the k' will decrease. Typically, for a HPLC column, k' starts to decrease once the load exceeds 100 µg/mL column volume (*see* Fig. 2). Comparable results are observed with gel media.

4.3. Selectivity

Selectivity is a companion concept to capacity factor. It refers to the relative retention volumes of two solutes (*see* Fig. 5, which depicts the calculation of the selectivity factor, or α) It is often possible to modify slightly the composition of a mobile phase so as to effect one of two components that are partially resolved. This could be achieved by adding a complexing reagent, modifying pH adding an ion pairing reagent to the mobile phase. It is a "fine tuning" of the mobile phase to optimize the separation of components.

4.4. Resolution

The factors of efficiency, capacity factor, and selectivity combine to express the resolution of a pair of solutes in a chromatographic system. Algebraic combination of N, k', and α yields the resolution equation:

$$R = \frac{\sqrt{N}}{4} \left(\frac{\alpha - 1}{\alpha} \right) \left(\frac{k'}{k' + 1} \right)$$

The physical and chemical parameters of a separation can be adjusted so as to effect the resolution.

Figure 6 conceptually illustrates the manipulation of each of these factors to effect resolution of a pair of eluitos.

It was noted that increased solute loads increased the HETP and decreased k'. Both of these trends have the effect of decreasing resolution as sample load is increased.

5. Isocratic and Gradient Separations

An isocratic separation refers to a separation where the mobile phase composition is invariant during the course of the run. There are a number of situations where varying the mobile phase com-

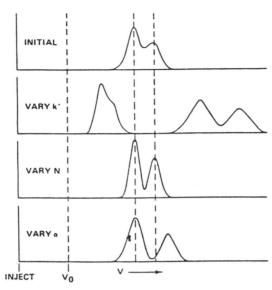

Fig. 6. Control of resolution by capacity factor, selectivity, and efficiency. Reprinted with the permission of MILLIPORE Corp.

position can increase the separation power of a system. One may, for example, load a sample on column under conditions such that certain components are strongly retained on the stationary phase at the column inlet, while other components pass through with little retention. Peptides eluted from an electrophoretic gel preparation are such an example. If one loads a reverse-phase column with such a sample and uses a mobile phase low in organic solvent, the peptide molecules will be strongly bound to the packing, and will concentrate as a saturated band at the head of the column. Electrophoretic buffer components will pass freely through the column. After passage of two or three column volumes of this mobile phase, the mobile phase is switched to one containing a higher content of organic solvent. This is referred to as a step gradient. During this step, the retained peptides can be resolved.

A step gradient is also used frequently at the end of a chromatographic separation. There may be some sample components that are strongly bound to the column. After separation of the components of interest, the mobile phase is switched to one that is a strong general eluant for all species. This step has the function of regenerating the column packing, sweeping off any strongly retained solutes.

Continuous gradients are employed to improve the resolution of poorly resolved peak groups, and to compress a separation of materials with widely varying capacity factors into a reasonable time frame. Increasing the k' of the early peaks (by decreasing the solvation power of the initial mobile phase) can effect a better separation of the early components. It also effects the later eluting components, with the result that the chromatographic run would be unduly long. The late eluting components might not even come off column.

With a gradient elution one continuously varies the mobile phase composition during the course of the separation. The solvation power of the mobile phase is low at the initial portion of the run. This increases the k' of the early eluates, allowing them to resolve. The increasing solvency of the mobile phase effects elution of the more strongly bound components.

Gradients are formed by use of mobile phase delivery systems that can mix two or more mobile phase solvents in varying proportions during the course of the separation. Early gradient systems used flow arrangements of solvent vessels in series to effect a continuous nonlinear change of composition as a run proceeded.

These arrangements have largely given way to multipump systems, with microprocessor-based flow proportioning, that can generate gradients of a variety of profiles, and with a high degree of repeatability.

6. Preparative Separations

There is a continuum of applications that range from the pure acquisition of information (analytical separations) to isolation and production of kilograms of material. Many of the preparative applications of chromatography encompass surprisingly small quantities of materials. Peptide sequencing can often be achieved with microgram quantities of pure material, and consequently can be obtained from the same equipment utilized for analytical separations.

There is a fundamental difference in the objective of an analytical and a preparative separation. In an analytical separation, one wishes to achieve complete resolution of all components, with accurate measurement of elution time and good quantitation of components. The objective of the preparative separation is to obtain a quantity of a component, or components, purified to some stipulated level.

These differing objectives generate different approaches to chromatography. In the analytical separation, the smallest practical packings are used. Load is kept as low as consistent with detector sensitivity. Flow rates of mobile phase are reduced to increase achievable column plate count. In the preparative separation, one maximizes the load and sacrifices resolution. As load is increased resolution is diminished, but it is still possible to obtain pure material by judicious fraction cutting of partially resolved peaks. Techniques of preparative and process scale chromatographic purification are addressed in elsewhere in this book. Suffice to say here that an analytical protocol can often be used to obtain milligram quantities of a solute by overloading a separation and taking a selected fraction cut. If this cut is not quite up to purity expectations, it is practical to rechromatograph the cut and shave off any "shoulder" from the resultant eluant. The much lower mass load of this single component will generally increase resolution enough to make this an expedient way to obtain adequate material for further procedures. The sequencing of peptide fragments from a protein digest is an example of such a micropreparative separation.

7. Modes

7.1. Gel Preparation Chromatography

GPC separations are effected solely on the basis of the effective molecular diameter of solutes. GPC packings are formed so as to have a controlled pore size. These can be either rigid structures, or crosslinked swollen gels, where pore dimensions are defined by crosslink density.

A chromatographic column is comprised of three elements of volume. There is the interstitial volume (V_i) which is the total volume of mobile phase that occupies the space between the particles of stationary phase.

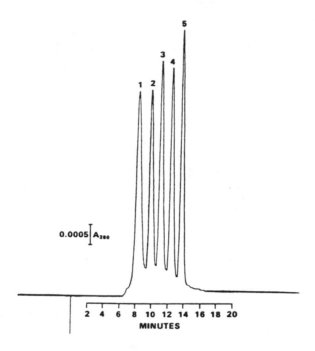

Fig. 7. Resolution of various mol wt proteins by GPC. Reprinted with the permission of MILLIPORE Corp.

The major fluid volume in a column consists of the fluid existing within the packing particles (V_p). Finally, there is the volume occupied by the packing material itself. Together V_i and V_p make up the total void volume (V_o) of the column. In a GPC packing the pore size distribution is carefully controlled. Molecules larger than the pore size cannot enter the pores, but remain in the interstitial volume, and elute from the column at V_i. Very small molecules, by contrast, are free to diffuse into the pore structure. Consequently, they move through the column more slowly, eluting at V_o.

Those molecules that are somewhat smaller than the effective pore size of the packing will have partial access to the pore volume, depending on their size and configuration; and this forms the basis of the GPC separation. Figure 7 depicts GPC column retention for various molecular weight proteins.

There are several limitations to GPC as an analytical tool for proteins. First, exclusion is a function of not only molecular weight, but also of steric factors. To obtain separation based on molecular weight, it is necessary to denature a protein, and eliminate disulfide bonds.

Second, all of the separation must occur within the range of $V_i < V_x < V_o$, where V_x is the elution volume of a particular protein. Achievable resolution is limited.

Finally, the range of molecular weights that a column can separate is constrained. All species larger than the pore size elute at V_i, and there is similar lack of separation of particles much smaller than the pore size. Fortunately, one can overcome this dynamic range limitation by placing GPC columns of graded pore size in series.

Notwithstanding the resolution limitations of GPC, it is frequently useful as a confirmatory technique that allows one to examine homogeneity and proximate mol size of a fraction obtained by another method.

7.2. Hydrophobic and Reverse Phase Chromatography

Chromatographic separation that exploits the hydrophobic character of the proteins has become the most widely used analytical chromatography mode. With the peptides and high molecular weight proteins, differences in the hydrophobic character of regions of the solute molecules form the basis of separation. In the case of amino acids, one can prereact the sample to form derivatives that have hydrophobic properties.

The factors that have driven the use of hydrophobic (reverse phase) separation are; high resolving power, and a broad application range.

Reverse phase packings have hydrophobic hydrocarbon groups bonded to the packing surface. The hydrophobic regions of a polypeptide will interact with the packing and adsorb to this surface. Elution results from the competition between the mobile phase and the stationary phase for the polypeptide. The more polar of the polypeptides will be readily eluted from a reverse phase column, whereas the less polar molecules will elute more slowly. Reverse phase chromatography exploits the small differences in hydrophobicity of the solutes. It achieves resolution by virtue of the thousands of interactions of solute molecules with the packing as they pass through the column.

There are a variety of reverse phase packings that are available, and possess varying degrees of hydrophobicity. The C18, or octadecyl silane bonded packings, are quite hydrophobic. These long chain groups project from the packing surface, presenting and oil-like hydrophobic surface. A variety of shorter chain and phenyl group packings provide a spectrum of hydrophobicity. The wide selection of reverse phase packings makes it possible to address a broad spectrum of solutes.

The packing substrate for reverse phase packings is silica. This is a highly polar material, and exposed silica surface would compound the hydrophobic adsorption mechanism of separation by binding the polar regions of a solute. In order to avoid such a "mixed mode" behavior, the packing manufacturer first bonds the functional hydrophobic group to the packing, via a silane chemistry. Following this, a second methyl silane reaction (called capping) ensues. The function of the capping silane is to occupy any exposed silanol surfaces and eliminate undesired polar surface characteristics of the packing.

Reverse phase packings are available in a range of particle sizes and pore sizes. Generally, the smaller pore sizes are used for low molecular weight solutes. These packings possess more capacity (owing to a higher surface area per volume of packing) and higher mechanical strength. Larger proteins may be blocked from access to small pore surfaces, and so packings of pore size >100 Å are recommended for large proteins.

A reverse phase separation results from a competition of the stationary phase and the mobile phase for the solute molecules. The mobile phase is an aqueous system, modified with miscible organics, typically acetonitrile, and/or low molecular weight alcohols. Increasing the hydrophobicity of the stationary phase (i.e., packing selection) has the effect of increasing solute retention. Increased organic content of the mobile phase has the opposite effect. The availability of a range of hydrophobic packings coupled with the ability to markedly alter the polarity of the mobile phase allows reverse phase separations to address a broad array of solutes.

The most frequently used mobile phases for RP-HPLC are acetonitrile/water systems. Phosphate buffers are generally employed to control pH and increase retention. Phosphate has the

Table 1
Capacity Factors with Different Ion Pairing Systems

Peptide	HPO$_4$	Hexyl sulfonate	Camphor-10 sulfonate	Dodecyl sulfate	Dodecyl ammonium phosphate	Tetrapropyl ammonium phosphate
RF	0.20	0.46	0.37	8.8	0.18	0.22
RFA	0.22	0.37	0.30	8.2	0.18	0.24
MRF	0.48	0.71	0.60	>16	0.32	0.24
MRFA	0.28	0.58	0.42	13.0	0.22	0.22
LW	0.48	0.70	0.72	-	0.20	0.25
LWM	0.77	0.92	0.81	-	0.18	0.28
LWNR	0.34	0.69	0.52	3.8	0.35	0.60
LWMRF	1.69	2.81	2.04	11.0	0.90	1.40

A = alanine, F = phenylalanine, L = leucine, M = methionine, R = arginine, W = tryptophan.

benefit of being non-UV adsorbing, and is thus compatible with the UV detectors commonly used. Volatile buffer systems (acetic acid, trifluoro acetic acid, triethyl amine) are used where fraction recovery is required.

Ion-pairing reagents can be used to adjust selectivity between specific solutes. These reagents interact at ionizing sites on the peptide (amino groups and/or carboxyls), permitting subtle adjustment of solute polarity. Table 1 above shows the effect on capacity factor of small peptide by various ion pairing reagents.

Ion pairing can often be used to resolve diastereomers. The pairing generates difference is presented hydrophobic surfaces, to form the basis of resolution.

It is often the case that an isocratic separation fails to resolve all components of a sample. Conditions required to resolve early eluting components may fail completely to elute strongly bound species. The use of a stronger solvent system will result in the early eluting solutes coming off in the void volume. In this situation gradients are used to resolve these complex samples. The shift in solvent composition profoundly effects the solute/mobile phase/stationary phase equilibria. This allows one to maximize resolution of individual solutes, and at the same time increases the dynamic range of separation. Figure 8 shows reverse phase separations of a

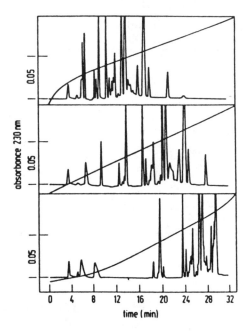

Fig. 8. Use of gradient profile to control resolution of hemoglobin A tryptic digest.

tryptic digest of hemoglobin A. The continuous curves plot the profile of the organic phase concentration. Note that with the convex-upward profile, the early eluters are inadequately resolved. Late eluting components are well resolved. The opposite situation obtains with the concave-upward profile.

For complex samples that are difficult to resolve, gradients are generally used. In most situations, it in not necessary to resolve all components, and one can tailor a gradient to maximize resolution of specific components. Complex samples will frequently employ a combination of step gradients (for components not of interest) with a shallow linear gradient designed to maximize resolution in a specific region of interest. Such gradient systems are routinely used in separation of amino acid derivatives, protein digests, peptide synthesis products, and so on. where the sample is complex and contains many similar components.

Reverse phase chromatography is frequently used as a preparative tool for peptide fractions. If a peptide has been isolated by HPLC or by another technique, such as ion chromatography or electrophoresis, it may be necessary to remove salts, glycine, or SDS. Rather than attempting dialysis on small samples, one can use an analytical RP-HPLC to effect a simple and rapid cleanup. The fraction is loaded onto the column in an organic-poor solvent, such that it binds strongly to the hydrophobic packing. The bound column can be washed with several volumes of mobile phase to remove undesired components. Following this, the peptide is eluted with a step-gradient of organic rich mobile phase. This has the effect of frontal displacement of the peptide, which elutes as a sharp peak of high concentration. Most often the same column that is used for isolation of a fraction is used for this cleanup technique.

7.3. Hydrophobic Interaction Chromatography

A variant of reverse phase chromatography is known as hydrophobic interaction. Reverse phase packings are used, but the mobile phase organic modifier is replaced with a salt. Retention in hydrophobic interaction is inversely related to the salt concentration, and the salt species can be used to control the selectivity.

7.4. Ion Exchange Chromatography

Ion exchange chromatography is based on the formation/dissolution of ion pair bonds between the stationary phase and the solute. The polypeptides contain both acidic and basic sites, and so can be chromatographed on both anionic and cationic packings. In most ion chromatography procedures, a sample is adsorbed onto the column, and then eluted by means of a gradient that progressively decreases charge interaction. This is achieved with pH adjustment and/or increasing salt concentration. An acid solute is eluted by decreasing pH, whereas a basic solute elutes with increasing pH.

There are a variety of ion exchange packing functionalities that are employed. Ion exchangers are categorized into strong acid/base and weak acid/base packings. Examples of strong exchangers would be a quaternary amine anion exchanger or a sulfonate cation exchanger. Secondary amines and carboxylic acids typify the corresponding weak exchangers. In general one chromatographs a strong acid/base solute with a weak exchanger and vice versa.

In developing a separation it is most helpful to have some knowledge of the titration curve (molecule net charge vs pH) of the solutes of interest. One can then select a pH range for separation that exploits the greatest difference of charge of the solutes.

Elution of solutes is effected both by changes in pH and in ionic strength of the mobile phase. Increasing ionic strength decreases the effective hydrated ionic radius, thus favoring elution of the solute. Elution is influenced not only by molarity of the mobile phase, but also by the ionic species. The retention sequence for anion exchange resins is as follows:

$$\text{Citrate} > SO_4^{-2} > \text{Oxalate} > I^- > NO_3^- > CrO_4^{-2} > Br^- > SCN^- > Cl^- >$$
$$\text{Formate} > \text{Acetate} > OH^{-4} > F^-$$

Cation exchangers exhibit the following retention sequence:

$$Ba^{+2} > Pb^{+2} > Sr^{+2} > Ca^{+2} > Ni^{+2} > Cd^{+2} > Cu^{+2} > Co^{+2} > Zn^{+2} > Mg^{+2} >$$
$$UO^{+2} > Tl^1 > Ag^+ > Cs^+ > Rb^+ > K^+ > NH_4^+ > Na^+ > H^+ > Li^+$$

The addition to the mobile phase of organic additives, and materials that can generate hydrophobic interactions or polar attractions will also influence a separation.

Figure 9 shows an anion exchange separation where all factors were held constant, save for the anionic species in the mobile phase. Note the marked shifts in selectivity factors of these proteins with the differing anions.

In some situations, the desired resolution of components is achieved by altering both the pH and the ionic content during gradient elution.

Figure 10 shows the chromatography of a monoclonal antibody preparation, first by a change (decrease) in pH with a binary gradient. Note the increased resolution achieved with a ternary

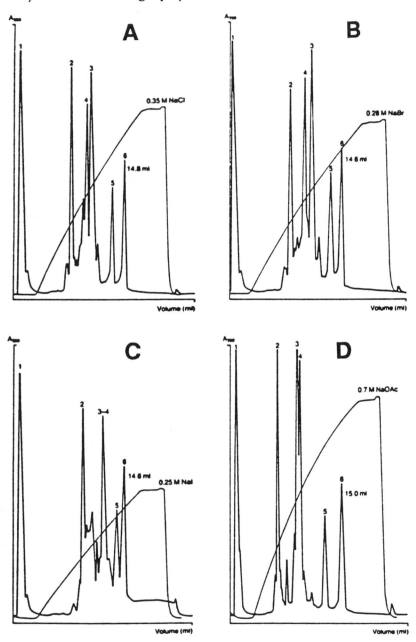

Fig. 9. Effect of varying the counter-ion species in ion chromatography.

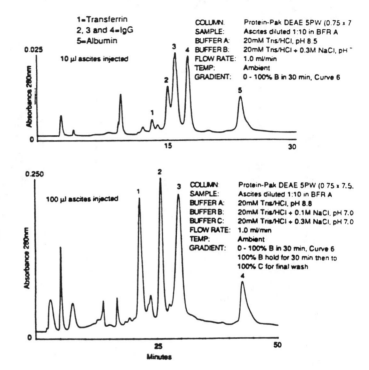

Fig. 10. Use of simultaneous pH and salt concentration dual gradients to effect resolution of a monoclonal antibody. Reprinted with the permission of MILLIPORE Corp.

elution system that simultaneously decreases pH while increasing the ionic strength.

Ion chromatography is a powerful tool for the resolution of the polypeptides, and can be used with undenatured species. Although the technique rests on simple fundamental principles, the design of a chromatography protocol can be complex.

7.5. Affinity Chromatography

Affinity chromatography is essentially a fixed bed adsorption/ desorption process that is controlled by the selectivity of the stationary phase. Most of the concepts of control of chromatographic resolution have little applicability to an affinity process. An affinity matrix consists of a stationary phase to which is affixed a ligand that binds specifically to some class of molecules. As such, affinity separations can be highly specific; or directed at broad classes of solutes. A monoclonal antibody affixed to a packing is an example of a highly specific affinity matrix, whereas a packing with Cibacron Blue dye attached binds to the broad group of proteins that contain nucleotide binding sites. The list below lists some affinity ligand systems commonly used.

Affinity Ligands

Ligand	*Group Specificity*
Cibacron Blue (triazine)	Nucleotide binding sites
p-aminobenzamidine	Trypsin family proteases
m-aminophenylboronic acid	Glycoproteins
Lectins	Glycoproteins, IgM
Protein A	Immunoglobulins
Protein G	Immunoglobulins
Heparin	Lipoproteins, lipases coagulation factors
Enzyme substrate analogs	Enzyme specific
Cofactors	Enzyme specific
Antigen	Antibody specific
Antibody	Antigen specific
Hydroxyapatite	Immunoglobulins

In an affinity purification, one slowly loads the sample onto the affinity column, adsorbing the specific substrate to the bound ligand. Load amount is determined by the binding capacity (mg/mL of column volume) of ligand on the column. After loading the column is washed with several volumes of a buffer or solution that will remove nonligated species. The column is then charged with an elution buffer that can disrupt the ligand–substrate bond, and the substrate is recovered in the eluant.

Note that affinity chromatography is also used to strip an undesired component from a preparation. Components isolated from the blood of the *Limulus* crab have the ability to bind lipopolysaccharide endotoxins. If this *Limulus* material is immobilized to a packing, this material can be used to effectively strip endo-toxins from a preparation. This can be a very useful technique to apply to biologics and recombinant products.

Affinity matrices include gel beads, various rigid packing materials, and high surface microporous membranes to which specific ligands have been covalently bound. A number of ligand matrices (triazine dyes, protein A, hydroxyapatite) are commercially available, but is quite common for the scientist to prepare his/her own affinity matrix by covalent bonding of a ligand to a solid substrate. Table 1 characterizes some of the linking chemistries frequently employed (2).

The various commercial suppliers provide a number of preactivated materials with which one can prepare a specific affinity matrix. Some of these preactivated matrices embody "spacer arms" as part of the linker chemistry. Spacer arms are helpful in some situations where a ligand must bind a high molecular weight substrate. Steric factors may prohibit adequate ligand-substrate binding, if a ligand is directly affixed to a solid surface. This problem is sometimes reuced if the ligand is coupled to a solid surface via a flexible carbon–carbon chain. Typical chain length is 8–12 carbon atoms (longer chains apparently permit folding of the ligand back to the solid surface, thus limiting accessibility) (Table 2).

Affinity purification is conceptually simple, and is widely applied as a preparative tool in purification of milligram quantities of materials. To date affinity chromatography has seen limited application as a process purification tool. The chapter on large scale processing methods details some of the complexities and limitations of large scale affinity purification.

7.6. Normal Phase Chromatography

The original chromatographic separations were performed on what is today called normal phase packing materials. The normal phase separation is the mirror image of reverse phase. The packing presents a highly polar surface, and the mobile phase is a nonpolar organic solvent. Separation is based on differences in the polar regions of solute molecules. The common normal phase packings are activated silica.

Table 2
Activating Procedures in Affinity Chromatography

Activating agent	Reagent toxicity	Activation time, h	Ligand coupling time, h	pH of coupling	Type of linkage	Stability of complex	Nonspecific interactions
Glutaraldehyde	Moderate	1-8	6-16	6.5-85	Michael's adduct Schiff's base	Good	–
CNBr	High	0.2-0.4 overnight at 4°C	2-4 h at 25-C, overnight	8-10.0	Isourea, imido, carbamate, N-substituted carbamate	Unstable at pH <5 or >10	Cationic
Bisoxiranes	Moderate	5-18	15-48	8.5-12.0	Alkylamide, ether, thioether	Excellent	–
DVS	High	0.5-2.0	Rapid	8-10.0	Michael's adduct	Unstable at high pH	–
CDI	Moderate	0.2-0.4 to 6 d	Overnight	8-9.5	N-substituted carbamate	Unstable at pH >10	–
Periodate	Nontoxic	14-20	Overnight	7.5-8.5	Alkylamine	Good	–
Trichloro-S-triazine	High	0.5-2.0	4-16	7.5-9.0	Triazinyl	Good	π-π type aromatic
Tresyl chloride	Moderate	0.5-0.8	Rapid	7.5-10.5	Alkylamine	Good	–
Diazonium	Moderate	–	0.5-1.0	6-8	Azo	Moderate	π-π type aromatic
Hydrazine	High	1-3	3-16	7-9	Amide	Excellent	–
NHS	–	2-4	Rapid	6.5-8.0	Amide	Hydrolysis, pH sensidve	–
4-Nitrophenyl chloroformate	–	0.5	Rapid to 2 d	–	Mixed anhydride, carbamatel urethane	Hydrolysis, pH sensitive	–

Normal phase separations are no longer widely used in protein chemistry. The nature of the mobile phases used is such that many samples are simply not soluble. There is much less control of capacity factor than one finds with the range of available reverse phase packings. At present normal phase separations are confined mainly to preoperative separations of synthetic peptides.

7.7. Chiral Chromatography

The synthesis and modification of peptides has created a need for chromatographic techniques that can resolve chiral variants of a molecule's structure. That class of stereoisomers that are "mirror image" molecules possess identical energy content for the isomeric forms. Because of this, these structures have virtually similar chemical and physical properties, and do not lend themselves to traditional chromatographic separation. Conformational and configurational stereoisomers, by contrast, may separate in traditional modes of chromatography.

A considerable body of methods has developed for the chromatographic separation of stereoisomers, with an extensive literature and comprehensive summaries (3). In general there are three approaches to developing a separation of chiral materials. In one fashion or another, they all address the formation of differential transient diastereomers between the solutes and the stationary phase to effect separation.

1. Chemically convert diastereomers by covalent chemical reaction with a reagent. The resulting reaction products will manifest different configurations, and can be separated on achiral stationary phases.
2. Perform chromatography on a stationary phase that has chiral selectivity.
3. Use a mobile phase that has chiral selectivity, with an achiral stationary phase.

Currently, there are many chiral columns commercially available, and the literature describes literally hundreds of mobile phase chiral additives for enantiomer separations. There is a like body of reagent systems to choose from for the conversion of enantiomers to diastereomers for HPLC analysis.

The table below provides an example of separation of some d,l α-amino acids on a C-18 column that as been coated with 1-hydroxy proline *(4)*. The eluant is a 10^{-8} *M* Cu(II) acetate in methanol/water 15/85 (v/v). Amino acids form complexes with divalent metal ions such as Cu, Zn, Ni, and so on. The competitive association of the solutes and the proline form the basis of chiral separation here.

Chiral Separation on a Proline Coated C-18 Column			
Amino acid	k_1'	k_d'	α
Alanine	0.58	0.91	1.56
Aspartic acid	0.08	0.10	1.17
Glutamic acid	0.13	0.17	1.33
Histidine	0.97	0.57	0.59
Serine	0.73	0.73	1.00
Threonine	0.83	0.85	1.02
Valine	1.63	7.42	4.55

More often than not, a chiral separation system is specific for a limited class of solutes. Some of the chiral separation systems are amenable to high loading (providing one has a sufficient α). In this situation it is feasible to do preparative separations.

8. Applications

There are many applications of chromatography of amino acids and polypeptides. The techniques used and problems encountered are a function of mol size, and is so divided in this chapter.

8.1. Chromatography of Amino Acids

Applications of amino acid analysis include compositional analysis of proteins, food analysis for amino acids content, clinical applications, peptide synthesis, and sequence analysis of proteins. In sequence analysis the individual amino acids are chromatographically identified as they are cleaved from the polypeptide being sequenced. Amino acid analysis of the unsequenced peptide fragment is also quite important, as the amino acid compositional analysis is a confirmatory check on the sequence data.

Some of the earlier methods of amino acid analysis include the use of paper electrophoresis (5), paper chromatography (6,7), and thin layer chromatography. These techniques, although providing satisfactory resolution, are slow and labor intensive. Gas chromatography is an option, but necessitates the preparation of volatile, stable derivatives of all the amino acids. Inasmuch as direct detection of low concentrations of amino acids in unfeasible, virtually all chromatographic methods require pre- or post-column derivitization. The combination of gas chromatography coupled with mass spectroscopy provides both excellent sensitivity and direct component identification.

At present rapid automated liquid chromatographic methods hold sway. The first generation amino acid analyzers were ion exchanges chromatography systems for the separation step. The first systems (8) were based on ion exchange chromatography and required 24 h/cycle, but subsequent improvements made dramatic reduction in analysis times. Various post-column derivatization chemistries were introduced, and these were integrated into automated systems that greatly reduced the direct labor of analysis. Some of the early systems utilized the Ninhydrin reaction to place a chromophore on column eluants. This procedure gave way to the use of fluorescent reagents such as fluorescamine and orthophthalaldehyde (OPA) (see Fig. 11). Integrated chromatography-reaction systems provided accurate, well resolved amino acid separations in a 1–2 h time frame.

Notwithstanding, ion exchange chromatography for amino acid analysis presents several limitations. Post-column derivatization equipment adds to system complexity and reduces resolving power. A fluor such as OPA requires that proline be first converted to a primary amine.

These amino acid analyzers have given way to methods that prederivitize the amino acids, and utilize reverse-phase HPLC. The derivitization chemistries provide two functions; they provide chromophores for UV and/or fluorescent detection of species, and they reduce the highly polar nature of the amino acids, and allow separation by reverse-phase chromatography.

There are a number of derivatization chemistries that can be successfully employed. The most common derivatization utilizes

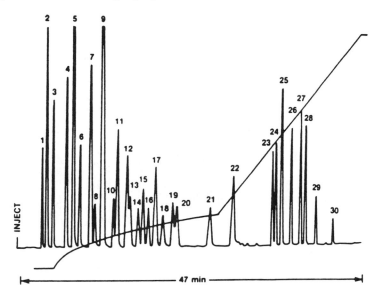

Fig. 11. Separation of OPA derivatives of an amino acid mixture. Reprinted with the permission of MILLIPORE Corp.

phenylisothiocyanate (PITC), the reagent used in the Edman method of peptide sequencing.

In sequencing, PITC is coupled with the peptide to form a N-terminal phenylthiocarbamyl (PTC) peptide. This is then cleaved with TFA to yield an anilothiazolinone (AZT) amino acid residue. The cleaved derivative is unstable, and is converted to the more stable phenylthiohydantoin (PTH) derivative using aqueous or methanolic HCl.

The complete reaction sequence in Edman sequencing entails reagent coupling, residue cleavage, and hydrolysis. When preparing PTH derivatives of free amino acids, the cleavage chemistry is not needed. If chromatography takes place immediately prior to derivitization with PITC, the PITC derivative amino acids can be separated and detected in the HPLC system. Separation can be achieved in less than 10 min, and detection sensitivity is such that a few picomoles of an amino acid can be measured.

Figure 12 shows the chromatography of a PITC derivitized amino acid mixture. Chromatography can be carried out on a vari-

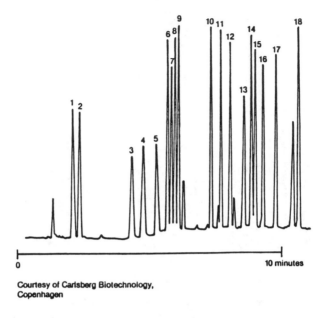

Courtesy of Carlsberg Biotechnology,
Copenhagen

Fig. 12. PITC derivatized amino acids. 1. ASP; 2. GLU; 3. SER; 4. GLY; 5. HIS; 6. ARG; 7. THR; 8. ALA; 9. PRO; 10. TYR; 11. VAL; 12. MET; 13. CYS; 14. ILE; 15. LEU; 16. NLE; 17. PHE; 18. LYS. Reprinted with the permission of MILLIPORE Corp.

ety of reverse phase packings. The mobile phase system is commonly a buffered acetonitrile gradient. Because of the need for high volume amino acid analysis that is entailed in protein sequencing, the suppliers have developed highly automated derivitization and chromatography equipment for this application.

8.2. Chromatography of Peptides

Polypeptides are amenable to virtually the whole range of chromatographic modes. In fact chromatography of peptides is not simply an analytical tool; it is one of the primary separation techniques used for preparative and production processing applications of peptides.

Applications include the analysis of fragments from protein digests and isolation of such fragments in preparation for sequencing. Present day sequencer mass requirements are so small that analytical column isolates provide adequate amounts of peptide for sequencing.

In many instances, HPLC is used in conjunction with electrophoresis to isolate a peptide for sequence analysis. If a fraction is excised from an electrophoresis gel, it is necessary to remove glycine and other electrophoretic reagents prior to sequencing. This can be done by first separating peptide from the electrophoresis gel (see elsewhere in this volume for various methods of isolation). The eluted peptide is then adsorbed onto a reverse phase column, and washed with several volumes of low organic content mobile phase. he washed peptide can then be eluted with a step gradient high of high organic phase concentration.

Chromatography is routinely used in the analysis of therapeutic peptides produced by bulk chemical synthesis, extraction of natural source materials, and recombinant expressed peptides.

The peptides are amenable to separation by most chromatography modes. When using HPLC packings it is preferable to utilize large-pore size packings (300Å) for high molecular weight peptides.

The most commonly employed mode for the peptides is reverse-phase. RP-HPLC provides excellent resolution, and is capable of high loads when a fraction is to be isolated. Mobile phase systems commonly are buffer/acetonitrile, or buffer/alcohol (methanol, ethanol, propanol). Peptides and proteins are soluble in relatively high concentrations of a buffer/acetonitrile system.

Peptide chromatography of protein digests yields a mixture of many components, as does the resolution of peptide synthesis products. The k' of peptide separations is quite sensitive to slight changes in mobile phase organic content, a few percent shift can cause major changes of k'. Gradient separations are generally developed to optimize component separation, with the organic phase concentration being increased by only 10–20% over the course of the separation.

Phosphate buffers are commonly used for peptide separations. Retention is generally increased at a pH below the pI of the peptide. Where fractions are to be recovered, volatile buffers such as trifluoroacetic acid, triethylamine, and triethylammonium acetate are employed.

Both retention time and selectivity are adjusted by the molarity and solvating strength of buffer salts employed. Ion pairing can also be employed to control the selectivity of a separation.

Chiral chromatography proves useful, particularly for synthetic peptides where racemic amino acids might be included into the peptide. Diastereomers are frequently resolvable because of differences in the presented hydrophobic regions of the molecules. The range of chiral derivatizations, columns, and mobile phase chemistries mentioned earlier are applied to the resolution of small peptides. As molecular weight (and the number of chiral centers) increases, chiral chromatography becomes less valuable.

8.3. Chromatography
of High Molecular Weight Proteins

This chapter arbitrarily defines a molecular weight region of 20–40 kDa as the boundary between a peptide and a protein. As the mol size of the poypeptide chain is increased, other factors come into play, which influence separation characteristics.

Chromatography of high molecular weight proteins poses particular issues. One may wish to retain the native state of the protein, rather than a denatured structure. This places some limits on the constituents and composition of mobile phases. Chromatography involving the hydrophobic sites of a protein is more difficult with high mol weight proteins. Because of the multiple hydrophobic regions of the proteins, they prove difficult to elute from a hydrophobic packing, to the extent that the elution conditions may completely denature the protein. The organic phase strength required may lie outside the solubility range of the protein. In such a situation one must use a less hydrophobic packing. Hydrophobic interaction chromatography is often employed. Here one uses a weakly hydrophobic packing surface, and controls the separation by altering the solubilizing power of the mobile phase. Typically high concentrations of salt are incorporated into the mobile phase.

In contrast to the peptides, where reverse-phase separations are predominant, large proteins are most frequently resolved by ion exchange columns. The packings traditionally used are the gels. They possess pore sizes able to aaccommodate the large proteins. The gels can be derivitized with a broad range of binding site chemistries. Because these are hydrogels, there is a minimum of nonspecific binding to the protein solutes. The gels will tolerate a broader pH range than the silica based packings, extending the range of mobile phase compositions available, and permitting more vigorous column cleaning/regeneration agents.

The original dextran-based gels have distinct operating limitations. These gels are soft and readily deformed. Consequently, columns must be operated at very low pressure differentials (to avoid bed collapse) and flow velocities. The long separation times can pose stability problems when separating labile proteins, and it may be necessary to perform chromatography in a cold room. The need to maintain low pressure differentials in turn makes impractical the use of small diameter packing particles with their higher inherent efficiencies.

The limitations of soft gels have been addressed with several improvements in packing technology. Higher modulus gel materials have been developed. These gels are based on materials such as agarose, acrylamides, and combinations thereof. The control of crosslink density in such gels often permits an increase in modulus, while still retaining sufficient permeability.

Silica-based packings are now available, where the porous silica particles have a coating, a few angstroms thick, of a hydrogel. This coating covers all surfaces of the porous structure, with the result that a protein moecule interacts with a gel surface similar in character to that of the gel packings. These packings do not possess the pH range stability of the gels, but they do operate over a broader pH range than conventional silane derivitized silica. The coating methods employed include adsorption coating with soluble polymers such as dextran, covalent coupling of such polymers, and *in situ* polymerization of branched hydrogels such as polyacrylates, on the silica.

More recently, a variety of rigid high strength polymeric packings have come into use. Several years ago, Pharmacia introduced their FPLC™ packings, based on rigid small diameter particles. These were made available in different surface chemistries, and provided HPLC resolution and speed to the modalities provided by gel chromatography. Presently there are a number of macroreticular packings available made of rigid polymer materials. These include styrenes, methacrylates, cellulosic derivatives, and other proprietary polymer systems. Columns packed with these materials are able to operate in the HPLC domain of several thousand psi. Porous carbon packings are also available. Protein chromatographic methods will be strongly influenced by the availability of these HPLC materials.

High resolution protein chromatography has given rise to new chromatography equipment. The soft gel apparatus used glass and plastics as the contact surfaces for solutes and mobile phases. HPLC equipment by contrast used stainless steel for wetted surfaces. This posed two issues: one was a concern that proteins might be inactivated by contact with metal surfaces. More significantly, the stainless steel has poor resistance to the salts (notably chlorides) used in protein mobile phases. As a consequence, HPLC apparatus has been developed for protein separations in which all stainless steel wetted surfaces have been replaced by polymers and/or ceramics.

The table below illustrates retention time of various proteins as a function of their nonpolar amino acid content (9).

Protein	Retention time, min	Hydrophobic amino acids
Lactalbumin	∞	35.8%
Elastase	∞	0 32.5
Myoglobin	45.0	31.3%
Lysozyme	37.5	26.4%
Cytochrome C	35.0	25.0%
Ribonuclease	27.5	21.8%
Neurotoxin 3	24.0	21.1%

As stated above, ion exchange chromatography is the most commonly used mode for protein separations. The advent of the high resolution packings described above has increased IEC utility.

Mobile phase systems must be designed so as to maintain the solubility of the proteins. Zwitterions may be incorporated to reduce solute mol interaction and aggregation. This is commonly used with the glycoproteins. Detergents can increase protein solubility. For IEC, nonionics can be used. If ionic detergents are to be used, they should match the charge sign of the stationary phase. Adjust surfactant concentration to avoid micelle formation.

The requirement for aqueous mobile phases precludes the use of normal phase chromatography. Reverse phase separations are often employed on proteins solubilized by denaturing agents, with hydrophobic interaction chromatography being employed for native proteins.

GPC is frequently employed: Although it has a limited resolution capacity, it provides an independent estimate of molecular weight of a fraction, and is often used as a confirmatory method, or as a check for homogeneity. Another common use of GPC is in the desalting of a protein fraction.

Affinity chromatography is very often employed with proteins, as an isolation or preparative tool for milligram quantities. There are various affinity packings applied in protein isolation (see table in Section 7.5.).

Protein A specifically binds immunoglobulin IgG, and is regularly used for antibody isolation. There are a number of Protein A packings and columns commercially available, as well as membrane-based affinity harvesting devices.

More commonly one will need to prepare an affinity column specific to a particular solute. There are a range of affinity packing substrates available that possess linking chemistries. One can covalently bind a specific ligand to one of these materials, pack a small bed with the derivitized packing, and achieve an affinity based isolation of a protein.

Immunoproteins are one of the frequently used applications of protein chromatography. IEC and affinity are regularly employed, there have been some specialized chromatography media developed for immunochromatography. The mineral hydroxyapatite exhibits some specificity for the immunoglobulins, and is used as a packing to isolate/purify them. A packing, whose trade name is Bakerbond ABx™ has been developed for antibody purification. It operates in a "mixed-mode" fashion by a combination of anion and cation exchange sites, coupled with hydrophobic interaction between solute and packing. In most chromatography applications one avoids mixed-mode separations. They render useless principals that guide the development of a protocol. In this defined antibody separation, however, the particular properties of the packing re optimized for a singular application. The indicator dyes frequently used in monoclonal antibody cell culture reactors interfere with proper separation on IEC packings, but do not so effect the ABx separation.

Bibliography

1. Horvath, C. (1983), *High Performance Liquid Chromatography,* vol. 3, Academic, p. 120.
2. *Affinity Chromatography Supports* Narayanan, Sunanda R., Crane, Laura, and Tibtech, A., eds., vol. 8, (Jan 90) , pp. 12–16.
3. *Chromatographic Separations of Stereoisomers* (1985), Souter, R . W ., ed., CRC, Boca Raton, FL.
4. Davankov, V. A., Bochkov, A. S., Kurgahov, A. A., Roumeliotis, P., and Unger, K. K. (1980), *Chromatographia* **13,** 677.
5. Gross, D. (1955), *Nature* **184,** 1298.
6. Consdon, R., Gordon, A. H., and Martin, J. P. (1944), *Biochem. J.* **8,** 224 .
7. Dent, C. E. (1948), *Biochem. J.* **43,** 169.
8. Spackman, D. H., Stein, W. H., and Moore, S. (1958), *Anal. Chem.* **31,** 1190.
9. Horvath, C. High Performance Liquid Chromatography, vol. 3, Academic, p. 55.

Four

Internal Structure and Organization

Relationship to Function

Felix Franks

1. Structural Hierarchy of Proteins

There are several levels of protein structure and different types of interactions that maintain such structures. Thus, the primary structure is the amino acid sequence in the covalently linked peptide chain. It contains all the information required for subsequent chemical and conformational changes that render the protein biologically active; its relationship to higher order structures and function is still somewhat obscure.

The secondary structure is produced by intrachain hydrogen bonding, to form sequences and domains of high internal symmetry. The energies required ($<10\,\text{kJ/mol}$) are small compared to those associated with covalent bonds. The tertiary structure arises from the spatial reorganization produced by the folding of these ordered domains relative to each other. Dominant interactions are believed to include hydrophobic, electrostatic, van der Waals, and hydration effects (and configurational entropy).

Quaternary structures are formed by the spontaneous, but ordered, aggregation of individual peptide chains (identical or different) to form multisubunit structures that are stabilized by noncovalent forces.

From: *Protein Biotechnology* • F. Franks, ed.   © 1993 The Humana Press Inc.

1.1. Experimental Approaches
to the Characterization of Protein Structure

Experimental techniques can be divided into chemical and physical methods.

1.1.1. Chemical Methods

Chemical methods include the following:

1. Amino acid sequence, location of —S—S—bonds, and quantitative carboxyl and amino terminal analysis define the primary structure and the number of peptide chains.
2. Chemical modification with monofunctional and polyfunctional reagents identifies "reactive" and "masked" residues in the peptide chain, which themselves are determined by the microenvironment (secondary and/or tertiary structure).
3. Susceptibility to proteolytic enzymes and other chain cleaving reagents (e.g., cyanogen bromide). Native (functional) proteins are resistant or almost resistant to proteolytic attack, whereas denatured (inactive) proteins tend to be rapidly hydrolyzed.

1.1.2. Physical Methods

Hydrodynamic properties monitor the size, shape, and charge of the macromolecule in solution. These are discussed further elsewhere in this book.

Scattering and absorption of radiation:

1. Small angle scattering by macromolecules (X-rays, neutrons, visible light) is a function of size and shape.
2. Absorption (visible and UV) and fluorescence measurements provide information about the microenvironment of chromophores (usually tyr, trp, and phe).
3. Optical rotatory dispersion (ORD) and circular dichroism (CD) properties are characteristic of particular ordered (secondary structure) domains.
4. Nuclear magnetic resonance (NMR) provides both structural (short-range) and dynamic information.

5. X-ray and neutron diffraction of *crystalline* proteins provides the complete three-dimensional structure in terms of the atomic coordinates. It requires previous knowledge of the amino acid sequence. X-ray diffraction relies on electron density and cannot, therefore, "see" hydrogen atoms, whereas neutron techniques can locate hydrogen (preferably deuterium) atoms.

2. Primary Structure

2.1. General Principles

Proteins are assembled on the ribosome. The primary amino acid sequence therefore contains *all* the basic information that is required for the subsequent conformational rearrangements leading to structures that convey biological activity to the polypeptide chain. Because the amino acid sequence is the most important property of the protein, in principle it should be possible to calculate a protein's higher-order structures from the primary sequence. Despite many attempts, this has not yet been achieved. In a few cases the primary structure can be related to function, e.g., casein, which behaves as a giant surfactant molecule, with phosphoserine residues (charged) at one end of the chain and a long hydrophobic sequence at the other. This arrangement promotes the formation of micelles.

The only known covalent cross links within a peptide chain are the disulfide bonds between cys residues. Disulfide bond stability and the role of such bonds in contributing to protein stability are subjects of lively debate. Attempts have been made to increase protein stability by the addition of —S—S— bonds (1), but the relationships are not clear.

2.2. Nomenclature

The following abbreviations for the amino acids are commonly used. The one-letter code is gaining in popularity.

Amino Acid Abbreviations

Amino acid	Conventional code	One-letter code
Valine	val	V
Leucine	leu	L
Isoleucineile	I	
Methionine	met	M
Phenylalanine	phe	F
Asparagine	asn	N
Glutamine	gln	Q
Histidine	his	H
Lysine	lys	K
Arginine	arg	R
Aspartic acid	asp	D
Glutamic acid	glu	E
Tryptophan	try or trp	W
Tyrosine	tyr	O
Proline	pro	P
Hydroxyproline	hyp or hypro	
Cysteine	cys	C
Serine	ser	S
Threonine	thr	T
Glycine	gly	G
Alanine	ala	A
Asparagine or aspartic acid	asx	B
Glutamine or glutamic acid	glx	Z

Typical examples of primary structures of proteins are described below.

2.3. Ribonuclease

The amino acid sequence of ribonuclease is shown in Fig. 1. This important protein consists of a single peptide chain containing 124 residues, including eight cysteine residues. Chemical analysis indicates disulfide links between the following pairs of cys residues: 4–5, 1–6, 3–8, and 2–7. The corresponding positions in the peptide chain are: 65–72, 26–84, 58–109, and 40–95.

2.4. Insulin

Insulin in the active state is composed of two peptide chains that are linked by disulfide bonds in two positions, as shown in

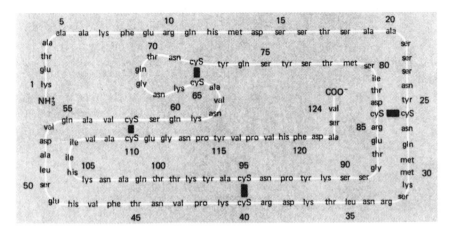

Fig. 1. Primary structure of bovine pancreatic ribonuclease, showing the positions of the four S—S bridges (from F. Wold, in *Macromolecules: Structure and Function* (1971) Prentice-Hall, New Jersey).

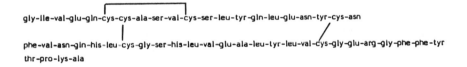

gly-Ile-val-glu-gln-cys-cys-ala-ser-val-cys-ser-leu-tyr-gln-leu-glu-asn-tyr-cys-asn

phe-val-asn-gln-his-leu-cys-gly-ser-his-leu-val-glu-ala-leu-tyr-leu-val-cys-gly-glu-arg-gly-phe-phe-tyr

thr-pro-lys-ala

Fig. 2. Primary structure of human insulin.

Fig. 2. The A chain of 21 residues also contains an intrachain disulfide link between residues 6 and 11. The B chain is composed of 30 residues (*see* also proinsulin).

2.5. Tissue Plasminogen Activator (tPA)

tPA, a fibrinolytic agent, is a derivative of the stoichiometric lys-plasminogen-streptokinase activator complex (2). The molecular weight is approx 131,000 Da. The catalytic site is blocked by a chemical group, e.g., *p*-anisoyl, which is attached to a serine residue (740) by an ester linkage. In aqueous solution, this protecting group is removed by hydrolysis. The molecule possesses five looped structures (kringles), held together by triple disulfide bonds. The kringle domains (lysine binding sites) are located in different parts

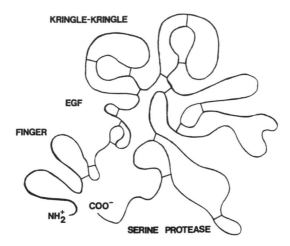

Fig. 3. Tissue plasminogen activator molecule, showing the distinct structural and functional domains and the positions of disulfide linkages: 41-71, 69-78, 86-97, 91-108, 110-119, 127-208, 148-190, 179-203, 215-296, 236-278, 267-291, 299-430, 342-358, 350-419, 440-520, 444-519, 476-492, 509-537.

of the plasminogen molecule from the catalytic site. They contain the elements by which the activator complex binds to the fibrin matrix that holds together the blood clot. Figure 3 is a schematic representation of tPA. Plasminogen is a good example of a single peptide chain protein that exhibits several distinguishable domains, each associated with a specific function and a specific structure.

2.6. Homology

The relationship between primary sequence, higher order structure, and function are again somewhat obscure. For instance, all globins have identical tertiary structures but a wide variety of compositions: 44 out of 54 hydrophobic residues are conserved, whereas none of the 21 polar residues is fully conserved. This illustrates a general principle: Buried hydrophobic residues tend to be more fully conserved than exposed residues.

Some proteins exhibit species differences in the amino acid composition, although *not* in the function of the particular protein. Table 1 summarizes the amino acid replacements in positions 8, 9, and 10 of the A chain of mammalian insulins.

Table 1
Amino Acid Substitutions in A Chains
of Insulins from Different Species

	Position		
	8	9	10
Beef	ala	ser	val
Pig	thr	ser	ile
Sheep	ala	gly	val
Horse	thr	gly	ile
Sperm whale	thr	ser	ile
Human	thr	ser	ile
Dog	thr	ser	ile
Rabbit	thr	ser	ile

Table 2
Amino Acid Differences in Cytochrome *c*
Between Different Species

Number of differences	
Human-monkey	1
Human-horse	12
Human-dog	10
Pig-cow-sheep	0
Horse-cow	3
Mammals-chicken	10–15
Mammals-tuna	17–21
Vertebrates-yeast	43–48

Vertebrate cyrochrome c contains 100 residues, of which only 36 are completely conserved in all species. These include a complete sequence of 11 residues in positions 70–80 (—NPKKOIPGTKM—). Table 2 shows amino acid differences between species. Despite the large degree of chemical variability, the protein fulfills the same function in all species. Table 3 gives details of the degree of conservation of the rarer amino acids that are thought to be important to the structure and function of cytochrome *c*.

Table 3
Conservation of Rarer Amino Acids
in Cytochrome *c*

Amino acid	Position
phe	10, 82
cys	14, 17
his	18
lys	27, 72, 73, 79, 87
pro	30, 71, 76
tyr	48, 67, 74
trp	59

In general, if a protein plays a central role in many different forms of life (e.g., glycolytic enzymes), then complete homology is observed. Thus, nucleohistones are 100% conservative.

In some cases single amino acid substitutions can give rise to pathological conditions, e.g., sickle cell hemoglobin. On the other hand, mutant proteins with minor amino acid changes can enable organisms to survive under adverse conditions (4), e.g., the alcohol dehydrogenase enzymes in halophilic bacteria contain more acidic residues (glu and asp) than do the corresponding mesophiles. This enables the bacteria to survive under conditions of high salt concentrations, as in the Dead Sea. In fact, being obligate halophiles, they cannot exist under normal salt conditions (0.15M).

With the increasing use of enzymes as industrial catalysts for biotransformations, attention is being focused on the properties of extremophiles, in particular, thermophiles, acidophiles, and alkalinophiles (4).

2.7. Homology and Database Searching

As more sequences are established, the literature can be searched for sequence and structure motifs and analogies/similarities established (5) that might in future help in the design of mutants or "new" proteins. In this way groups of proteins can be identified that have similar structures but quite different functions, and vice versa (*see* Table 4). Databases can also be searched for sequence motifs derived from homologous sequences. Structure-

Table 4

Some Heterofunctional Protein Families

Discovered by Database Searches (from ref. *(5)*)

Serpins[a]		Lipocalins[b]
α_1-antitrypsin	leuserpin-2	retinol-binder
α_1-antichymotrypsin	ovalbumin	purpurin
α_2-antiplasmin	angiotensinogen	complement C8, gamma
antithrombin 3	thyroxine-binder	α_1-microglobulin
protein C inhibitor	corticosteroid-binder	β-lactoglobulin
monocyte PA inhibitor	endosperm protein Z	orosomucoid
C1 inhibitor		

β-2 Glycoprotein-type[c]	EGF-Containing[d] proteins	
cartilage proteoglycan	thyroid peroxidase	LDL receptor
complement B reg	complement C9	sea urchin reg protein
complement C2	factor VII	factor IX
complement C4-binder	protein S	urokinase
factor XIIIb	thrombomodulin	Drosophila notch
complement factor H	cartil matrix protein	vaccinia virus
decay accelerating factor	factor X	lin 12, *C. elegans*

[a]Serpins are named thus because several are known to be serine protease inhibitors.

[b]Lipoalins bind lipophilic agents.

[c]The β-2 glycoprotein unit is about 60 residues long and contains a disulfide bond.

[d]The EGF-unit, first observed as epidermal growth factor, consists of 40–45 residues with three disulfide bonds.

related sequence motifs are sequence motifs that are reliably related to common structures; several examples are shown in Table 5 *(5)*. Common structural motifs can recur with quite different sequences. Searching can be performed in many different ways, e.g., by sequence, torsional angles (*see below*), solvent accessibility, tertiary interactions, or disulfide bridges. Database searches are particularly useful to gain an understanding of how mutations can affect (or not affect) certain structural and functional features.

Interesting relationships have recently been described. For instance, striking sequence similarities have been found between sarafotoxins (snake toxins) and endothelins (natural mammalian peptides involved in blood pressure control) *(6)*. Both groups have

Table 5
Examples of Protein Motifs*

Sequence motifs	
(GPP)	Collagen triple helix (multiple repeats)
FDTGS	Acid proteinase cleavage site
G.G..G	Nucleotide binding motif (simplified)
Structure-derived sequence motifs	
AA..K	Peptide containing at least two helical residues
GG.L	Structure XXββ
Structural motifs	
Helix; strand, turn	Secondary structure
β-hairpin, βαβ-unit, Greek key, 4α-bundle	Supersecondary structure
Jelly roll, TIM barrel	Tertiary structure

*A dot indicates no conservation. The code for structure is β = β-strand; X = not α, β, or turn; see ref. (5).

two —S—S— bonds linking cys-1 to cys-15 and cys-3 to cys-11. Both groups have 21 residues with C-terminal trp and the following identical residues: asp-8, glu-10, his-16, and ile-20.

3. Secondary Structure

Compared to the covalent bond energies involved in the primary structure (>200 kJ/mol), the secondary structure relies on weak, hydrogen-bonding interactions (≈10 kJ/mol). It originates from the stereochemistry of the peptide bond, shown in Fig. 4. The six atoms within the shaded area lie in one plane because the C—N bond has some double bond character. Rotation can occur about the C—C and C—N bonds, and successive residues in the polypeptide chain are characterized by a pair of torsional angles, as shown in Fig. 5. By convention, ø = 0 when the four atoms C—N—C_α—C lie in one plane, and ψ = 0 when N—C_α—C—N are coplanar. If successive residues take up similar orientations (same

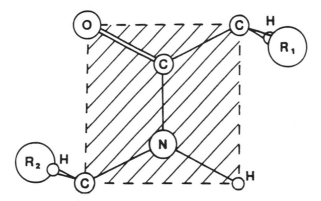

Fig. 4. Stereochemistry of the *trans*-peptide bond, as found in proteins.

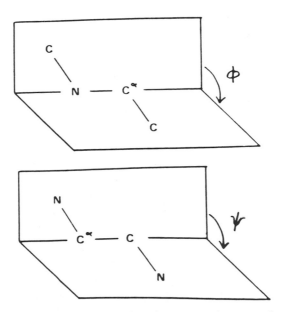

Fig. 5. Torsional angles ø and ψ that determine the secondary structures of proteins (from F. Wold, in *Macromolecules: Structure and Function* (1971) Prentice-Hall, New Jersey).

values for ø and ψ) to one another, a symmetrical structure results. Such symmetry is most likely in cases in which the amino acid sequence shows repeating patterns.

3.1. The α-Helix

Apart from possible interactions between amino acid side chains, the α-helix is the form of lowest free energy. It therefore forms spontaneously, especially in fibrous proteins (keratins in hair, wool, feather, and skin), which usually contain repeating patterns in their primary sequences.

The helix is described in terms of a symmetry index η_m, where η is the number of residues that form a perfect repeat after m turns. For the α-helix, $18_5 = 3.6$ peptides per turn. The helix is right-handed with a pitch (rise per turn) of 0.54 nm and an average rise per residue of 0.15 nm. The representation of the α-helix in Fig. 6 shows that hydrogen bonds are formed between NH groups and the CO groups of the *fourth* residue along the chain. The hydrogen bonds are almost parallel to the helix axis; this makes for maximum bond strength and macroscopic elasticity. The intrinsic stabilizing forces can be enhanced or diminished by interactions between amino acid side chains. Proline and hydroxyproline are incapable of inclusion in the helix because they do not contain NH groups. Glycine is also considered to be an α-helix structure breaker.

3.2. The β-Pleated Sheet

Another high-symmetry form is the so-called β-structure, which is adopted by peptide chains that cannot easily form an α-helix. It is an extended, corrugated structure with $\eta_m = 2_1$ in which alternate side chains project above and below the axis of the chain, respectively. The chains are linked laterally by hydrogen bonds to form sheets, as shown in Fig. 7. Two possible configurations exist, depending on whether neighboring chains run parallel or antiparallel. In fibrous proteins, only antiparallel structures are observed. This structural arrangement provides high tensile strength, but renders the fiber brittle.

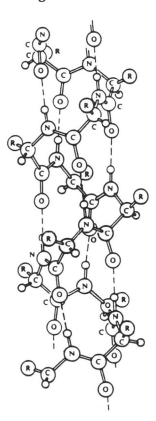

Fig. 6. The α-helix. Hydrogen bonds are indicated by broken lines. Amino acid side chains are designated by R (from B. W. Low and J. T. Edsal, in *Currents in Biochemical Research* (1968) (D. E. Green, ed.) Wiley Interscience, New York).

3.3. Other Regular Structures: Polyproline Helices

Although secondary structures found in proteins are generally confined to α-helices and β-sheets, other regular peptide structures are possible; some are described in Table 6. Of these, polyproline is particularly important, because it occurs in collagen, the most abundant protein in nature. Collagen contains a high proportion of pro and hypro residues and is therefore unable to form regular arrays of intramolecular hydrogen bonds. Two

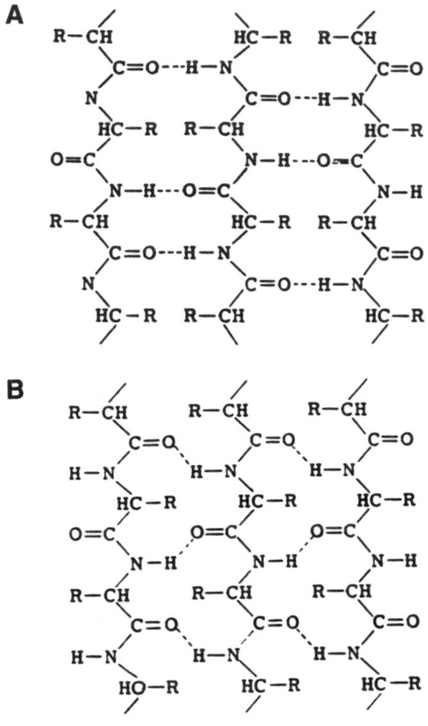

Fig. 7. The β-sheet structures commonly found in proteins: (a) antiparallel, and (b) parallel.

Table 6
Regular Secondary Peptide Structure,
with Physical Parameters

Structure	Symmetry index, η_μ	Peptides per turn, η /m[a]	Residue repeat, nm
α-Helix	18_5	3.60	0.15
β-Sheet (parallel)	2_1	2.0	0.325
β-Sheet (antiparallel)	2_1	2.0	0.35
ω-Helix	4_1	−4.0	0.1325
Polyglycine II	3_1	±3.0	0.31
Polyproline II	3_1	−3.0	0.312
Polyproline I	10_3	3.33	0.19

[a]Positive values for right-hand, negative values for left-hand helices.

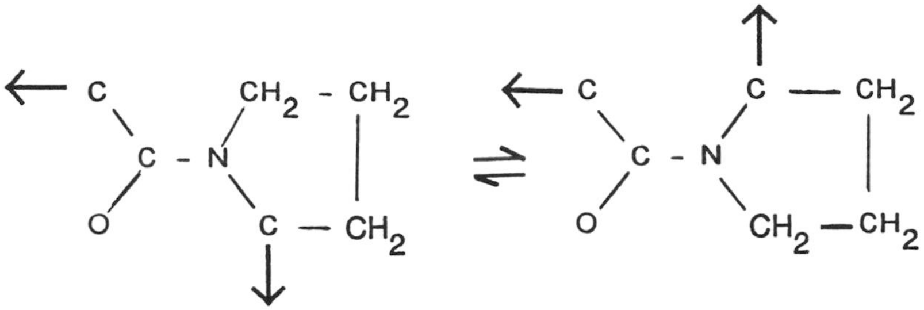

Fig. 8. *Cis-trans* proline isomerism.

symmetrical structures exist: polyproline I with $\eta_m = 10_3$, a right-handed helix in which successive amino groups are in the *cis*-position, and polyproline II, with $\eta_m = 3_1$, a left-handed helix with *trans*-imino groups. It should be noted that the possibility of *cis-trans* isomerism also exists for the peptide bond. However, the *trans*-form is much more stable, so that *cis*-isomers have never yet been detected.

The free energies of the two diproline isomers are very similar, so that they are both capable of existence; the isomerization equilibrium is shown in Fig. 8. In the absence of stabilizing intrachain hydrogen bonds, the helices are presumably stabilized by solvation interactions, and the choice of solvent determines the particular isomer that will be stable. The isomerization is thus solvent-dependent; it is also slow and can readily be followed experimentally. *Cis-trans* proline isomerization is believed to be implicated in denaturation and renaturation processes (7).

3.4. Steric Maps

By plotting the torsional angles (*see* Fig. 5) for successive pairs of residues along the peptide chain, an energy map can be constructed by calculating the energies between all pairs of atoms as functions of their distance of separation as the angles are altered and summing over all atomic pairs. This requires assumed potential functions for atom–atom interactions (*see later*). Contours of equal energy can then be drawn on a (ø,ψ) map. A potential energy map for dialanine is shown in Fig. 9. For the α-helix: ø = 110°, ψ = 135°. For the β-sheet, α = 45° and ψ = 290° (parallel) and

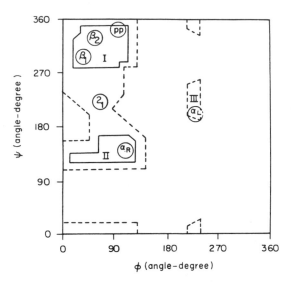

Fig. 9. Hard-sphere steric map for dialanine. Broken lines: allowed regions for outer contact distances; solid lines: regions for normal contact distance.

330° (antiparallel). Similarly, the other regular structures in Table 6 can be identified. Figure 10 shows such a map for lysozyme, each dot representing the sequential $(\varnothing,\psi)$ configurations of the dipeptides in the chain. It is seen that high concentrations of dots are found in the α-helix and β-sheet regions but somewhat displaced from the actual $(\varnothing,\psi)$ positions shown in Fig. 9; this is believed to be owing to the influence of side chain and/or solvation interactions, which would tend to distort (and destabilize?) the idealized regular structures. The map in Fig. 10 also reveals a number of dipeptides that fall well outside the range of low energy configurations; they correspond to loop sequences, often associated with glycyl residues that connect the secondary structure elements. The potential energy contours are drawn at 4.2 kJ/mol intervals, between zero (broken contour) and −12.6 kJ/mol. The exact shapes of the contours are determined by the atom–atom potential functions chosen for the energy minimization.

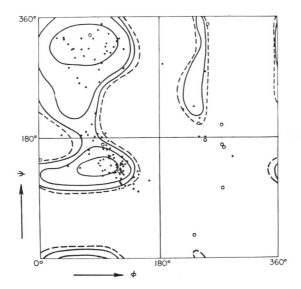

Fig. 10. Torsional angle map of lysozyme (dots) with energy contours of alanine: — zero contour; — contours at intervals of 1 kcal/mol (4.18 kJ). From Ramachandran and Sasisekharan, *Adv. Protein Chem.* **23,** 238 (1968).

4. Nonstructured Elements

Loop regions, also referred to as reverse turns, connect regular helix and β-structures. In most proteins, loops account for >30% of the structure. They lie mainly on the protein surface, are flexible and form binding and recognition sites (e.g., immunoglobulins and DNA-binding proteins). Turns can be very short, involving a single residue, or extended over a sequence of several residues; this is illustrated in Fig. 11 *(8).*

5. Tertiary Structure

The tertiary structure is the spatial disposition of various ordered domains with respect to each other; it is also the origin of the *native* or *functional* state of the protein in vivo. The interactions that maintain tertiary structures are weak (<5 kJ/mol), short-range and sensitive to enviromnental factors, such as pressure,

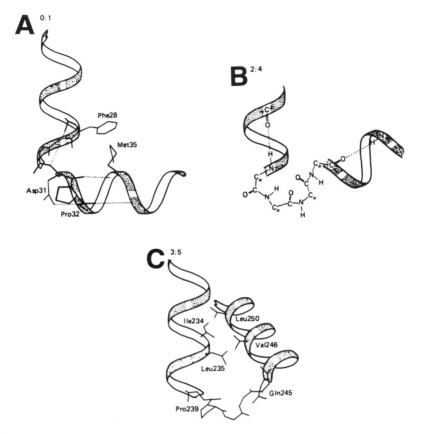

Fig. 11. Loop structures in subtilisin, an α-helical protein: (a) 1-residue loop, only a single residue has non-α angles; (b) 2-residue loop in which the second residue is often glycine; (c) 3-residue loop. Reproduced with permission from ref. *8*.

temperature, pH, ionic strength, and the composition of the solvent medium. The tertiary structure is therefore labile and easily disrupted.

5.1. Classification of Tertiary Structures

Many attempts are on record to classify known tertiary structures and then to calculate the generation of the tertiary structure from the primary sequence. Tertiary structures can be grouped in terms of secondary structure sequences in the peptide chain, as

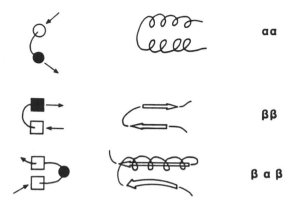

Fig. 12. Classification of tertiary structures according to secondary structure sequences.

shown in Fig. 12, where α-helices are represented by circles and β-structures by squares. The structural elements are connected by loops. The arrows show the direction of the peptide chain, from the amino to the carboxy terminal. Figure 13 shows how these structure elements are formed during the folding process. We distinguish between sequences of α-helices, sequences of β-structures, and alternating α- and β-structures. In all α- or all β-structures, structural domains are antiparallel. In α/β structures, the β-sheets tend to lie parallel to each other (*see* Fig. 7), separated by α-helices that run in the opposite direction. Examples of proteins based on these structure sequences are given in Fig. 14, and Table 7 provides a summary of proteins belonging to each of the four groups.

5.2. Spatial Disposition of Structured Domains

X-ray and neutron diffraction techniques provide ever more detailed information on three-dimensional structures of proteins in the crystalline state, and it is generally believed that the secondary and tertiary structures in solution resemble those in the crystal in the essential features, i.e., those that determine the biological effectiveness of the molecule.

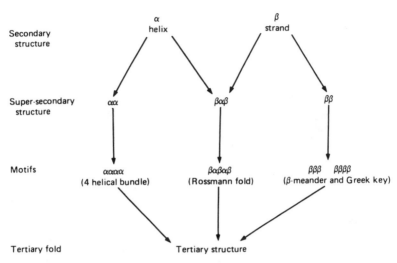

Fig. 13. Hierarchy of structures, leading to folded tertiary structures and giving rise to different loop types. Reproduced, with permission, from ref. 8.

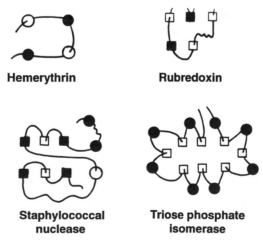

Fig. 14. Examples of different types of secondary structure sequences.

Table 7

All-α	All-β	α + β	α /β
Haemerythrin	Rubredoxin	Insulin	Flavodoxin
Myogen	Immunoglobulins	PTI	Dehydrogenases
Globins	Prealbumin	Ribo-nuclease	Triose-phosphate isomerase
	Superoxide dismutase	Lysozyme	Phospho-glycerate kinase
	Concanavalin	Papain	Hexokinase
	Chymotrypsin	Staph. nuclease	Carboxypeptidase
		Thermolysin	Subtilisin

The projections in Fig. 14 can now be further elaborated to illustrate the topology of the molecule as a whole. Examples are shown in Fig. 15. Hemerythrin and rubredoxin are simple examples of the αα and ββ types, respectively. In both cases, the structural units are arranged in pairs. Staphylococcal nuclease is an example of the αα + ββ type, in which two triplets of antiparallel β-sheets are joined to three helical segments. In the α/β arrangement of triose phosphate isomerase, alternating α and β sequences resemble a barrel arrangement in which the parallel β-strands run in one direction and the α-helices in the opposite direction. The folding of the protein gives rise to a hollow cylinder.

The relationship between a primary sequence and the resulting folded tertiary structure is not well established, but often the primary structure, when, folded into the α-helix, provides a cylinder with hydrophobic surfaces or "patches" that can serve as the nucleation sites during folding or assembly. This is illustrated in Fig. 16 for a helical sequence (148–165) of apolipoprotein A-1 in the form of a projection (9). Bearing in mind the symmetry index of the helix, 18_5, then starting with met-148 and counting four residues, it is found that ala-152 lies almost directly below met-148; similarly, val-156 is below ala-152, and so on. The helix thus contains a

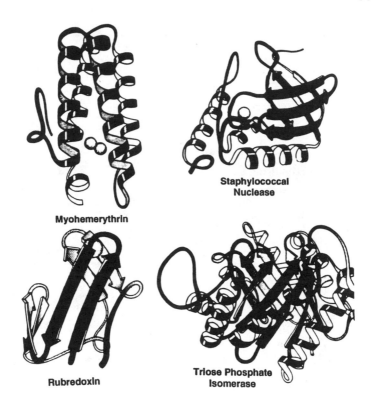

Fig. 15. Different types of secondary structure sequences, folded to adopt tertiary structures (from C. Chothia and A. M. Lesk, in *Protein Folding* (1980) (R. Jaenicke, ed.) Elsevier, Netherlands).

hydrophobic face which, during folding or aggregation with other subunits, becomes a preferential contact site. This conclusion could be tested by the synthesis of a model 22-residue peptide, containing only three amino acids: lys, glu, and leu, *see* Fig. 17. It was found to possess considerable biological activity, of the type associated with apo-A1 lipoprotein.

5.3 Tertiary Structure and Homology

Natural selection produces protein families with very different amino acid sequences that nevertheless have very similar tertiary suctures. For example, the globins occur in many widely

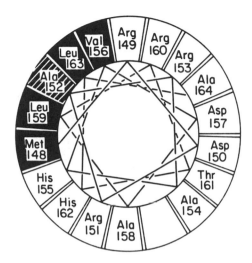

Fig. 16. Helical wheel projection of the sequence 148–165 of apolipoprotein A-1. Hydrophobic residues are shown in black. Reproduced, with permission, from ref. 9.

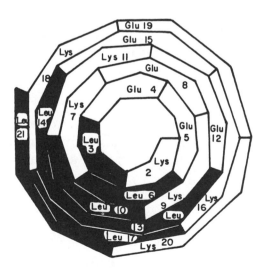

Fig. 17. Axial projection of the helical 22-residue peptide model for apolipoprotein A-1, *see* text. Reproduced, with permission, from ref. 9.

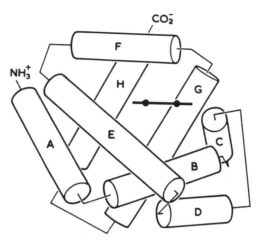

Fig. 18. Spatial disposition of the α-helices in globins; the heme group is shown as a bold line (from J. S. Richardson, in *Protein Folding* (1980) (R. Jaenicke, ed.) Elsevier, Netherlands).

differing organisms, where they store and transport oxygen. They contain a heme group in which the iron can reversibly bind oxygen. Each molecule (or subunit) contains 136–154 amino acid residues, of which only 16% are conserved. Despite this variability in composition, the tertiary structures are remarkably similar: Seven or eight α-helices are assembled, as shown in Fig. 18. In an extreme case, a comparison of nine different globins has established that the same folded structure is obtained, but only two residues are conserved!

On the other hand, a comparison of pancreatic trypsin inhibitor (PTI), which contains 58 residues with a number of different toxins having the same number of residues, but totally different functions, shows that although only 19 residues are the same, yet the folding pathway and folded structure of the different proteins is identical. Finally, a survey of several plastocyanins (small Cu-containing proteins) indicates that of the 21 amino acid residues buried in the interior of the proteins, 20 are conserved, whereas none of the 46 surface exposed residues is fully conserved. It appears that the crucial feature for identical tertiary structures is the maintenance of the hydrophobic interior of the protein, whereas a greater degree of variability can be tolerated in those residues that lie on the periphery, exposed to the solvent. This is not to say, however, that *all* mutant proteins have the same degree of biological activity.

5.4. Recent Developments in Determinations of Complex Structures

Apart from neutron diffraction, the most significant advance has been the development of synchrotron X-radiation methods. High resolution diffractograms can now be obtained fram small crystals with short exposure times (ca. 500 ns). The methods are being used to study proteins the sequences of which are still unknown (10). The technique has been applied to obtain high resolution structures of large enzymes, e.g., β-lactamase (11), and even of enzyme–substrate complexes, such as the glycogen phosphorylase-b/maltoheptanose complex.

5.5. Hydration Structures

Protein crystals contain up to 70% water. Its involvement in determining and stabilizing specific structures has long been a subject of study and speculation. With the advent of neutron diffraction it has become possible to assign coordinates to those O and H atoms that are sufficiently well localized, or where certain sites have a high degree of occupancy (probability scattering intensity), i.e., fast exchange. Water molecules are found preferentially at the following sites:

1. Coordinated to metal atoms as ligands, e.g., carboxypeptidase: $Zn(his)_3H_2O$;
2. Coordinated to backbone -NH and >C=O groups, forming connecting links between different residues, involving one, two, or three water molecules; and
3. Linking backbone groups to side chain groups.

Complex water geometries have been observed, i.e., distortion of hydrogen bond angles and a distribution of hydrogen bond lengths. Although the exact role of hydration interactions as a contributing factor to protein structure and stability is still a subject of intense speculation, there is general agreement that a number of water molecules must be regarded as performing an essential role in the maintenance of the native protein structure. Indeed, the term "protein" implies peptide + water. A fuller discussion of the contribution of water and hydration to the stabilization of native proteins will be found in Chapter 12.

6. Quaternary Structure

Most important large proteins cannot be crystallized, so that higher order structures cannot be determined by high resolution diffraction methods. The mode of aggregation and self-assembly must then be established by macroscopic methods, e.g., electron microscopy, hydrodynamics, chromatography, electrophoresis, small angle scattering, or spectroscopy. The structural information so obtained is somewhat rudimentary and confined largely to general supramolecular shapes and relative dispositions of subunits within the assembled structures.

As summarized elsewhere in this volume, many proteins in their active state are composed of aggregates of identical or different peptide subunits. Aggregation numbers range from 2 to several thousands, e.g., 2100 for tobacco mosaic virus coat protein. The supermolecular structure is formed by a process known as self-assembly, from the correctly folded individual subunits. The driving forces for self-assembly are believed to be mainly hydrophobic interactions, although contributions from hydrogen-bonding and charge-charge effects cannot be ruled out.

Self-assembly is highly specific: If two teramic enzyme complexes (aldolase and glyceraldehyde-3-phosphatase) in a mixture are dissociated and then allowed to reassemble, this process results in a complete regeneration of the two individual enzymes, despite the fact that many random collisions must occur between peptide chains belonging to the *different* enzyme. There is no evidence that dissociation and reassembly occur via the same pathway.

The complexity and specificity of self-assembly is well illustrated by the construction of T-even bacteriophages, as portrayed in Fig. 19. Three different, but converging, pathways have been identified by which the head, the tail, and the tail fibers are assembled. For instance, in the construction of the tail the first event

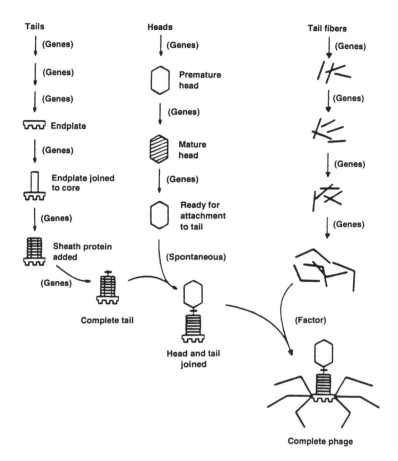

Fig. 19. Self-assembly scheme of T-even bacteriophage (from R. N. Perham, in *Characterization of Protein Conformation and Function* (1979) (F. Franks, ed.) Symposium Press, London).

is the aggregation of subunits to form the base plate. The tail tube is then built up with an α-helical arrangement of subunits. Around this tail tube a second array of polypeptides that acts as the contractile element in the sheath is then built up. Finally, the tail assembly is completed. Experiments on mutants have shown that there is a sequential information transfer during the assembly process, so that subunits cannot become misassembled before the right structure has appeared for their incorporation.

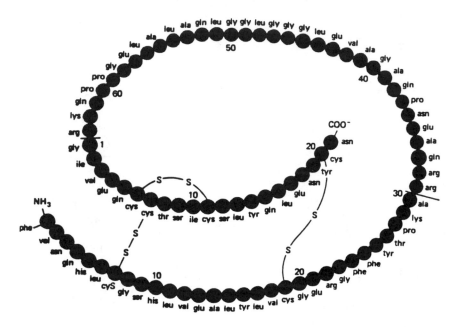

Fig. 20. Primary sequence of proinsulin. Compare with the active hormone sequence in Fig. 2.

This assembly information is contained in so-called propeptides (also known as signal peptides or teleopeptides), short amino acid sequences that are essential for the correct folding/ assembly of the subunits, but that are subsequently cleaved off by a proteolytic enzyme. It is thus impossble to dissociate such a structure in vivo and to reassemble it.

One of the simplest cases of spontaneous in vivo folding and assembly is provided by insulin. The protein that is synthesized on the ribosome is proinsulin, the primary structure of which is shown in Fig. 20. It consists of a single chain and has no hormonal activity. It is activated by trypsin to produce insulin and a peptide of molecular weight 4000 that contains residues 31–63 of the original molecule. The propeptide is therefore required to ensure the correct folding of the chain, but until the removal of part of the chain, the protein remains inactive. The active protein now has two

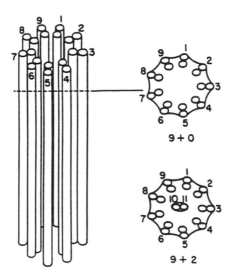

Fig. 21. Schematic diagrams showing the pairwise organization of micro-tubules in a flagellum of eurkaryotes (from G. H. Brown and J. J. Wolken, in *Liquid Crystals and Biological Structures* (1979) Academic, Florida).

chains, but once the tertiary structure is disrupted, it cannot refold correctly because the necessary information has been removed by the cleavage of the C-chain.

6.1. Motile Structures

Cilia and flagella are the simplest motile organelles. A characteristic structure in eukaryotic cells involves 11 protein filaments of which nine (paired microtubules) are peripherally located, whereas the other two are usually found at the center of the flagellum, as shown schematically in Fig. 21. The filaments are composed of the globular protein tubulin, which is the basic subunit of microtubules. The flagella or cilia of prokaryotes are composed of flagellin that resembles tubulin, but is of a simpler, helical design. For instance, bacterial flagellin protein takes up an α-helical pattern, as shown in Fig. 22.

Fig. 22. Flagellin (related to tubulin) α-helical assembly, as found in the flagella of prokaryotes. The protein forms a single strand, compared to the double-stranded tubulin assembly (from G. H. Brown and J. J. Wolken, in *Liquid Crystals and Biological Structures* (1979) Academic, Florida).

Compared to the flagellum, muscle is a vey complex proteinaceous contractile assembly of specialized fibers (myofibrils). The two major components, actin and myosin, self-aggregate and then associate with one another to form actomyosin, the active contractile system. Two additional globular proteins, tropomyosin and troponin, are associated with the actomyosin structure and regulate the contractile motion. The total supermolecular assembly is shown in Fig. 23. The myofibril is composed to a hexagonal array of these assemblies.

6.2. Structural Proteins

Collagen is one of the most studied proteins. It forms the connective tissue of the bulk of metazoan animals, but also occurs in immunoglobulins, bone, skin, and cornea of vertebrate eyes. The collagen molecule is a single polypeptide chain:

$$NH_2—(20\ \text{residues})—(gly—X—Y)_{300}—(20\ \text{residues})—COOH$$

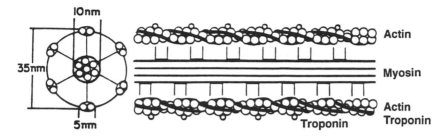

Fig. 23. Protein components and quaternary structure of muscle acto-myosin fibrils (from G. H. Brown and J. J. Wolken, in *Liquid Crystals and Biological Structures* (1979) Academic, Florida).

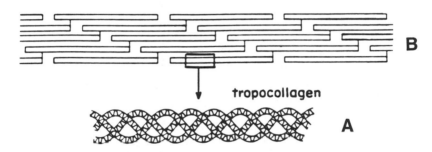

Fig. 24. Three-stranded tropocollagen molecule (A) and their method of assembly in the collagen fiber (B), showing overlap and gap regions (from G. H. Brown and J. J. Wolken, in *Liquid Crystals and Biological Structures* (1979) Academic, Florida).

where 30% of the X or Y residues are pro or hypro. The conformation resembles the left-handed polyproline-II helix, and three molecules aggregate to form a tropocollagen triple-stranded structure, as illustrated in Fig. 24A. The individual triple structures are laid down in the collagen fiber to produce the staggered array shown in Fig. 24B. Their exact positions in the fibril are determined by the 20-residue propeptide sequence that is subsequently removed, leaving the 67 nm displacements. The regular occurrence of glycine allows the three chains to approach sufficiently closely to produce the collagen superhelix. The molecules are also linked by covalent bonds and hydrogen bonds, all of which enhance the tensile strength of the fibrils.

Table 8
Predominant Amino Acid Residues
(Chemical Rhythm) of Fibrous Proteins

Protein	Source	Predominant residues, percent of total residues
Fibroin	Silk moth	gly 45, ala 29, ser 12
Fibroin	*Chrysopa* egg stalk	ser 43, gly 25, ala 21
Elastin	Lung tissue	gly 32, apolar 60
Resilin	Insect exoskeleton	gly 38, ala 11
Keratin	Whole wool	ser 11, cys/2 11, glu 11
Myosin	Rabbit muscle	glu 18, lys 11
Actin	Rabbit muscle	glu 11
Fibrinogen	Human blood	glu 15, asp 13

Table 8 summarizes other quaternary protein assemblies that give rise to load-bearing structures. A common feature is the chemical periodicity of certain amino acids, usually glycine. An interesting variant of the common helical and β-sheet structures is the protein found in the egg stalk of the lacewing fly *Chrysopa flava*. The stalk has to support the weight of the egg and must therefore possess a rigid structure, unlike most other fibrous proteins that fulfill functions in which high tensile strength is required. The egg stalk protein is characterized by high proportions of serine (43%), glycine (25%), and alanine (21%). The rigidity is achieved by short stretches of β-structures that double back on themselves by means of reverse turns, with a glycine residue at each corner. The stucture is referred to as a cross-β-structure.

6.3. Immunoglobulins

Immunoglobulins are multisubunit proteins responsible for providing adaptive immunity in vertebrates. The unique characteristics of the adaptive immune response are specificity, dissemination, amplification, and memory.

All immunoglobulins have a common basic structure consisting of two "heavy" and two "light" polypeptide chains linked by —S—S— bridges, as shown in Fig. 25. The general formula for all immunoglobulins is therefore $(H_2L_2)_n$. Their diversity derives from the different types of heavy and light chains.

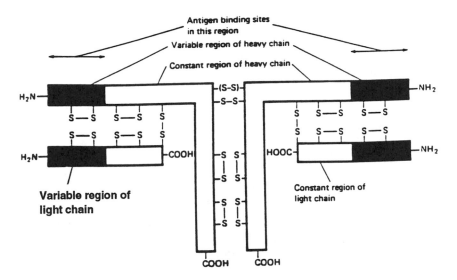

Fig. 25. Basic structure of immunoglobulin molecule, showing the two light and two heavy peptide chains and constant and variable regions.

6.4. Multisubunit Enzyme Complexes

The aggregation of several different subunits facilitates the performance of a series of sequential reactions, in which the product of one such reaction serves as the substrate of the next. The reaction sequence can thus be performed without the need for random diffusion of intermediates and the possibility of decomposition of reactive species. An example is provided by yeast pyruvate dehydrogenase. Under anerobic conditions, pyruvate is decarboxylated, eventually to yield ethanol. In aerobic yeast, pyruvate is converted to acetyl-CoA by a complex system of three individual enzyme functions, each associated with a different polypeptide chain. The three enzymes have molecular weights of 100,000, 80,000, and 56,000, respectively, and the particle of the multisubunit complex has a diameter of 30–40 nm. The complex is held together by noncovalent forces and can easily be dissociated. Enzyme E2 is itself composed of 24 subunits, clustered in trimers at the corners of a cube. In the active complex, the polypeptide chain stoichiometry (E1:E2:E3) is 2:1:1, which indicates a total of 96 peptide chains and a molecular weight of 8×10^6. The subunits will self-assemble in vitro, but in the absence of the peptide chains belonging to E2, no assembly occurs. E2 forms the core to which E1 and/or E3 can be added independently. E2 therefore has two roles: It has an enzyme function and it also forms the structural core around which the complex is built.

All enzymes that perform complex chemical functions, such as synthases or kinases, are multisubunit structures of greater or lesser complexity.

6.5. Nuclear Proteins

Higher cell DNA is organized in a specific manner into chromosomes of a fibrillar appearance (chromatin fibers) that contain equal amounts by weight of DNA and proteins (nucleohistones). The histones themselves are complexes of different types of subunits of the types H2A.H2B (tetramers) and H3.H4 (dimers). The subunits themselves contain no cysteine residues and they exist in the random coil configuration. They resemble giant surfactants in that they are composed of large (structured) hydrophobic domains and flexible polar end sequences. Conformational transitions occur at high ionic strength in the central and C-terminal regions to yield a structured state. Structured apolar regions then self-aggregate to form large complexes with structured cores and unstructured C- and N-terminal regions (largely basic).

Subunit H1 does not form oligomers or complexes. It undergoes an intramolecular transition with the conserved (central) section taking on a globular structure, as the ionic strength is increased.

Histones form complexes with DNA, especially H3 and H4, which are thought to be the major structural proteins in the nucleosome. They provide the core around which DNA is wound (84 base pairs per turn). H1 is bound to the edge of the particle and seals off two turns of DNA (168 base pairs). The low resolution model of a nucleosome, as obtained from small angle neutron scattering data, is shown in Fig. 26.

6.6. Blood Clotting

Hemostasis is associated with a large number of individual steps, whereby the soluble protein fibrinogen is converted into the insoluble final product fibrin, by the action of the enzyme throm-

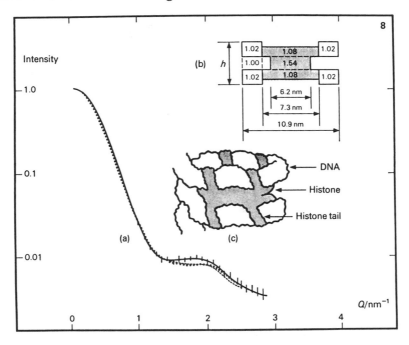

Fig. 26. The model for the nucleosome particle that provides the best agreement between the calculated and experimental neutron scattering data. Calculated curves are for the structures with heights of 5.0 nm (. . . . .) and 5.3 nm (—). Reproduced, with permission, from D. M. J. Lilley and J. F. Pardon, *Chem. Britain* **15,** 182 (1979).

bin. Fibrin is a crosslinked protein that provides a protective barrier to bleeding. Its precursor, fibrinogen, is rich in aspartic acid and exists as a dimer with each monomer composed of three peptide chains of molecular weights:

Aα: 67,000
Bβ: 58,000
gamma: 47,000

Its schematic structure is shown in Fig. 27, but the 3D-structure is still uncertain; in the electron microscope it appears as three globules linked by a thread. The Aα and gamma chains of each dimer are linked by five permanent S–S bonds near the N-terminal; there are also four labile S–S bonds within each monomer. The

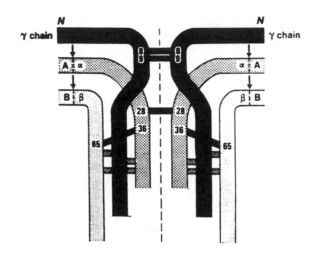

Fig. 27. Assembly of the fibrinogen dimer, showing permanent and labile S—S linkages.

production of the fibrin clot involves a reaction of thrombin with the A and B chains at arg–gly linkages. Simultaneously fibrinogen is modified by yet another protein, plasmin, to produce crosslinked fibrin by an as yet unidentified mechanism.

6.7. Extracellular Proteins

Laminin is a multidomain protein that binds to collagen IV and heparin sulfate. It can be degraded with elastase to yield nine fragments, of which some are α-helical and some aperiodic. Fraction 3 is mainly a β-structure and binds heparin. Fraction 1 binds to cells.

Fibronectin has two major domains, linked by two —S—S— bridges. Each domain consists of looped structures held by —S—S— bridges.

7. Structure Calculations

Protein structures are required mainly to gain a better understanding of assembly processes and stability, and to alter functional properties, such as stability toward temperature or pH, binding capability or specificity, or a catalytic property of an enzyme. All these modifications imply a change in the free energy of the (mutant) protein relative to the wild type and/or a change in the chemical environment of the particular protein. Since the number of possible sequence isomers is extremely large, it is usually the aim of structure and energy calculations to arrive at the lowest free energy state that the protein can adopt, because this is by many believed to correspond to the equilibrium conformation, an assumption that, however, still remains to be proved.

Theoretical approaches face two major problems. First, the free energy space available to a polypeptide chain is very large. A small protein of, say, 100 residues possesses approx 100 torsional angles, each of which might have access to three minimum energy states. The polypeptide then has $3^{300} \approx 10^{100}$ conformations. It will not become feasible in the forseeable future to cope with energy searches of this magnitude. The second problem concerns the realistic description of atom–atom interactions on which the energy calculations are based. For large molecules, the employment of *ab initio* quantum mechanical methods is out of the question. Instead, one has to resort to the additive sum of various classical interaction energy terms, such as covalent bonds, bond angles, torsional angles, van der Waals interactions, hydrogen bonds, and electrostatic energies. The parameters associated with all these contributions to the global (free) energy must be obtained from *ab initio* calculations on small model compounds or from experimental data.

At present, the prediction of *de novo* peptide structures is out of the question, unless some drastic simplifying assumptions are introduced into the calculations, e.g., regarding a protein as a set of sequential secondary structures that are then assembled into a plausible folded, native structure. On a more limited scale, theoretical methods can assist in the prediction of structures and properties of proteins that are locally modified, e.g., by single point mutations. On a more ambitious level, it is sometimes possible to predict the properties of homologous structures in which only few amino acid residues are conserved but that exhibit identical folding patterns.

7.1. Normal Mode Analysis

Normal modes (<300 cm^{-1}) correspond to motions, e.g., bond vibrations, on a picosecond timescale. They result from the general motion of the whole molecule (elongation, twisting). Such motions are experimentally determined by infrared spectroscopy. The calculations assume simple harmonic (parabolic) potential energy functions U(r) for such motions and only require force constants as parameters. The theory is based on vacuum vibration potentials. The calculated vibration amplitudes usually agree well with those obtained from crystal diffraction data.

7.2. Computer Simulation Techniques (12)

7.2.1. Static Modeling

The starting point is a fixed structure, if possible, the known crystal structure for which the global potential energy is calculated. A residue substitution is then performed and the change in energy calculated, on the assumption that the overall structure has not been changed. The potential function used in such a calculation may incorporate some or all of the contributing interactions mentioned above and be of the form

$$U(\mathbf{r}) = U(\mathbf{r}_1, \mathbf{r}_2, \ldots \ldots \mathbf{r}_N)$$

and that describes the potential energy of the molecule in terms of the positions $\mathbf{r}_i$ of the N atoms. Changes in the atomic positions are then introduced and the potential is recalculated. If a change has reduced $U(\mathbf{r})$, then it is accepted, otherwise it is rejected. A configuration corresponding to a minimum potential energy can thus be arrived at which is assumed to correspond to the native, equilibrium state.

7.2.2. Molecular Dynamics

The static simulation provides information about the equilibrium structure, but the search of configuration space can be extended to provide dynamic information. The same potentials are then used in a calculation of the time evolution of the native protein structure. This is achieved by numerically solving Newton's equations of motion for the system of N atoms of masses m_i and configurations $\mathbf{r}_i$ over time intervals, usually of the order of 10^{-15} s. The method is referred to as Molecular Dynamics (MD) and usually yields a lower energy minimum than that predicted by static methods.

MD techniques, although more expensive in terms of computer time, provide valuable dynamic information, such as possible dynamic equilibria between alternative conformations, binding rates, bond flexibility, and temperature effects on structures. For example, in BPTI (bovine pancreatic trypsin inhibitor), two distinct conformational states have been identified with mean lifetimes of 75 and 110 ps, respectively, that interconvert, and MD searches of the hormone somatostatin at elevated temperatures have revealed a number of low energy states able to interconvert *(13)*.

7.3. Approximations, Limitations, and Applications

Probably the most serious limitation of present simulation methods is that, whereas potential energies are calculated, protein stability is a *free energy* function (Gibbs or Helmholtz) that differs from the potential energy by the entropic term TΔS. In the case of proteins, the entropic contribution to the free energy is large and includes solvent effects and hydrophobic "interactions." (The latter are frequently, but mistakenly, treated as van der Waals type interactions.) For that reason, they cannot be incorporated into the potential energy function. Over the past few years, methods have been developed to calculate *relative* free energy changes and they can be applied to protein calculations, e.g., for the determination of changes in the free energy of an enzyme during substrate binding.

Neither protein–water nor water–water interactions can be included in the calculations in any but a very approximate form. Even for the "simple" water dimer alone (20 electrons, 6 nuclei), more than 20 U(r) functions have been proposed since the first water simulations were performed during the early 1970s; none of them has been found to be entirely satisfactory in accounting for the physical properties of both liquid water and water vapor over the whole temperature range *(14)*. A more detailed discussion of the complex problems posed by protein hydration is presented elsewhere in this volume.

To estimate loop structures, ø, ψ values are required for all amino acid main chain and side chain angles; e.g., for a loop of 6 residues, there are 12 main chain angles plus approx 10 side chain angles, requiring 10^{18} energy calculations. In some of the simpler simulation treatments only main chain atoms are included, a somewhat heavy-handed approach to a subtle process.

A major conceptual problem centers on the validity of the assumption of pairwise additivity of atom interactions (the superposition principle) in the energy minimization. Protein stability is more likely to depend on many-body forces. At present there is a lack of statistical mechanical theories that would permit the calculation of many-body and cooperative interactions.

Despite the above (and other) limitations, protein structure calculations are finding increasing applications in several areas, e.g., in helping with the interpretation of complex NMR spectra. A particularly useful application is the combination of thermodynamic information with calculations for nonchemical changes that are not accessible to experiment. Consider the binding of a substrate S to two mutant enzymes E_A and E_B. The relative binding constant is obtained from the thermodynamic cycle

$$E_A + S \rightleftharpoons (E_A \bullet S)\ldots K_1$$
$$\Updownarrow \qquad \Updownarrow$$
$$E_B + S \rightleftharpoons (E_B \bullet S)\ldots K_2$$
$$\vdots \qquad \vdots$$
$$K_3 \qquad K_4$$

The horizontal processes are obtained by experiment but could not be simulated, because the complex changes in hydration states could not be incorporated in such calculations in any realistic manner. However, the two vertical, nonchemical processes, i.e., the mutations, are not easily accessible to experiment but can be calculated. The relative binding constants are given by $K_2/K_1 = \exp[-(\Delta G_2 - \Delta G_1)/RT]$ and, since the process is cyclical, then

$$\Delta G_2 - \Delta G_1 = \Delta G_4 - \Delta G_3$$

so that the calculated and experimental values can be checked against one another. If the simulation technique is considered to be reliable, then the effect of making a single mutation on the binding equilibrium can be obtained without the necessity of actually preparing the mutant.

8. Synthetic Proteins *(15)*

8.1. Partial Enzymes

The first attempts to create synthetic proteins involved the selective removal of portions of an enzyme and testing for activity, as a basis for eventual peptide synthesis. Thus it was found that RNase, when "trimmed" to 63 residues, retains 15% of its activity and 100% of its specificity *(16)*.

8.2. β -Structures

By searching databases for sequences that produce pure β-structures, it has been possible to synthesize so-called betabellins (bell-shaped) as synthetic β-structures joined by pro-asn loops *(17)*. The most successful attempts to date to produce structures which fold spontaneously have been performed with the following sequence:

HTLTASIPDLTYSIDPPNTATCKVPDθTLSIGB

where B = β-alanine and θ = iodo-phe, as a potential heavy atom derivative. This sequence does not correspond to that of any known protein.

8.3. Helical Structures

Four helix (FELIX) bundles, similar in structure to hemerythrin have been produced. They contain 79 residues and 19 different amino acids, but the sequences used do not exhibit any homology with any known protein. Unlike "natural" proteins, the artificial analogs require a disulfide bond for the teriary structure to be rendered stable.

8.4. Template-Assisted Synthetic Protein (TASP)

Peptide side chains have been grafted onto a peptide backbone at lys residues placed close together to constrain the side chains to close proximity. Side chain sequences are chosen for their propensity to form a-helices (*18*). Side chains can also be engineered to form β-structures. Thus, a 4-β version uses the sequence (val-lys)$_4$gly; it shows a pure β-type CD spectrum, but only at pH 12, presumably because of charge repulsions. All TASPs are quite soluble. This type of chain branched structure exhibits a high degree of stability and would appear to show promise as a synthetic peptide-based catalyst, provided that specific binding and active sites can be introduced.

Bibliography

1. Creighton T. E. (1988), *BioEssays* **8**, 57–63.
2. Castellino F. J. (1981), *Chem. Rev.* **81**, 431–446.
3. Hecht K. and Jaenicke R. (1989), *Biochemistry* **28**, 4979–4985.
4. Ng T. K. and Kenealy W. R. (1986), in *Thermophiles* (Brock T. D., ed.), Wiley, New York, p. 197.
5. Thornton J. M. and Gardner S. P. (1989), *TIBS* **14**, 300–304.
6. Kochva E., Wollberg Z., and Bdolah A. (1991), *Chem. Britain* **27**, 132–134.
7. Krebs H., Schmid F. X., and Jaenicke R (1983), *J. Mol. Biol.* **169**, 619–635.
8. Thornton J. M., Sibanda B. L., Edwards M. S., and Barlow D. J. (1988), *BioEssays* **8**, 63–69.
9. Kaiser E. T. (1987), in *Protein Engineering* (Oxender D. L. and Fox C. F., eds.), Liss, New York, p. 193.
10. Garner C. D. and Helliwell J. R. (1986), *Chem. Britain*, p. 835.
11. Samraoui B. et al. (1986), *Nature*, **320**, 378–380.
12. van Gunsteren W. F. (1988), *Protein Engineering* **2**, 3–9.
13. DiNola A., Berendsen H. J. C., and Edholm O. (1984), *Macromolecules* **17**, 2044–2050.
14. Finney J. L., Quinn J. E., and Baum J. O. (1985), *Water Sci. Rev.* **1**, 93–170.
15. Richardson J. S. and Richardson D. C. (1989), *TIBS* **14**, 304–309.
16. Gutte B. (1975), J. Biol. Chem., 250, 889–904; (1977), *J. Biol. Chem.*, 252, 663–670.
17. Kaiser E. T. and Kezdy F. J. (1985), *PNAS* **80**, 1137–1143.
18. Mutter M. (1988), *TIBS* **13**, 260–265.

Five

Solution Properties of Proteins

Felix Franks

1. Application of Physicochemical Techniques

Physicochemical measurements find application principally in the determination of size, charge, shape, conformation, interactions (e.g., binding, hydration), and dynamics (diffusion, reaction rates). Most techniques are limited to dilute solutions; some require optical transparency.

2. General Properties of Macromolecules

The shape and behavior of macromolecules in solution are determined by

1. Intramolecular effects (size, charge, conformation);
2. Solvation interactions;
3. Interactions *between* macromolecules (subunits);
4. External variables (temperature, pressure, pH, and so on).

Solvation (hydration) is also sensitive to the same external variables.

Changes in function or activity can be induced by changes in the environmental variables and the composition of the solvent medium. Technological performance critically depends on conformation, hydration (water holding capacity), and solubility. Biological performance is normally restricted to narrow physiological ranges of pressure, temperature, and medium composition. There are many examples, however, of evolutionary adaptation to

From: *Protein Biotechnology* • F. Franks, ed. © 1993 The Humana Press Inc.

extreme physical and chemical environments: thermophiles, psychrophiles, halophiles, acidophiles, and so on (1). Such adaptations are usually achieved through minor changes in the amino acid composition (2).

Like other macromolecules, proteins can be characterized by their chain conformations that, in turn, are described by the values adopted by the torsional angles. The chain properties of most synthetic polymers in solution, such as end-to-end distance, molecular weight, radius of gyration, and so on, are expressed in terms of average values that can be calculated from the monomer dimensions, the flexibility of the main chain bonds, the degree of polymerization, and the nature of the side chains. By contrast, the methods of polymer statistics cannot be applied to native proteins, because they possess *specific* conformations under physiological conditions, all molecules being identical in every respect.

3. Solubility

For the in vivo functioning, the solubility of a protein is unimportant. For the isolation and purification of proteins, solubility is one of the most important attributes. Proteins are polyelectrolytes, hence their solubility is governed mainly by electrostatic interactions.

3.1. Electrostatic Factors: pH Titrations

In detemining charge/charge interactions, the pK_a and pK_b values of the individual amino acids play an important role. They are summarized in Table 1. Only those amino acids with acidic or basic side chains affect the solution behavior (except the end groups). Histidine is the only amino acid with a buffering capacity in the physiological pH range of 6–8.

A pH titration of a protein provides a useful fingerprint of the number of ionic groups that are accessible to the solvent medium. Dissociation equilibria are expressed in terms of $\bar{\imath}$, the number of dissociated protons per molecule (3). Figure 1 is the titration profile of ribonuclease that should be read in conjunction with the amino acid composition/sequence. Not all of the 36 potentially

Table 1
Dissociation Constants (pK_a) of Amino Acids

Amino acid, or residue	pK, —COOH free acid	pK, —NH_2	pK, Side chain free acid	pK, Side chain in protein
Glycine	2.3	9.6		
Alanine	2.3	9.7		
Leucine	2.4	9.6		
Serine	2.2	9.2		
Threonine	2.6	10.4		
Glutamine	2.2	9.1		
Aspartic acid	2.1	9.8	3.9	4
Glutamic acid	2.2	9.7	4.3	4
Histidine	1.8	9.2	6.0	6
Cysteine	1.7	10.8	8.3	10
Tyrosine	2.2	9.1	10.1	10
Lysine	2.2	9.0	12.5	10
Arginine	2.2	9.0	12.5	12
Terminal —NH_2		8		
Terminal —COOH	3			

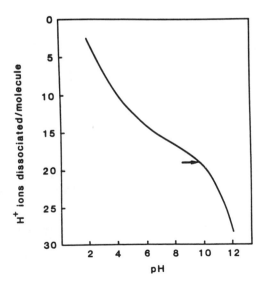

Fig. 1. Titration curve of ribonuclease at 25°C, ionic strength = 0.01.

titratable groups are in fact titrated in the range $2 < pH < 12$. Also the shape of the curve depends on the ionic strength (I), reflecting electrostatic effects on ion binding/dissociation.

Individual groups cannot be resolved from the titration profile, but the curve can be seen to be divided into several main regions, depending on the respective pK values of the amino acids. More detailed information can be obtained from a spectrophotometric titration curve of, say, tyrosine at 295 nm. In the case of ribonuclease this indicates that of the six tyr residues, three titrate normally, with $pK \approx 10$, but three have abnormally high pK values.

A semiempirical, but very useful method for analyzing titration profiles makes use of an "intrinsic" constant, $K_{i,in}$ and an electrostatic free energy of interaction $\Delta G°(\bar{z})$, such that

$$K_i = K_{i,in} \exp[-\Delta G°(\bar{z})/RT] \tag{1}$$

where $\bar{z}$ is the average charge on the protein molecule. It can then be shown that for any single type of residue i,

$$pH - \log \frac{x_i}{1 - x_i} = pK_{i,in} - 0.868w\bar{z} \tag{2}$$

where x_i is the fraction of class i that have been titrated, and w is an empirical constant (3).

The linearity increases as I is reduced. Equation (2) also gives the intrinsic pK of the protein (the pH value for which $\bar{z} = 19$). Note that the isoelectric point, pI, is a function of the ionic strength, the nature of the buffer used, and any other solutes present. It is *not* dependent on protein concentration. Table 2 provides a summary of pI values and ranges for a variety of proteins. It must be emphasized that even at the isoelectric point the protein molecule carries a charge and it is only the *net* charge that is zero. Since proteins are polyelectrolytes, many of their properties depend on their net charge, the charge distribution, ionic strength of the medium, and its dielectric permittivity.

Table 2
Isoelectric Points (pI) of Proteins[a]

Protein	pI	Protein	pI
Salmine	12.1	Collagen	6.7
Thymohistone	10.8	Gelatin	4.7–5.0
Ovalbumin	4.7	α-Casein	4.5
Conalbumin	7.1	Lysozyme	11.1
Serum albumin	4.9	Myoglobin	7.0
Myogen A	6.3	Hemoglobin (human)	7.1
β-Lactoglobulin	5.0	Hemocyanins	4.6–6.4
Livetin	4.9	Hemerythrin	5.6
γ$_1$-Globulin (human)	6.6	Cytochrome c	10.6
γ$_2$-Globulin (human)	8.2	Rhodopsin	4.5
Myosin A	5.3	Chymotrypsinogen	9.5
Tropomyosin	5.1	Urease	5.0
Thyroglobulin	4.6	α-Lipoprotein	5.5
Fibrinogen	4.8	Bushy stunt virus	5.3
α-Crystallin	4.8	Vaccinia virus	5.3
β-Crystallin	6.0	Prolactin	5.7
Arachin	5.1	Insulin	5.3
Keratins	3.7–5.0	Pepsin	1.0

[a]Values given refer to measurements at low ionic strength (< 0.01).

3.2. Classification of Proteins by Solubility

A purely phenomenological classification system makes use of the solubility:

Globulins	Insoluble in water, but dissolve in aqueous salt solutions.
Albumins	Soluble in water and in salt solutions.
Prolamins	Insoluble in water, but dissolve in 50–90% aqueous ethanol. Prolamins have only been detected in plants.
Glutelins	Insoluble in all the above solvents, but dissolve in dilute solutions of acid or base; they only occur in plants and are not well characterized.
Scleroproteins	Insoluble in most ordinary solvents; they swell in salt solutions.

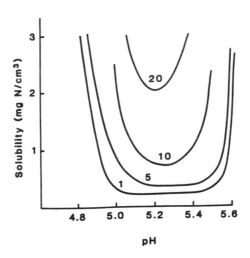

Fig. 2. The effects of pH and NaCl concentration on the solubility of β-lactoglobulin; concentrations in mol/L.

3.3. Solubility as Index of Viability/Denaturation

Technological performance often depends on the complex interplay of several physical properties. They determine technological attributes such as storage life, moisture loss, and response to additives, such as salts.

Solubility is a good index of the extent of denaturation, and the pH/ionic strength/solubility profile is often the best practical index of technological usefulness. Figure 2 shows the effects of pH and NaCl concentration on the solubility of β-lactoglobulin, and Fig. 3 shows the solubility of isoelectric carboxyhemoglobin as a function of I for various electrolytes. Two important principles are illustrated: First, the solubility undergoes a minimum in the neighborhood of the isoelectric pH, and second, the effect of salt on the solubility is complex with *qualitatively* similar behavior being produced by all salts. The quantitative effects differ widely between salts. Thus, besides the general ionic strength effect, there is another effect that is ion-specific. At low salt concentrations the ionic strength effect predominates, and *all* salts raise the solubility. At $I > 0.15$, ion-specific effects are first observed, with some ions reducing the solubility, and others enhance it further.

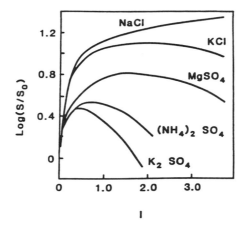

Fig. 3. Influence of different electrolytes at the same ionic strength on the solubility of isoelectric carboxyhemoglobin. Note the differences in salting in/out behavior, indicating that the electrolyte effects are not of a purely electrostatic nature.

The combined effects of salt concentration and pH on protein solubility are combined in Fig. 4. They are seen to be quite complex. The qualitative features are common for any protein/salt/acid system, but considerable differences become apparent when different combinations of salts and acids are compared, as discussed below.

3.4. Specific Ion Effects

At any given pH, the solubility S can be expressed by

$$\log S = \log S_o - K_s I \tag{3}$$

where S_o is the solubility at $I = 0$ and K_s is a constant, specific to the ions, but not to the nature of the protein. S_o is very dependent on the protein. At pH = pI, K_s becomes independent of temperature and pH.

Ion-specificity follows the Hofmeister series (lyotropic series) (4) that also controls micellization of surfactants, protein conformational stability, flocculation of colloids, and so on. It has been established that hydrophobic effects are implicated (5). The Hofmeister series of ion-specific effects is shown on the next page:

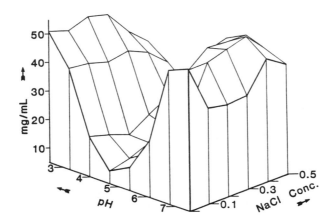

Fig. 4. Fababean storage protein solubility (mg/mL) as a function of pH and ionic strength.

salting-out   neutral   salting-in

$SO_4'' > HPO_4'' > OAc' > F' > cit^{3-} > Cl' < NO_3' < I' < CNS' < ClO_4'$
$Li^+ > Na^+ < K^+ < Rb^+ < Cs^+$
$Me_4N^+ < Et_4N^+ < Pr_4N^+ << Bu_4N^+$

Solubility determinations are complicated by the fact that ions also affect the stabilities of the native states of proteins and their degree of association. Thus, salting-out ions (and molecules) enhance the stability of native states, whereas salting-in ions (and molecules) give rise to denaturation (6,7).

3.5. Solubility and Processing Conditions

The total soluble nitrogen content of a protein, or its unit activity is markedly influenced by several factors related to the processing conditions, such as isolation, precipitation, and resolubilization. As a general rule, acid or salt precipitation is to be preferred over alkali or heat precipitation or freeze/thaw treatments.

3.6. Solubility and Nonelectrostatic Factors

Salting in/out can also be achieved by nonelectrolytes. Thus, carbohydrates salt-out (8–10); K_s is proportional to the number of –OH groups. Urea salts-in (11). Its solubilizing action is frequently employed in the processing of protein inclusion bodies. Alcohols exhibit an extremely complex behavior: e.g., ethanol salts-out at low concentrations and low temperatures; at higher temperatures it salts-in at all concentrations (12). Despite many claims to the contrary, these effects do not depend on polarity (dielectric constant) but on the eccentric properties of water–alcohol mixtures themselves (13).

The nonelectrolytes can be fitted into the Hofmeister series of ion-specific effects. There is as yet no convincing explanation for such effects. They cannot be accounted for by conventional electrostatic theories that treat the solvent medium as a dielectric continuum and ionic effects only in terms of concentrations and valences (ionic strength). On the other hand, the series has been "rediscovered" at regular intervals during the past hundred years.

4. Solution Properties of Large Molecules

Apart from the charge, the main determinants of the solution behavior of large molecules are size and shape. Since most measurement techniques are based on solution thermodynamics, we summarize briefly the main features of binary solutions.

4.1. Thermodynamics
of Ideal and Nonideal Solutions

The chemical potential μ_i of component i in a solution is referred to the ideal solution (Raoult's law) by

$$\mu_i - \mu_i^o = RT \ln x_i \qquad (4)$$

where x_i is the mol fraction concentration, and μ_i^o is the chemical potential of the *pure* component $i(x_i = 1)$. More practical units of concentration are mol/L or g/L (c_i). In the latter case, μ_i^o refers to $c_i = 1$ g/L.

Nonideal behavior is taken into account by the activity coefficient f_1 by which the concentration must be multiplied for Eq. (4) still to hold. For a binary solution in the limit of infinite dilution $f_1 \to 0$ as $x_1 \to 1$ (or $c_2 \to 0$). Usually $f_1 < 1$. Also in a binary solution ($x_1 - x_2$) and $\ln(1 - x_2) = -x_2 - x_2^2/2...$, so that

$$\mu_i - \mu_i^o = -RT(x_2 + ...) + RT \ln f_1$$

and $x_2 = c_2 V_1^o/M_2$, where V_1^o is the molar volume of the solvent (water) and M_2 is the molecular weight of the solute. Making this substitution for x_2, we obtain

$$\mu_i - \mu_i^o = -(RTV_1^o c_2)/M_2 + RT \ln f_1$$

Expressing $\ln f_1$ in terms of a power series in c_2:

$$\ln f_1 = -(\alpha c_2^2 + \beta c_2^3 + ...)$$

so that Eq. (4) now becomes

$$\mu_i - \mu_i^o = -RTV_1^o\,[c_2/M_2) + Bc_2^2 + ...] \qquad (5)$$
$$= -RTV_1^o c_2/M_2(1 + BM_2 c_2 + ...) \qquad (6)$$

B is known as the second viral coefficient and serves as a convenient measure of nonideality. Equation (6) serves as the basis for the determination of molecular weights.

4.2. Solution Properties and Molecular Shapes

The second viral coefficient B and many other solution properties of macromolecules are sensitive to the shape (conformation) of the large molecule. A linear polymer in which there is little or no resistance to rotation around the bonds adopts a *random* conformation (random coil), the average dimensions of which can be calculated *(14)*. The *average* radius of gyration, $<r_g>$, is defined in terms of the *average* squared distance $<r_i^2>$ of residue i from the center of mass, by

$$r_g^2 = \left(\Sigma <r_i^2>\right)/N \qquad (7)$$

where N is the number of residues. For a true random coil, r_g is proportional to $N^{1/2}$. The structural proteins in which the predominant secondary or quaternary structure is helical adopt a rod-like conformation in solution. On the other hand, globular proteins can often be quite adequately described as spheres or ellipsoids. Nonideal behavior in solutions of macromolecules arises primarily from their size. The center of each molecule is excluded from a volume determined by the volumes occupied by all the other molecules. B therefore depends on this *excluded volume (u):*

$$B = Nu/2M_2^2 \tag{8}$$

where N is the Avogadro number. The actual value of u is calculated from the molecular dimensions:

for spheres: $u = 8M_2v/N$ and $B = 4v/M_2$
for rods: $u = 2LM_2v/Nd$ and $B = Lv/dM_2$

where v is the specific volume of the polymer and L and d are the length and diameter of the rod.

      The excluded volume effect always reduces μ_1, i.e., it tends to push molecules apart. It is used to effect separation, purification and concentration of proteins. An example is given in Fig. 5: The excluded volume effect causes two polymers of different shapes (PEG and Dextran) to separate spontaneously into two phases. The protein to be purified will become concentrated in one of these phases. The fractionation can be further improved by the aid of salting-out.

4.3. Quality of Solvent

      The thermodynamics of mixing a solute and a solvent are determined by the balance of the interactions between solvent molecules, solute molecules, and cross-interactions between solute and solvent. If the sum of the self-interactions balances the cross-interactions, the solvent is known as a theta (θ) solvent and the solution exhibits a pseudoideal behavior, with $B = 0$. When preferential solvation occurs (good solvent), $B > 0$, whereas in a poor solvent the polymer molecules tend to aggregate and $B < 0$. In the

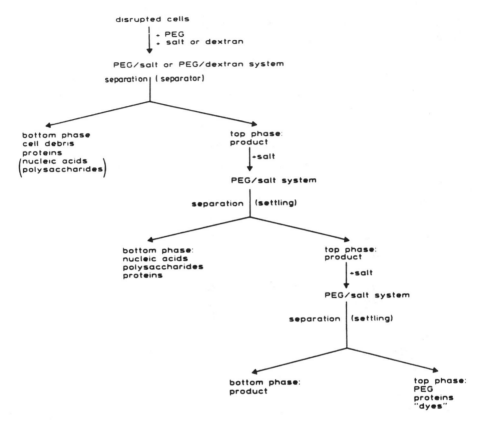

Fig. 5. General scheme of extractive protein purification (three-step process).

extreme case the polymer will precipitate. The concept of good and poor solvents is related to the phenomenon of salting in/out, already discussed in Section 3.5. Table 3 summarizes some calculated nonideality parameters for solutions of macromolecules at a concentration of 5 mg/cm³, a typical protein concentration in solution experiments (15).

5. Membrane Equilibria

Semipermeable membranes find multiple applications in biochemistry and biotechnology, e.g., dialysis, membrane filtration, osmosis, and reverse osmosis. The membrane acts as a barrier that

Table 3
Solution Nonideality Corrections, as Expressed in Eq. (6)[a]

Particle	B	$(1 + BM_2c)$
Sphere, $M = 10^5$	3×10^{-5}	1.015
Rod, $L/d = 1000$, $M = 10^5$	7.5×10^{-4}	1.375
Random coil, $M = 10^5$	5×10^{-4}	1.250
Random coil, θ solvent	0	1.000

[a]Units: B in $cm^3/g^2/mol$, $c = 5$ mg/cm^3.

retains some substances and allows others, of a lower molecular weight, to pass. In practice, cellophane membranes that are commonly used have a cutoff of $M < 10,000$, but membranes can be prepared that select at much higher or lower molecular weights.

5.1. Osmotic Pressure

The separation of a protein solution from the pure solvent by means of a semipermeable membrane leads to a flux of solvent into the solution that gives rise to an excess pressure (osmotic pressure). Equation (5) can then be written as

$$\mu_1 - \mu_1^o = - RTV_1^o(c_2/M_2 + Bc_2^2 + \ldots) + V_1^o\pi \qquad (9)$$

where π is the osmotic pressure. Equation (9) applies at atmospheric (low) pressure and low c_2. Now

$$\pi = RT(c_2/M_2 + Bc_2^2 + \ldots) \qquad (10)$$

In the limit of $c_2 \to 0$

$$\pi = RTc_2/M_2 \qquad (11)$$

analogous to the ideal gas equation, from which the molecular weight of the protein can be obtained. In practice, molecular weights are calculated from the relationship

$$\pi/c_2 = RT/M_2 + BRTc_2 + \ldots \qquad (12)$$

Table 4
Molecular Weights Obtained
from Osmotic Pressure Measurements

Polymer	M_n
Ovalbumin	44,600
Hemoglobin	66,500
Aldolase	156,500
Aldolase subunit	42,400
Amylose	32,000–150,000

by a plot of π/c_2 vs c_2 and extrapolation to $c_2 = 0$. The method has several disadvantages. (1) The possibility of protein aggregation/denaturation while the determination is in progress, (2) the measurement should be performed at net zero charge, but the pI may not be known, (3) the method gives no indication of the purity of the sample, i.e., a *mean* molecular weight may be obtained. This is a well-known effect in polymer chemistry.

Table 4 summarizes some molecular weights obtained from osmotic pressure measurements, and Fig. 6 is a typical osmotic pressure plot for native aldolase and its dissociated subunits. The slopes of the lines are proportional to B, according to Eq. (12); note the larger B for the denatured subunits.

5.2. Equilibrium Dialysis

Dialysis can be used to purify proteins in solution by the removal of low molecular weight impurities, but also to measure the binding of ions or molecules. The protein solution is contained in a membrane bag (phase α) that is suspended in a solution containing the low molecular weight substance component 3 (phase β). Component 3 will distribute itself between phases α and β, until, at equilibrium, $\mu_3^\alpha = \mu_3^\beta$. Neglecting pressure differences, then

$$RT \ln c_3^\alpha f_3^\alpha = RT \ln c_3^\beta f_3^\beta$$

or

$$c_3^\alpha f_3^\alpha = c_3^\beta f_3^\beta$$

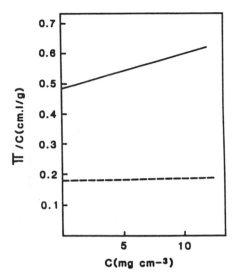

Fig. 6. Osmotic pressure data for native aldolase at neutral pH (broken line) and the individual subunits of the protein in 6M GuHCl (solid line).

If we now assume that both bound and free component 3 exist in phase α, and that the environments on both sides of the membrane are sufficiently similar, so that $f^{\alpha}_3 = f^{\beta}_3$, then c^{α}_3 (free) $= c^{\beta}_3$ and

$$c^{\alpha}_3 \text{ (bound)} = c^{\alpha}_3 - c^{\beta}_3 \qquad (13)$$

Unless binding is strong, the results obtained from dialysis equilibrium measurements must be treated with caution.

5.3. Donnan Equilibrium

When the macromolecule carries a charge (e.g., protein), the membrane equilibrium is further complicated *(16)*. Let the protein be designated by PX_z, which dissociates according to

$$PX_z \rightarrow P^{+z} + zX^-$$

where X^- is a monovalent counterion. In the presence of a salt BX, which dissociates into B^+ and X^-, and at equilibrium, we have

$$c_\beta^a \, c_X^a = c_B^\beta \, c_X^\beta$$

In addition, electroneutrality demands that

$$c_\beta^a - c_B^\beta = -z c_p^a \, c_\beta^a / \left(c_B^\alpha - c_B^\beta \right) \tag{14a}$$

$$c_X^a - c_X^\beta = z c_p^a \, c_X^a / \left(c_X^\alpha + c_X^\beta \right) \tag{14b}$$

In other words, at equilibrium, the counterion X will be more concentrated on the polymer side of the membrane, whereas the concentration of the cation B will be lower. These concentration differences will contribute to the osmotic pressure, and it can be shown that

$$\pi = RT \left\{ \frac{c_p}{M_p} + \frac{z^2 c^2 p}{4 M_p^2 c_{BX}} \right\} \tag{15}$$

In other words, the solution behaves as if it were nonideal, with a virial coefficient proportional to the charge/mass ratio. This effect is minimized at the isoelectric point, but can become large at pH values far removed from pI.

It is evident from Eq. (15) that π will approach its ideal value as c_{BX} becomes large. Analogs of the Donnan effect also exist in other physical measurements (sedimentation, light scattering), so that it is advisable to use high salt concentrations for molecular weight determinations, especially when measurements are made at pH values far removed from pI.

6. Diffusion, Sedimentation, Viscosity

Like the equilibrium properties, the dynamics of macromolecules in solution are also sensitive to their sizes, shapes, and charges. Several experimental techniques make use of the responses of such molecules to applied fields, such as centrifugal, electrical or shear forces, or combinations of the above.

6.1. Diffusion and Sedimentation

The basic equation governing the free diffusion of particles in a continuous medium is

$$(\partial c/\partial t) = D(\partial^2 c/\partial r^2) \text{ (Fick's law)} \qquad (16)$$

where the left-hand side is the rate of change of concentration and $\partial c/\partial r$ is the concentration gradient; D is the proportionality constant (diffusion coefficient). By imposing certain boundary conditions, Eq. (16) can be integrated:

$$(\partial c/\partial t)_x = [c_o/2(\pi Dt)^{1/2}]\exp[-x^2/4Dt] \qquad (17)$$

where c_0 is the initial concentration. In practice, D is measured by the broadening of an initially sharp interface, as detected by a suitable optical device. Figure 7 is a schematic representation of the diffusion broadening, from which D can be obtained as

$$A/H = 2(\pi Dt)^{1/2} \qquad (18)$$

where A is the area of the diffusion peak and H its height. Provided the molecular weight is known, D can provide information about the shape of the protein in solution through the introduction of f, the frictional coefficient, which is a measure of the asymmetry and hydration of the protein molecule:

$$f/f_o = (f/f_o)_{\text{hydration}} (f/f_o)_{\text{sym}} \qquad (19)$$

The frictional coefficient f_o is that of an unhydrated sphere of the same mass and is given by

$$f_o = 6\pi\eta r \qquad (20)$$

in terms of the viscosity of the solvent (η) and the radius of the sphere (r). The axial ratios, f/f_o, diffusion coefficients, and molecular weights of proteins are summarized in Table 5. If the shape of the molecule is known from some independent measurement, then f/f_o permits the evaluation of the degree of hydration. On the other

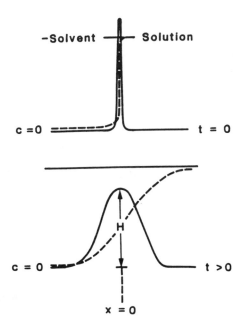

Fig. 7. The broadening of an initial sharp boundary ($t = 0$) as a function of time (t) during free diffusion. The broken line represents the concentration as a function of the position in the cell (x). The solid line shows the gradient, as given in Eq. (17).

hand, assuming zero hydration, f/f_o for an ellipsoid can be calculated from its dimensions. Thus, $(f/f_{o\ Asym}) = 1.5$ corresponds to a prolate ellipsoid with an axial ratio of 10:1. The axial ratio is best determined from scattering measurements (*see* Section 7.).

The development of the analytical ultracentrifuge has facilitated the determination of sizes and shapes of macromolecules with a high degree of precision. Two distinct techniques find application: (1) sedimentation equilibrium and (2) sedimentation velocity. In the former method the speed of rotation is adjusted so that the centrifugal force just balances diffusion and there is no net solute migration. M is then given by

$$M = \{RT[d(\ln\gamma)/dc]/r\omega^2(1-\bar{v}\rho)\},(1/c)(dc/dr) \qquad (21)$$

Table 5
Diffusional Properties and Molecular Weights of Proteins

Protein	Molecular weight	Diffusion coefficient, $D_{20,\omega} \times 10^7$	Frictional ratio, f/f_o	Sedimentation coefficient $s_{20,\omega}$
Cytochrome c	13,370	11.4	1.19	1.17
Ribonuclease	13,683	11.9	1.14	
Lysozyme	14,100	10.4	1.32	
Chymotrypsinogen	23,200	9.5	1.20	2.54
β-Lactoglobulin	35,000	7.82	1.25	2.85
Ovalbumin	45,000	7.76	1.17	
Serum albumin	65,000	5.94	1.35	4.6
Hemoglobin	68,000	6.9	1.14	4.46
Catalase	250,000	4.1	1.25	11.3
Urease	480,000	3.46	1.20	18.6
Tropomyosin	93,000	2.24	3.22	
Fibrinogen	330,000	2.02	2.34	7.63
Collagen	345,000	0.69	6.8	
Myosin	493,000	1.16	3.53	6.43
Tobacco mosaic virus	40,590,000	0.46	2.03	198

where ω is the angular velocity, $\bar{v}$ the specific volume, and ρ the density. Values of c and dc/dr are obtained from the experimental measurements, which are extrapolated to infinite dilution, where the activity coefficient term $1 + [d(\ln\gamma)/dc]$ approaches unity.

This is the most precise method for the determination of M, and has been further refined to reduce the time required for equilibrium to be established.

In the sedimentation velocity method the centrifugal force on the macromolecule is increased until it just balances the frictional force opposing migration. The solute then moves down the tube at a constant velocity. The molecular weight is obtained by the use of the Svedberg equation

$$M = RTs/D(1 - \bar{v}\rho) \qquad (22)$$

where s is known as the sedimentation constant and is given by

$$s = (dx/dt)/\omega^2 x \; s^{-1}$$

x being the distance from the center of the rotor. The basic unit is taken as $10^{-13}s^{-1}$ (one Svedberg). Sedimentation constants for most proteins fall in the range 1–50 Svedbergs. Table 5 includes some sedimentation constant data.

Both s and D depend on the concentration via the frictional coefficient, f, already referred to. The experimental data must therefore be extrapolated to $c \rightarrow 0$. They must also be referred to some standard condition, e.g., water at 20°C. If the measurements are performed in a buffer, b, at temperature, T, then

$$s_{20,\omega} = [(1 - \bar{v}\rho)20_{,\omega}\eta_{t,b}/(1 - \bar{v}\rho)_{T,}\eta_{20,\omega})] \; s_{T,b}$$

it being assumed that f is accurately proportional to the viscosity η.

Various refinements have been developed to increase the resolving power of the ultracentrifuge. In particular, the introduction of a density gradient with the aid of an "inert" substance, such as sucrose, makes possible the separation of the components in a mixture of proteins, and the s value of each fraction can be calculated.

6.2. Viscosity

Viscosity data supplement the information derived from diffusion and sedimentation measurements. The contribution of a macromolecule to the viscosity of a solution depends mainly on its volume, in contrast to the frictional coefficient, which depends on the linear dimensions.

A fluid is termed Newtonian when its viscosity is constant, independent of the shear stress. For a dilute solution of solute concentration ϕ (volume fraction)

$$\eta/\eta_0 = 1 + k\phi \tag{23}$$

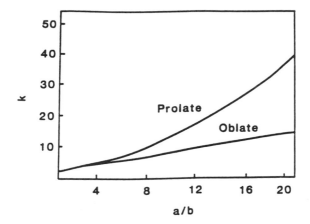

Fig. 8. Constant k in the viscosity Eq. (23) for ellipsoids of different axial ratios.

where η_0 is the viscosity of the solvent and $k = 2.5$ for spherical particles. For other shapes $k > 2.5$, as shown in Fig. 8 for ellipsoids of different axial ratios. Equation (23) is usually extended to take account of solute–solute interactions at higher concentrations:

$$\eta_{sp} = \frac{\eta}{\eta_o} - 1 = \frac{\eta - \eta_o}{\eta_o} = k\phi + k'\phi^2 \dots \tag{25}$$

where η_{sp} is known as the specific viscosity and is dimensionless. In most cases it is more convenient to express concentration in terms of g/cm^3. Equation (24) then becomes

$$\eta_{sp}/c = kv + k'v^2c \tag{25}$$

The limit of η_{sp}/c as $c \to 0$ is called the intrinsic viscosity $[\eta]$. It depends on the properties of the isolated protein molecule, and from Eq. (25)

$$[\eta] = kv$$

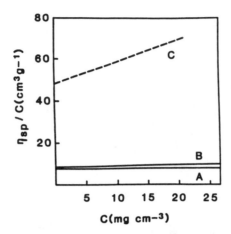

Fig. 9. Determination of the intrinsic viscosity of crab hemocyanin. (A) Native protein ($M = 10^6$); (B) Subunits ($M = 7 \times 10^4$); (C) Denatured (GuHCl) protein.

In other words, $[\eta]$ is a function only of the shape (through k) and the volume of the protein, and $k \geq 2.5$. For many proteins the *anhydrous* specific volume is set at ~0.75, so that the minimum value of $[\eta]$ is approx 2 cm³/g. The $[\eta]$ data included in Table 5 illustrate that this value is sometimes approached. The slope of the η_{sp}/c vs c plot gives an indication of the protein–protein interactions. Figure 8 shows that neither intact molecules of hemocyanin nor the dissociated, but native, subunits interact strongly. Pronounced aggregation does however take place between the unfolded subunits. Also from Fig. 9 we see that the unfolded subunit appears to be much larger than its native, compact counterpart. A useful expression for *random-coil* polymers (including proteins) is

$$[\eta] = KM^{a} \tag{26}$$

where $a \geq 0.5$. For a θ-solvent, $a = 0.5$. K is an empirical constant that depends on the degree of hydration of the macromolecule. For compact folded proteins there is no direct relationship between $[\eta]$ and M.

When the viscosity is a function of the shear rate, the fluid is said to be non-Newtonian. Two types of behavior can be distinguished: shear thinning and shear thickening. The former is more common and is observed for elongated molecules (fibrous proteins, nucleotides) that can be aligned by the flow gradient. Non-Newtonian flow can provide problems in the experimental determination of [η]. For instance, the shear rate in an ordinary capillary viscometer is very high, so that artificially low [η] values are indicated. Viscosity measurements, especially on concentrated solutions, should normally be performed over a range of shear rates and extrapolated to zero shear rate.

7. Scattering of Radiation

If an assembly of molecules is irradiated with light of wavelength λ, their electronic charge distribution is affected, resulting in a dispersion of the radiant energy in directions other than that of the incident radiation. This is the common basis for *all* scattering processes (X-rays, light, or neutrons).

For very dilute solutions of macromolecules in which the solute molecules behave as independent point scatterers and are small compared to λ, the scattering intensity I_θ at an angle θ is expressed by

$$R_\theta = \frac{r^2}{1 + \cos^2\theta}$$

$$= \left[\frac{2\pi^2\eta_0^2(dn/dc)^2}{N\lambda^4}\right] cM = KcM \tag{27}$$

where R_θ is the Rayleigh scattering ratio, I_0 is the intensity of the incident beam, n is the refractive index, and N is the Avogadro number. For an ideal solution, Eq. (27) reduces to

$$Kc/R_\theta = M^{-1}$$

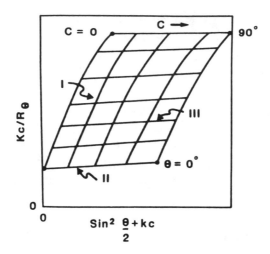

Fig. 10. Typical Zimm plot, as obtained from scattering data at different scattering angles and concentrations (for details, *see* text).

For a nonideal solution, a second viral coefficient B is introduced, analogously to Eq. (10):

$$Kc/R_\theta = M^{-1} + 2Bc \qquad (28)$$

When the greatest dimension of the scattering particle exceeds the wavelength of the radiation, then the scattering experiment becomes more complex, but also more informative. In that case

$$\frac{K_c}{R_\theta} \sim \left\{ \frac{1 + 16\pi^2 r_g^2}{3\lambda^2} \ sin^2 \ \frac{\theta}{2} \right\} \{M^{-1} + 2Bc\} \qquad (29)$$

It is now necessary to extrapolate Kc/R_θ to $c \to 0$ *and* $\theta \to 0$. This is achieved by means of a Zimm plot, as shown in Fig. 10. The points along lines such as that marked I refer to measurements at constant c, but different θ. The value of Kc/R_θ that will be approached at $\theta = 0$ and at the particular c value is given by the point at which line I has the abscissa value Kc, where K is some arbitrary constant. The equation of line II is

$$\frac{K_c}{R_\theta}\Big|\theta = O = M^{-1} + Bc$$

If a number of determinations is made at constant θ, but varying concentrations (line III), the extrapolation to $c \to 0$ will yield

$$\frac{K_c}{R_\theta}\bigg|\, c = O = M^{-1}\left[1 + \frac{1 + 16\pi^2 r_g^2}{3\lambda^2}\; sin^2\; \frac{\theta}{2}\right]$$

Both procedures therefore yield an estimate of M and the radius of gyration r_g is also obtained from the experiment. Its interpretation in terms of the dimensions of the particle does, however, depend on the shape of the particle. For instance, for a sphere of radius R, $r_g = (3/5)^{1/2} R$, whereas for a very long rod of length L, $r_g = L/(12)^{1/2}$.

7.1. Small Angle X-Ray Scattering

Although light scattering is adequate for the determination of M, the method fails when the dimensions of particles with $r_g < 10$ nm are to be estimated. Unfortunately, most materials absorb strongly in the wavelength region <300 nm, but they become transparent again in the X-ray region. However, the scattering is almost entirely confined to a narrow angular region. Experimentally this does not matter, because reliable extrapolations to q = 0 can be performed. For the low angle range we can write

$$-h^2 r_g^2/3 = \frac{-16\pi^4 r_g^2 sin^2(\theta/2)}{3\lambda^2} \tag{30}$$

with $h = (4\pi/\lambda)\sin(\theta/2)$. A linear relationship exists between ln $i(\theta)$ and $sin^2(\theta/2)$, with a slope $-16\pi^4 r_g^2/3\lambda^2$ and an intercept ln $i(0)$. Even the smallest protein molecules can be investigated by this method.

8. Optical Spectroscopy

In biochemistry optical spectroscopy is usually confined to the observation of absorption spectra (except for fluorescence). The requirement for the absorption of energy is that the energy available must be related to the difference between energy levels, according to Planck's law, $\Delta E = h\nu$, where ν is the frequency and h is Planck's constant. Three important spectral features provide the desired molecular information:

1. Frequency: corresponds to the energy associated with the transition ΔE.
2. Intensity: corresponds to the concentration of the species responsible for the transition.
3. Band shape (usually width at half-height): provides dynamic information (diffusion), environment (hydrogen bonding, and so on), and interactions (binding, denaturation).

Spectroscopic measurements are usually also confined to dilute solutions for which the Beer-Lambert law governs the absorption process:

$$A = \log(I/I_0) = \varepsilon l c \qquad (31)$$

where A is the absorbance, I and I_0 are the intensities of the transmitted and incident beam, l is the optical path length, c the concentration (usually in mol L) and is the extinction coefficient, with the units L mol^{-1} cm^{-1} in Eq. (31).

Experimental data are sometimes expressed as percent transmission, $100I/I_0$, and other concentration units can also be used, e.g., g/100 cm^3. This is useful when the molecular weight of the absorbing species is not known.

The types of absorption spectra and the molecular information that can be obtained in different parts of the frequency spectrum are summarized below:

Far infrared	>10,000 nm	pure rotation
Near infrared	1000–10,000 nm	bond vibrations
Visible/UV	200–800 nm	outer shells electrons
X-ray		inner shell electrons

Experimentally, the determination of absorption spectra requires a source of radiation of the requisite frequency range, a monochromator, the sample, and a suitable detection system. Most instruments in common use also include a beam splitter, so that the beam that passes through the sample is automatically compared with the beam that is made to pass through a reference standard of some kind. The output is then passed through a difference circuit and an amplifier, and the spectrum is either recorded or the data collected on disk for future processing.

8.1. Infrared Spectroscopy

Pure rotation spectra can only be obtained from dilute gases, where collisions between molecules are negligible. Protein spectra are obtained from solutions, and therefore the near infrared region becomes of interest. Unfortunately there is such a wealth of detail in the spectra that it is often difficult to assign observed bands to definite bond vibrations. It must be remembered that a nonlinear molecule with n atoms has $3n-6$ fundamental vibrational modes. A complete analysis of the protein spectrum is therefore out of the question. Nevertheless, certain groups display characteristic vibrational transitions, which are also sensitive to the environment and to hydrogen bonding, and this enables useful information to be derived from such spectra. The >CO group has a fundamental stretching vibration near 1700 cm^{-1} and- the >NH group stretch vibration is observed near 3400 cm^{-1}. (Note that the frequency expressed in wave numbers, per cm, is the reciprocal wavelength, i.e., the number of wavelengths cm^{-1}.) Both the >CO and >NH vibrations are very sensitive to hydrogen bonding. Their involvement in hydrogen bonding shifts the observed frequencies to lower values. One troublesome interfering factor is the strong absorption of water that occurs in the two regions 3500 and 1500 cm^{-1}. This makes the direct observation of fundamental stretching vibrations difficult, though not impossible with modern instrumentation.

In practice, the two most informative infrared bands are the so-called amide I and amide II bands (17). The former is the >CO band, referred to above, and the latter is a combination of the >NH deformation and the CN stretching vibration. The amide I and II band frequencies are diagnostic of the protein secondary structure, as shown in Table 6. A further series of amide bands (III–VII) has also been identified and assigned to specific vibrational modes involving the peptide groups. Because of the sensitivity of the vari-

Table 6
Infrared Frequencies of Amide I and II Bands
of Polypeptides in Various Secondary Structures (cm^{-1})

Conformation	Amide I	Amide II
Random coil	1656 (s)	1535 (s)
Helix	1650 (s)	1516 (w)
	1652 (m)	1546 (s)
Pleated sheet (parallel)	1645 (w)	1530 (s)
	1630 (s)	1550 (w)
Pleated sheet (antiparallel)	1685 (w)	1530 (s)
	1632 (s)	

ous amide bands to hydrogen bonding interactions, they can be used to monitor the interaction of water with the polar residues of the protein. Thus, infrared spectroscopy has become a popular technique for the study of protein hydration.

8.2. Visible and UV Spectroscopy

Visible and near UV spectra originate from relatively low-energy electronic transitions. Two main groups of chemical species are involved: compounds containing metals (especially transition metals), such as metalloproteins, and large aromatic ring structures or conjugated double-bond systems. Electronic transitions are usually very broad because they include a large number of closely spaced bands, each corresponding to a vibrational energy change.

Among the proteins, most of the measured absorption in the 260–280 nm region results from the aromatic amino acid residues phe, trp, and tyr. Figure 11 shows the absorption spectra of these three residues. Farther in the UV, other groupings begin to absorb. Thus, at ≤230 nm other amino acid side chains absorb, but also the electron displacements in the peptide backbone produce absorption spectra. A strong band is observed at 200 nm and a weaker one at 225 nm. Below 180 nm oxygen, water and all other solvents absorb strongly, so that only dry materials can be studied under vacuum conditions; hence the term vacuum UV. As is the case for most spectroscopic techniques, the details of visible/UV spectra

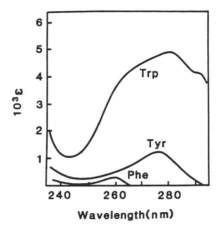

Fig. 11. Absorption spectra of aromatic amino acids.

are sensitive to the environment of the chromophore and can therefore be used to monitor conformational changes, during which the absorbing residues become more or less able to perform diffusional motion or to interact with water.

UV and visible spectra are particularly useful as fingerprinting methods; devices such as intensity ratios and/or derivative and difference spectroscopy are informative. Both tyr and trp have derivative peaks at 284 nm, but only trp has a peak at 290 nm. Figure 12 is a "calibration" curve of the intensity ratio 284/290 nm for tyr/trp mol ratios shown. It is a useful tool in amino acid analysis. Other intensity ratios, e.g., 280/260 or 280/320 nm, are used routinely for the detection of nonprotein impurities. Maxima in the first derivative spectra of trp, tyr, and phe are summarized in Table 7.

8.3. Fluorescence Spectroscopy

In the case of absorption spectroscopy, one measures energy absorbed during the excitation of a molecule from the ground state to some excited state, but without considering the eventual fate of the excited state. In a few cases, however, energy is reradiated at a different frequency from that of the exciting radiation. This process is called *fluorescence*. The fluorescence spectrum is independent of the wavelength of the exciting radiation, but it appears at a

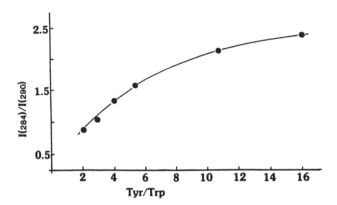

Fig. 12. Ratio of the first derivative peak intensities for solutions containing various molar ratios of tyrosine and tryptophan in 0.5M NaCl. Values along the ordinate axis are the measured ratios of the "284" derivative peak to the "290" derivative peak.

Table 7
Maxima for First Derivatives of UV Absorbance Spectra
for Phenylalanine (1.70 mM), Tyrosine (0.223 mM),
and Tryptophan (2.23 mM)

Amino acid	Major maxima, nm	Minor maxima, nm
Phenylalanine	267	242
	263.5	247
	257.5	
	252	
Tyrosine	284.5	
	277	
Tryptophan	290.5	272.5
	283	

higher wavelength, as seen in Fig. 13 for the case of tyrosine. The fluorescence spectrum is a sensitive fingerprint for a compound and, since very low intensities of emitted radiation can be detected, it serves as a very sensitive analytical technique.

If the exciting radiation is polarized, then the degree of depolarization of the fluorescent radiation provides a measure of the Brownian diffusion of the chromophoric species.

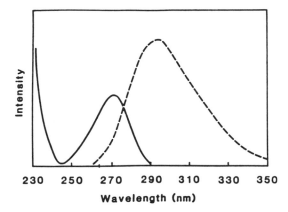

Fig. 13. Absorption (solid line) and fluorescence (broken line) spectra of tyrosine.

Under favorable conditions, excitation energy can be transferred from one chromophore to another, so that the fluorescence spectrum of the latter is observed. The efficiency of such an energy transfer is a function of r^{-6}, where r is the distance between the two chromophores involved. For proteins it is usually found that even when excitation is in the phe or tyr bands, the trp fluorescence spectrum is in fact observed.

Valuable information can be obtained by excitation at several wavelengths, e.g., 270, 280, 290, 297 nm. This is illustrated by the following example: Assuming zero contribution from phe, then at 270 nm only tyr excites. If its emission is efficiently relayed to trp, then only a trp emission is observed. The half-band width monitors the efficiency of the energy transfer. At 297 nm excitation, only trp is excited. In a polar environment its emission is at 350 nm. If trp is buried in a nonpolar region, then its emission is observed at 325 nm, i.e., unfolding (dissociation) can be monitored from shifts in the emission band. Also, if tyr becomes displaced from trp, the energy relay may be broken and at 280 and 290 nm excitation, both emission bands may be evident. Figure 14 shows the emission spectra of purified vicilin; the absence of trp is evident.

Fluorescent probes are molecules that are covalently linked to reactive amino acid side chains, e.g., lysine. They are useful tools for monitoring flexibility, the rate of conformational changes,

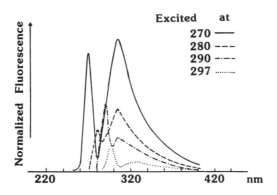

Fig. 14. Fluorescence emission spectrum of chromatographically purified fababean vicilin storage protein.

polarity, and dissociation constants. TNS (2-*p*-toluidinyl-naphtha-
lene-6-sulphonate) is a commonly used probe. The information is
most valuable when the exact location of the probe on the peptide
chain is known. It must also be remembered that the probe is bulky
compared to an amino acid residue, so that an observed fluorescence change may in fact be due to a perturbation caused by the
introduction of the probe into the protein.

8.4. Optical Activity, Circular Dichroism (CD), and Optical Rotatory Dispersion (ORD)

The responses of dissolved molecules to polarized radiation
depend explicitly on their structural asymmetry. Hence this
response is a useful identifying tool for the type of structure that
exists in a protein molecule.

Optical rotation is the consequence of a different refractive
index (n) for left and right circularly polarized light, whereas CD
results from a corresponding difference in absorption. CD is usu-
ally expressed as molar ellipticity [θ], which is equal to 3300 $\Delta\varepsilon$,
where $\Delta\varepsilon = (\varepsilon_L - \varepsilon_R)$ is the difference in the extinction coefficients. If
the dimensions of $\Delta\varepsilon$ are L(mol/cm), then [θ] is in deg cm^2/decimol.
The ORD is the wavelength dependence of the optical rotation.
The Kronig-Kramer equation expresses the relationship between
the optical rotation and CD at a particular wavelength λ:

$$[m']\lambda = 2.303 \frac{9000}{\pi^2} \int_0^\infty \Delta\epsilon\lambda' \frac{\lambda'}{\lambda^2 - (\lambda')^2} d\lambda' \qquad (32)$$

In proteins there are three distinct kinds of asymmery that can give rise to optical activity:

1. The primary structure may be asymmetric. The α-carbon atoms of most amino acids have four different substituents.
2. A helical secondary structure can result in optical activity for electronic transitions in the main chain or in helically organized amino acid side chains.
3. The folded, tertiary structure may lead to the introduction of a symmetric or weakly asymmetric group into an asymmetric environment. Thus, tyr, which in isolation produces a weak optical activity, may exhibit marked activity when the residue is buried within the folded structure.

The three main types of organization in proteins, namely α-helix, β-sheet, and "random" coil, have distinctive CD spectra, as shown in Fig. 15. CD band intensities at 207, 290, or 222 nm have been used to determine the percentage of helix in a protein. An alternative method is by the se of the Moffitt equation:

$$[m'] = \frac{a_0\lambda_0^2}{\lambda^2 - \lambda_0^2} + \frac{b_0\lambda_0^4}{\left(\lambda^2 - l_0^2\right)^2} + .. \qquad (33)$$

where a_0 and b_0 are constants and $[m']$ is the molar rotation. The importance of this equation lies in the fact that it predicts a functional dependence of the ORD of helical polymers on $(\lambda^2 - \lambda_0^2)^{-2}$, where λ_0 is the center of the ORD band. Figure 15 also shows how the "pure" spectra associated with specific secondary structures can be mixed to calculate the percent helix, β-sheet, and random coil in a globular protein. On the whole, the agreement with X-ray diffraction results is satisfactory (*see* Table 8). Some empirical working rules have been established that help in the intepretation of CD and ORD spectra:

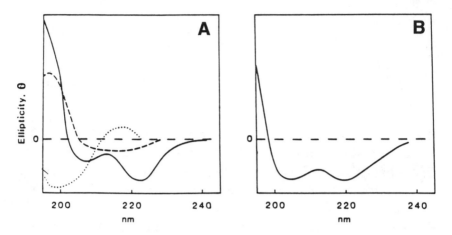

Fig. 15. (A) CD spectra of poly-L-lysine in the α-helical (solid line),
β-sheet (broken line), and random coil (dotted line) conformations. (B) Calculated CD spectrum of myoglobin, based on the model spectra in (A).

Table 8
Helix Content of Proteins:
Comparison of CD
and X-Ray Diffraction Results

	Percent α-helix	
Protein	CD	X-ray
Myoglobin	77	77
Lysozyme	29	29
Ribonuclease	18	19
Papain	21	21
Lactate dehydrogenase	31	29
α-Chymotrypsin	8	9
Chymotrypsinogen	9	6

1. The spectra are additive, i.e., they are the simple sum of the
    spectra of the individual chromophores.
2. The amplitude of the ORD curve or the rotational strength of
    a CD curve is a direct measure of the degree of asymmetry.

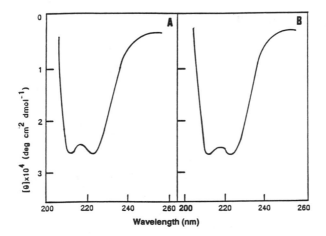

Fig. 16. CD spectra of (A) fibroblast HuIFN-β and (B) *E. coli* rHuIFN-β at 23°C and pH 6.8. After ref. *19*.

3. A chromophore that is symmetric can become optically active when placed in an asymmetric environment. This may or may not be accompanied by a change in λ_0.

4. The value of λ and the magnitude and sign of $\Delta\varepsilon$ at λ_0 allow the chromophore to be identified because λ_0 is always very near the absorption maximum of the molecule.

Although for small proteins the above rules have been found to be satisfactory, it must be emphasized that they are empirical and that care needs to be taken in the deconvolution of CD spectra, especially where "unexpected" secondary structures may exist. They have been found in glycoproteins where the bulky oligosaccharide residues force the peptide backbone into structures that are neither α-helical nor β-sheet; e.g., Antarctic fish antifreeze glycoprotein appears to have a left-handed helix because of the high degree of threonine O-glycosylation. This gives rise to a CD spectrum *(18)* that cannot be decomposed into the usual three components shown in Fig. 16.

CD spectra provide sensitive fingerprints for secondary structure; *see* Fig. 16 for a comparison of a mammalian and a recombinant protein *(19)*.

9. Magnetic Resonance

Two groups of techniques have in recent years been developed to the stage where they constitute powerful tools for the probing of macromolecular structures and interactions (20). Both of them are based on the fact that a spinning charged particle behaves as a magnet; it possesses a magnetic dipole moment, which, according to the laws of quantum mechanics, can adopt only certain values, measured as spin quantum numbers. If such a particle has a non-zero spin, then it will interact with an external magnetic field, and different energy levels can be distinguished. The energy levels involved are very closely spaced and correspond to radiation in the microwave region of the spectrum. In practice, the microwave radiation is kept at constant frequency and the magnetic field is swept so as to bring various transitions into resonance; hence the term magnetic resonance.

The two techniques referred to above are electron paramagnetic resonance (EPR), which relies on the existence of unpaired electrons for the nonzero spin, and nuclear magnetic resonance (NMR), in which nuclei with nonzero spin give rise to the spectrum. Magnetic resonance techniques have reached such degrees of refinement that only the briefest accounts can be given here. The interested reader is advised to consult one of the many specialist texts that already exist and more of which appear with regular frequency.

9.1. Electron Paramagnetic Resonance (EPR)

With the exception of very few molecular species (e.g., oxygen), unpaired electrons appear only in transition metals with incomplete d or f shells and in free radicals. Few biological molecules contain unpaired electrons, so that a popular device is to attach covalently a group of atoms that contains such an unpaired electron and is also chemically stable. The technique is known as spin labeling.

The frequency v of the radiation corresponding to the spin transition is

$$v = g\beta H/h \tag{34}$$

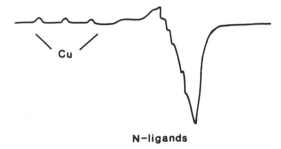

N—ligands

Fig. 17. EPR spectrum of the copper-conalbumin complex, showing the splittings caused by the copper *d*-electrons and the nitrogen ligands.

where β is the Bohr magneton, H is the magnetic field, h is Planck's constant, and the so-called g factor is the local magnetic field produced by the molecule that contains the unpaired electron. The measurement of g therefore identifies the source of the spectrum. In addition the magnetic moment of the electron interacts with the magnetic fields of nearby spinning nuclei, and this gives rise to the splitting of the line (hyperfine splitting). If the electronic spin is affected by more than one spinning nucleus, the spectrum can become quite complex. In principle, if the spin quantum number of the nucleus is I, then the epr spectrum has $(2I + 1)$ lines of equal intensity.

The spectrum in Fig. 17 illustrates the complex appearance that is the hallmark of many epr spectra. It is the spectrum of the copper–conalbumin complex and is composed of two distinct features: the splitting on the left (low-field region) is caused by the interaction between the d electron with the spinning Cu nucleus. The hyperfine splittings on the right result from the interactions of the electron with nitrogen ligands.

Apart from structural information, epr spectra also provide an indication of dynamic properties, e.g., rigidity or flexibility of a residue, and diffusion rates. By attaching a spin label to specific amino acid residues, it then becomes possible to monitor the motional freedom of that residue as affected by certain environmental factors.

9.2. Nuclear Magnetic Resonance (NMR)

All spinning nuclei that have a nonzero spin give rise to NMR signals. Nuclei that are frequently used in studies of proteins include ^1H, ^2H, ^{13}C, ^{14}N, ^{15}N, ^{17}O, and ^{31}P. Of these, the proton receives the most attention. As is the case with other branches of spectroscopy, there are several distinct features of the spectrum that provide different types of information. The frequency at which a signal appears is closely related to the energy of the process that gives rise to the signal. The intensity of the signal is a measure of the concentration of the species, and the band shape provides information about the type of interaction involved and its dynamics.

NMR techniques have been developed into powerful tools for the study of small and medium sized(?) proteins, protein fragments, or subunits (21). They are highly specialized and require a considerable depth of knowledge; the technical details are well beyond the scope of this book. The following account constitutes a summary of the varied types of molecular information that can be obtained from different NMR experiments.

In NMR it is the fact that the observed resonance frequency of a given nucleus is very sensitive to its environment that makes the technique so valuable. The shift in resonance frequency caused by the chemical environment is known as the *chemical shift*. Chemical shifts are usually measured as displacements from some reference standard and are measured in ppm:

$$\text{Chemical shift (ppm)} = \frac{\upsilon - \upsilon_{ref}}{\upsilon_{ref}} \times 10^6$$

Figure 18 summarizes some proton chemical shift values, referred to the protons in $SiMe_4$, a commonly used standard. Compared to other nuclei the proton shifts are small; they lie within a range of 0–15 ppm. By comparison, ^{13}C shifts cover a range of 350 ppm and ^{31}P shifts, a range of 700 ppm. Nevertheless, because of the common occurrence of hydrogen nuclei in organic compounds, proton resonance is the commonly applied technique.

Reference was made above to the sensitivity of the chemical shift to many environmental factors. Some of these factors are of special importance in protein characterization. *Intramolecular shield-*

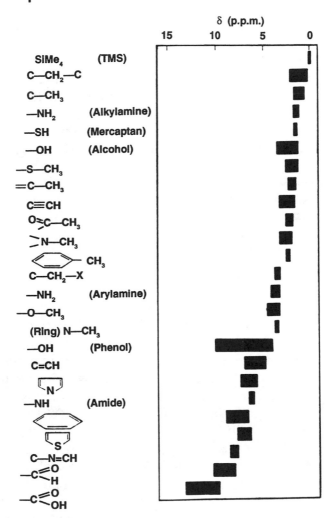

Fig. 18. Ranges of ¹H NMR shifts (referred to as SiMe₄) of functional groups that commonly occur in proteins (ref. 20).

ing can produce shifts of up to 20 ppm; it is caused by the movement of electrons by the external magnetic field, thus generating a field at the nucleus opposite in direction to that of the applied field. The effect is particularly pronounced in aromatic molecules and those containing conjugated double-bond systems. The direction of the shift produced (up-field or down-field) depends on the position of the nucleus with respect to the carbon atoms making up the aromatic system.

Paramagnetic shifts are caused by unpaired electrons. They can amount to 20 ppm and can be detected when the unpaired electron is as much as >1 nm distant from the nucleus producing the resonance. The effect is particularly useful because it is related to r^{-3} where r is the distance between the nucleus and the unpaired electron. Paramagnetic shifts therefore provide information of a structural nature.

Intermolecular effects occur if the electron distribution in one molecule affects the chemical shifts of nuclei in another molecule. In polymers such effects can also be intramolecular, i.e., residues in one part of the chain can influence the chemical shifts of nuclei in a distant part of the chain, if the two regions are placed in close proximity as, for instance, in the tertiary, folded stucture of a protein.

If a molecule undergoes chemical or physical changes of a reversible nature or is subjected to reversible changes of environment *(chemical exchange)*, the observed chemical shift depends on the rates of such processes. If exchange is very slow, two distinct signals are observed because the molecule will then be in either of the two states, rather than in transition between them. As exchange becomes more rapid, the two lines will broaden and eventually overlap and merge. In the limit of very rapid exchange, only one narrow signal is observed at a position that constitutes an average between those of the two shifts observed for slow exchange. Exchange broadening is a useful phenomenon because the areas of the two resonance lines are proportional to the number of nuclei producing the lines, and hence to the respective populations of the two states. Examples of chemical exchange effects include the binding of substrate molecules to enzymes, *cis-trans* isomerism, and other conformational changes.

There are yet two further important effects to be considered. The magnetic field produced by a spinning nucleus is affected not only by unpaired electrons, but also by the fields due to other, nearby nuclei. This gives rise to spin–spin interactions that result in line splitting (spin coupling). The appearance of multiplets can be of great help in the assignment of signals to specific protons.

The spacing between the components of a multiplet bears a relationship to the distance between the nuclei that give rise to the splitting. The spacing is expressed as a coupling constant, J, which is measured in Hz and is related to the torsional angle ø about the—CN bond in a peptide group by

$$J = 8.5 \cos^2 ø, \qquad 0 < ø < 90° \; cis$$
$$J = 9.5 \cos^2 ø, \qquad 90 < ø < 180° \; trans$$

The above equations (Karplus equations) are two examples of relationships between J and torsional angles. Similar equations have been derived for other types of chemical linkages, e.g., H—C— C—H, H—C—O—H. Their applications are subject to various assumptions and complications. Although much useful structural information can be derived from spin coupling interactions, these processes give rise to very complex spectra. This is the case especially where the various protons are chemically and environmentally very similar, i.e., they have similar chemical shifts. This is the case for carbohydrates. Figure 19a shows the 500 MHz spectrum of H/D-exchanged sorbitol in aqueous (D_2O) solution, together with the proton signal assignments. The hexitol molecule possesses eight [1]H-nuclei, which give rise to the observed spectrum. The complex nature of this spectrum is illustrated by the very large number of signals (owing to spin coupling), yet covering a very narrow range of chemical shifts, because the chemical environments of the hydrogen atoms are closely similar. The useful device of spin decoupling produces a simplification of the proton spectrum. The comparison of the spin-decoupled spectrum with the primary spectrum allows for a more reliable assignment of individual signals to individual protons. The correctness of the spectral assignments can then be verified by generating a computer-simulated spectrum, as shown for sorbitol in Fig. 19b. The complete analysis of the alditol spectrum and its comparison with similar spectra of its stereoisomer mannitol, in aqueous and organic solvents has produced valuable information about the solvent-dependent conformations adopted by such flexible, hydrophilic molecules (22). In particular, it has been established that the molecular conformations in solution show distinct differences from the configu-

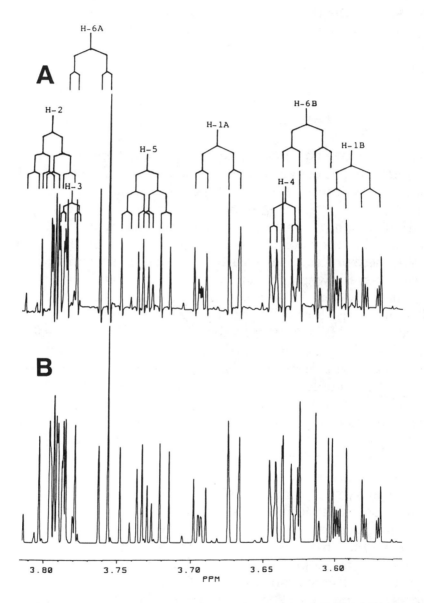

Fig. 19. 500 MHz proton nmr spectrum of sorbitol in D$_2$O: (a) experimental spectrum with assignments and (b) computer simulated spectrum. Reproduced from F. Franks and J. R. Grigera, *Water Sci. Rev.* **5**, 187–289 (1990).

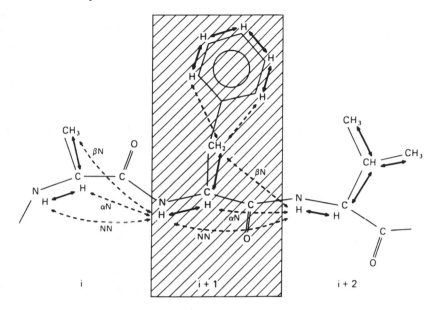

Fig. 20. A tripeptide segment of a protein chain. Protons with through-bond coupling are shown by solid arrows. Protons whose resonances can only be linked through space are shown by broken arrows. Three ways of making connections between the shaded phe residue and its neighbors are shown. Reproduced, with permission, from R. M. Cooke and I. D. Campbell, *BioEssays* **8,** 52–56 (1988).

rations adopted by these molecules in the crystalline state. Similar experiments on peptides could prove informative and would probably be somewhat simpler, because of the greater chemical variety of the various protons in such molecules.

The effects of spectral overlap can also be overcome by the application of advanced experimental procedures that allow two-dimensional spectra to be recorded. Such procedures, just like spin decoupling, simplify the spectrum, allowing reliable assignments to be made. Two types of experiment are usually carried out: 2D-COSY (correlated spectroscopy), which detects spin–spin interactions through chemical bonds (J-coupling), is shown by arrows in Fig. 20, and 2D-NOESY (nuclear Overhauser effect spectroscopy), which identifies through-space spin coupling, is shown by broken line arrows. Provided that the primary sequence is known, all protons in Fig. 20 can be correctly assigned and the conformation of the tripeptide established, because the COSY and NOESY spectra provide the necessary proton–proton distance information.

The final NMR effect to be considered in this abbreviated discussion is *nuclear magnetic relaxation*. In the absence of a magnetic field, the magnetic moments of a system of nuclei ($I = 1/2$) will have random orientations and the populations of the two magnetic energy levels are equal. When a field is applied, the nuclei must adopt one of two allowed orientations; the populations are given by the Maxwell-Boltzmann distribution law. The equilibrium is attained by a first-order rate process with a rate constant $1/T_1$, where T_1 is known as the *spin-lattice relaxation time*. It follows that when the field is removed, the nuclei will relax according to the same rate law. The important feature of T_1 is that it can be related to the Brownian diffusion of the molecule that contains the nucleus under study, or to motions of a segment of such a molecule, e.g., the aromatic ring in phenylalanine.

Depending on the plane of magnetization, another relaxation time, T_2, is defined. Here T_2 measures the decay of magnetization in the x and y axes, whereas the decay of magnetization along the z axis is measured by T_1. If the molecular motion is rapid and isotropic, then $T_1 = T_2$. This is the case for many small molecules, but not necessarily for polymers or for systems in which diffusion is inhibited. Two practical results are of importance: T_2 is related to the nmr line width by $\Delta v_{1/2} = 1/\pi T_2$, where $\Delta v_{1/2}$ is the line width at half-height and T_2, being directly related to molecular motion, becomes long with rapid motion.

As is the case with other spectroscopic techniques, NMR is subject to a number of technical problems and limitations. Many of them, e.g., the powerful water proton signal that dominates aqueous solution spectra, can now be overcome by the use of various refinements and modifications. Chief among them has been the introduction of Fourier transform techniques, although greatly improved spectral resolution resulting from the development of more powerful magnets and the diversification of the nuclei that can be studied have also contributed materially to the ever-growing popularity of NMR.

Certain useful rules can be applied to the interpretation of NMR spectra:

1. The proton shift of a particular group, e.g.,—OH, is affected by other groups in the same molecule, e.g., alcohol, amino acid;
2. Proton shifts are also affected by molecular weight and molecular complexity. Such effects derive from paramagnetic centers, ring currents, or electric fields produced by charged groups;
3. A change in the proton shift as a result of treating the protein with a chemical or physical agent is indicative of a change in the structure of the protein, so that the environment of the proton has changed. Such a change can be of a minor nature, e.g., small shifts in pH;
4. Covalent coupling of nuclear spins gives rise to splitting. The number of lines is symptomatic of the nature of the group in question. The coupling constant J is related to the through-space distance of the nuclei involved. In a protein, if some chemical treatment results in a change in J, then a conformational change is indicated;
5. In a system that is subject to chemical exchange, e.g., an enzyme–substrate interaction, the width of the line and/or the number of lines provide information about the population of the two states and the rate of exchange between the two environments;
6. The line width is a measure of the relative mobility of the nucleus. Note, however, that changes in band width can result from exchange phenomena (*see* rule 5 above);
7. The binding of a ligand produces changes in the spectra of the ligand and of the macromolecule with which it interacts, particularly of those nuclei that are close to the site of the interaction.

Both the position of the shift as well as the line width are usually affected. In the interpretation of protein spectra, the observed signals must first of all be assigned to the constituent amino acids. The latter have been well studied and tabulations exist of the proton and carbon resonances under various conditions of pH. Table 9 lists the ^{1}H resonances of three amino acids, each one characteristic of a different type: apolar, charged, and aromatic.

Table 9
Chemical Shifts (Hz) of ^{1}H Resonances in Amino Acids and Peptides;
Measured at 220 MHz with an Internal Standard $CH_3Si(Me)_2(CH_2)_3SO_2Na^+$

Hydrogen type	Equivalent H per residue	Compound	Chemical shift
Leucine	6	L-Leucine	208
CH$_3$		Gly-leu	197
		his-leu	193
β-CH$_2$ + γ-CH	3	L-Leucine	374
		Gly-leu	350
		His-leu	345
α-CH	1	L-Leucine	813
		Gly-leu	920
		His-leu	905
Lysine	2	L-Lysine	321
γ-CH$_2$	2	Poly-L-lysine	315
δ-CH$_2$	2	L-Lysine	375
β-CH$_2$	2	L-Lysine	412
ε-CH$_2$	2	L-Lysine	664
	2	Poly-L-lysine	660
α-CH	1	L-Lysine	821
	1	Poly-L-lysine	947
Tyrosine	1	Gly-tyr	628
β-CH$_2$	1	Tyr-ala	694
α-CH	1	Gly-tyr	972
	1	Tyr-ala	906
Aromatic *ortho* to OH	2	L-Tyrosine	1514
	2	Gly-tyr	1504
Aromatic *meta* to OH	2	L-Tyrosine	1583
	2	Gly-tyr	1572

When the primary sequence is known, fragments can then be obtained and purified by stepwise enzymatic cleavage of the protein. The determination of the NMR spectra of the fragments then allows certain residues to be assigned to a particular region of the macromolecule. However, the chemical shift of a given residue in the fragment is not always identical with that in the intact molecule.

A further device in the assignment of peaks makes use of the ability of certain residues (or groups of residues) to bind ligands. Partial deuteration, when possible, simplifies the spectrum to be analyzed, because proton resonances from the deuterated residues will vanish. Simple organisms can be grown in a completely deuterated medium to which a single protonated amino acid might be added, so that only signals due to this residue will appear in the spectrum. This method is powerful, but expensive. When a full structure is known from X-ray or neutron diffraction, the NMR spectrum can be compared with the predictions of the structure. Many mutant proteins have been found that differ from the mesophilic protein by only one amino acid residue, although performing the same function and, presumably, possessing the same three-dimensional structure. A comparison of the NMR spectra should make possible the identification of the particular residue, but spectral changes can also occur for those residues that interact with the mutant residue. Finally, spin–spin decoupling can be used to simplify the spectrum and help in the assignment of signals to specific residues and their positions in the macromolecule. In recent years, methods employing two-dimensional NMR have also been developed and are now being used extensively in conformational studies of macromolecules. Of all the instrumental techniques that have been developed and refined during the past two decades to probe the properties of macromolecules in solution, NMR is by far the most powerful. Developments in magnet design, microelectronics, and data processing equipment have further enhanced the potency and versatility of NMR techniques.

9.2.1. Examples of NMR Applied to Protein Chemistry

The following examples serve to illustrate just a few of the applications of NMR techniques to the study of various aspects of protein chemistry and technology.

1. The proton of the α-C atom of some amino acids is sensitive to the secondary structure to the extent that two characteristic chemical shifts are observed for the α-helix and the random coil. The fraction of the amino acids in the α-helical configuration can thus be calculated.
2. One of the most common applications of NMR is to the study of protein unfolding (denaturation), caused by thermal, chemical, or other means. Figure 21 is a diagrammatic proton spec-

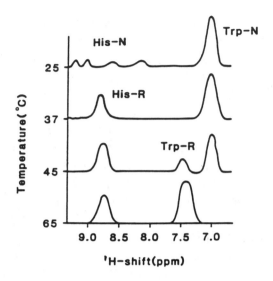

Fig. 21. Histidine and tryptophan proton NMR spectra of a hypothetical protein showing various stages of denaturation (N = native and R = random coil, denatured).

trum of the histidine and tryptophan residues in a protein. The subscripts N and R refer to the native and random coil denatured state of the protein. At 25°C the protein is in its native state and the four histidine signals indicate four distinct chemical environments. The intensity of the tryptophan signal is large compared to that of the individual histidine signals, indicating the presence of several trp residues in the same (average) environment. At 37°C the his signals have merged and the resulting signal is found in the position associated with the R state. However, the region of the protein in which the trp residues are buried is still in its native conformation. At 45°C the resonance of trp in the R state first appears and denaturation is complete at 65°C. The area of the subsidiary peak observed at 45°C is 1/6 of that of the total trp signal, indicating that one trp residue is affected by exposure to 45°C. By further refinement of the NMR studies it can be established which of the several trp residues is responsible

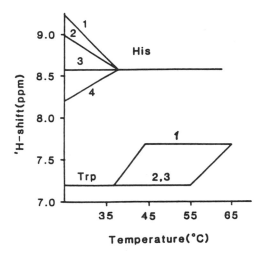

Fig. 22. Plot of the line positions shown in Fig. 21 as function of temperature.

for the signal at 7.5 ppm at 45°C. A plot of the line positions as a function of temperature, as shown in Fig. 22, provides detailed information about the thermal denaturation of the protein, especially if the peptide sequence and tertiary structure are known. From the various line widths, the relative mobilities of the residues can be estimated. Thus, the unchanged width of the trp residue before and after denaturation indicates that the trp residues are situated in a flexible region of the protein. Normally, a narrowing of the signal would be expected upon denaturation.

3. The active site of an enzyme can be probed by the change in the NMR signals of certain amino acids as substrate is added. Thus, Fig. 23 shows the effect of 3'-CMP concentration on the five histidine signals in ribonuclease (RNase); only three residues (12, 48, and 119) show significant effects. More extensive studies, using also the proton signals of the substrate, as well as imidazole instead of histidine and phosphate instead of 3'-CMP, have shown that his-119 is hydrogen bonded to the phosphate group of 3'-CMP. On the other hand, his-12 is

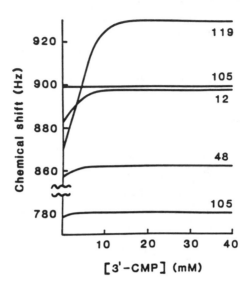

Fig. 23. Effect of 3'-CMP concentration of the proton NMR signals of histidine protons in ribonuclease (for details, *see* text).

not affected by the binding of inhibitors to the enzyme, indicating that any shift in the his-12 resonance is caused by a
more general evironmental effect, e.g., a change in pK, and
that it must be close to the active site, without taking a direct
part in the catalytic function.

4. Binding can be studied more directly by monitoring the signal of a ligand or its analog, e.g., ^{129}Xe, for hydrophobic interactions. Binding of Ca^{2+} to calmodulin has been studied by
^{113}Cd and ^{43}Ca, to show that calmodulin binds two Ca^{2+} ions
but that there are two distinct sites with different binding
affinities (23).

5. Great advances have been made in the quantitative determination of protein structure in solution with the aid of techniques such as paramagnetic shift reagents, spin decoupling,
and two-dimensional NMR. A classic study of lysozyme
illustrates some of the methods and conclusions. By taking a
difference spectrum of a solution with and without Gd(II), it
was possible to identify those amino acids that are close to the

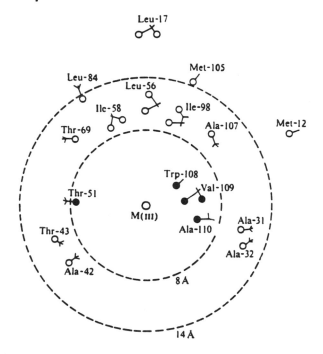

Fig. 24. Map of the metal binding site in lysozyme, as obtained from X-ray diffraction and ¹H NMR (ref. *18*).

metal binding site. Figure 24 is a map of the results obtained from X-ray crystallography, which is fully consistent with the NMR findings that three methyl signals and one aromatic signal are primarily affected: the Me signals belong to val-109 and ala-110, whereas the aromatic proton signal belongs to trp-108.

Although the early studies relied on a knowledge of the crystal structure, this is no longer necessary because with the aid of computers the various shifts can be fitted to a variety of possible structural models and refined.

6. The effect of point mutations on protein stability can now be studied with the aid of high field spectrometers (up to 600 MHz). Figures 25a and b show the spectra of wild type *Staph.*

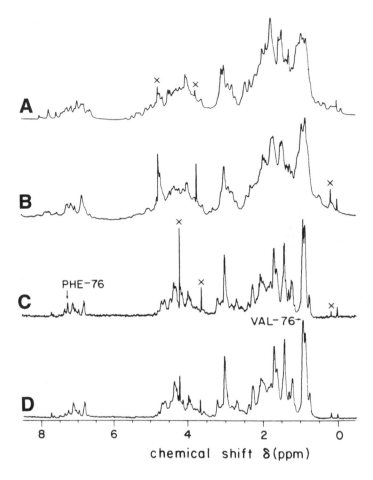

Fig. 25. Comparison of proton nmr spectra of two mutants of staph. nuclease at pH 7.6 and room temperature (A and B); for details, *see* text. Spectra C and D were obtained from the corresponding heat denatured proteins at 80°C. Reproduced, with permission, from ref. 24.

nuclease and a mutant phe 76 —> val at two temperatures (24). At 30°C the spectra of the native proteins show differences, but at 80°C (c and d) they are identical, except for the val/phe resonances. It is concluded that differences exist in the folded states. The val mutant is found to be less heat resistant, but its catalytic function is not affected.

9.3. Determination of Protein Structure (Conformation) by NMR (25)

The experiment is performed in two distinct stages:

1. The assignment of two protons in the spectrum, which involves the determination of the shifts of two specific protons i and j.

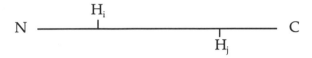

2. Determination of a proton–proton distance constraint to establish if these two protons are closer together *in space* than 0.5 nm.

This process is repeated for all protons in the molecule. A family of structures is then calculated, by using geometric constraints (experimental constraints plus covalent structure). Finally, these structures are refined by combining geometric constraints with potential energy functions (e.g., energy minimization or molecular dynamics). A complete three-dimensional structure can thus be constructed *(25)*.

Assuming that the primary sequence is known, the measurement of the 2D-COSY spectrum determines those pairs of protons that are close in a chemically bonded sense; *see* Fig. 26. The 2D-NOESY spectrum, coupled with distance geometry calculations shows those pairs of protons which are close in space, also shown in Fig. 27.

The combination of the two 2D-spectra then permits the construction of the tertiary (folded) structure of the protein. Figure 28 shows typical COSY and NOESY protein spectra *(21)*. They are

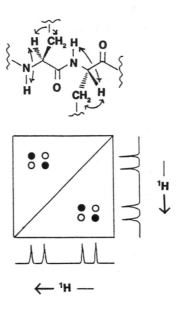

Fig. 26. 2D-COSY spectrum; coupling is shown by arrows.

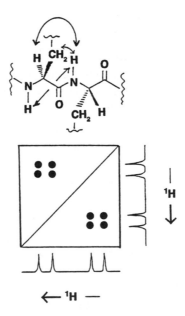

Fig. 27. 2D-NOESY spectrum; as for COSY + through space proton coupling, shown by arrows.

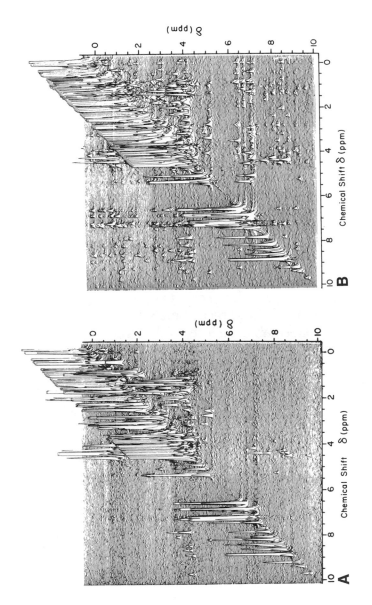

Fig. 28. 2-D nmr spectra of turkey ovomucoid third domain: (A) COSY and (B) NOESY spectra from the same sample. Reproduced, with permission, from ref. 21.

2D NOESY   3D NOESY

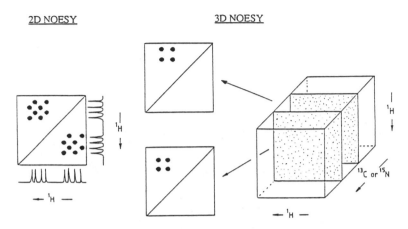

Fig. 29. 3D-NOESY spectrum, for details, *see* text.

complex, with some lines overlapping, which causes problems with the spectral assignments, such problems becoming more serious with increasing molecular weight (increased line widths). A useful device is to extend the scope of the experiment by isotopic enrichment with, e.g., ^2H, ^{13}C, or ^{15}N, and recording 3D-spectra *(26)*. Such spectra separate the information in a third dimension, according to the ^{13}C or ^{15}N shift of the atom attached to one of the ^1H nuclei involved (e.g., the peptide group); see Fig. 29. Each slice through the cube contains just a portion of the information in the 2D-spectrum, thus enhancing the resolution.

Bibliography

1.   Jaenicke R. (1980) *Protein Folding*, Elsevier, Amsterdam.
2.   Schellman J. A. and Hawkes R. B. (1980) in *Protein Folding* (Jaenicke, R., ed.) Elsevier, Amsterdam, p. 331.
3.   Tanford C. (1961) *Physical Chemistry of Macromolecules*, John Wiley & Sons, New York, Chap. 8.
4.   Hofmeister F. (1988) *Arch. Exptl. Pathol. Pharmakol.* **24**, 247–260.
5.   von Hippel P. H. and Schleich T. (1969) *Acc. Chem. Res.* **2**, 257–265.
6.   von Hippel P. H. and Wong K. Y. (1965) *J. Biol. Chem.* **240**, 3909–3923.
7.   Hamabata A. and von Hippel P. (1973) *Biochemistry* **12**, 1256–1261.
8.   Arakawa T. and Timasheff S. N. (1982) *Biochemistry* **21**, 6536–6544.
9.   Uedaira H. (1980) *Bull. Chem. Soc. Japan* 53, 2451–2455.
10. Gekko K. and Morikawa T. (1981) *J. Biochem.* **90**, 39–50.

11. Franks F. and Eagland D. (1975) *Crit. Rev. Biochem.* **3**, 165–219.
12. Brandts J. F. and Hunt L. (1967) *J. Amer. Chem. Soc.* **89**, 4826–4840.
13. Franks F. and Desnoyers J. E. (1985) *Water Sci. Rev.* **1**, 171–232.
14. Richards E. G. (1980) *An Introduction to Physical Chemistry of Large Molecules in Solution*, Cambridge University Press, Cambridge, UK.
15. van Holde K. E. (1971) *Physical Biochemistry*, Prentice-Hall, NJ.
16. Eisenberg H. (1976) *Biological Macromolecules and Polyelectrolytes in Solution*, Clarendon Press, Oxford.
17. Mahler, H. R. and Cordes E. H. (1966) *Biological Chemistry*, Harper & Row, New York.
18. Franks F. and Morris E. R. (1978) *Biophys. Biochim. Acta* **540**, 346–356.
19. Utsumi H., Yamazaki S., Hosoi K., Kimura S., Hanada K., Shimazu T., and Shimizu H. (1987) *J. Biochem.* **101**, 1199–1208.
20. Dwek R. A. (1973) *Nuclear Magnetic Resonance in Biochemistry* Clarendon Press, Oxford (1973).
21. Markley J. L. (1987) in *Protein Engineering* (Oxender D. L. and Fox C. F., eds.), Liss, New York, pp. 15–25.
22. Franks F., Dadok J., Ying S., Kay R. L., and Grigera J. R. (1991) *J. Chem. Soc. Faraday Trans.* **87**, 579–585.
23. Forsen S., Thulin E., Drakenberg T., Krebs J., and Scamon K. (1980) *FEBS Lett.* **117**, 189–194.
24. Markley J. L., Westler W. M., Chan T.-M., Kojiro C. L., and Ulrich E. L. (1984) *Fed. Proc.* **43**, 2648–2654.
25. Kaptein R., Boelens R., Scheek R. M., and van Gunsteren W. F. (1988), *Biochemistry* **27**, 5389–5395.
26. Oschkinat H., Griesinger C., Kraulis P. J., Sorensen O. W., Ernst, R. C., Gronenborn, A. M., and Clore, G. M. (1988) *Nature* **332**, 374–376.

Six

Posttranslational Processing of Proteins

Johannes M. F. G. Aerts
and André W. Schram

1. Introduction

To obtain their final structural features and related function, newly formed polypeptides undergo various forms of modification. The nature of these protein modifications is highly diverse, such as correct folding of the polypeptide, stabilization of conformation by generation of sulfide bridges, unfolding to a transport competent state, assembly into multi homo- or hetero-oligomers, acquisition of prosthetic groups, covalent attachment of phosphate, sulfate, fatty acid, complex lipid or sugar groups, and specific proteolytic modifications. The types of modification shown by an individual protein are not only dependent on its amino acid sequence but also on the modification machinery that is (sequentially) encountered during its life span.

A rapidly increasing number of intracellular proteins are known to be subject to phosphorylation and dephosphorylation of specific serine, tyrosine, or threonine residues. These particular modifications are catalyzed by specific protein kinases and phosphatases. Because of its diverse, complex, and reversible nature, this form of modification is not discussed in detail. Attention is focused on more permanent forms of protein modification, in particular specific covalent attachment of moieties to amino acids and specific proteolytic processing.

From: *Protein Biotechnology* • F. Franks, ed. © 1993 The Humana Press Inc.

Various forms of protein modification are restricted to specific cellular sites and some of these processes are directly related to targeting of proteins to specific locations within and outside cells. For this reason the different types of protein modification are described in this chapter in relation to routing of proteins to distinct intra- and extracellular locations.

1.1. Targeting of Proteins: Translation on Free or Membrane-Bound Ribosomes

In this section, some concepts and terminology regarding routing of newly formed proteins to their destinations are introduced.

Proteins are targeted to various locations with high specificity and efficiency by means of specific amino acid sequences ("signal sequences") in their polypeptide moieties. In eukaryotic cells, the endoplasmic reticulum, Golgi apparatus, secretory vesicles, specific storage granules, endosomes, lysosomes (mammalian cells), or vacuoles (plant cells, lower eukaryotes) are to be considered as elements of a large system of membrane-bound organelles for which the terms vacuome or vacuolar system are used. Transport between compartments of the vacuolar system is vectorial and occurs via membrane-flow, i.e., via fission and fusion of "container" transport vesicles. The plasma membrane is also part of the vacuolar system and consequently, material is exchanged between elements of the vacuolar system and the extracellular space without the necessity of translocation across membranes. In bacteria, the equivalent of the vacuome is the cytoplasmic membrane. In the case of Gram-negative bacteria, their outer membrane constitutes an effective impermeant bilayer boundary that separates the cytoplasmic membrane from the surrounding medium. Gram-positive bacteria lack an impermeant outer membrane and therefore, the cytoplasmic membrane is more or less in direct contact with the surrounding medium in these prokaryotes.

Five major targeting routes can be discerned. First, the so-called "secretory" route that comprises transport of proteins to membranes or the interior of the vacuolar system and transport of secretory proteins to the extracellular space. Second, there are four distinct routes to specific intracellular compartments: the mitochondria, the chloroplasts, the peroxisomes, and the nucleus.

Most proteins destined for the secretory route share the occurrence of a "secretory routing signal," being either a cleavable N-terminal signal peptide or a noncleavable internal signal sequence located near the N-terminus. This was the first targeting signal in proteins discovered *(1–5)*. Initially it was found that translation of proteins with this signal always occurred on membrane-bound ribosomes and was coupled to membrane translocation. For this reason, it is often generalized in textbooks that proteins destined for the secretory route show obligatory a tight coupling of translation and translocation, whereas this is not the case for proteins destined for other noncytoplasmic locations (mitochondria, chloroplasts, peroxisomes, nucleus). In other words, these two categories of proteins would differ fundamentally on the basis of cotranslational vs posttranslational translocation across the first membrane that is encountered during their life span. This generalization becomes increasingly questionable *(see below)*. However, it has to be stressed that in eukaryotic cells, most secretory route proteins indeed are cotranslationally translocated in contrast to the posttranslational translocation of proteins destined for other cellular compartments.

2. Proteins of the Secretory Route and Their Modification

2.1. Targeting of Proteins to the Secretory Route

2.1.1. Multiple Secretory Route Signals

The majority of proteins destined for the secretory route in eukaryotes and prokaryotes are initially synthesized containing a characteristic hydrophobic secretory route signal *(see e.g., ref. 6)*. However, there are indications that some proteins are alternatively targeted to the secretory route. A number of secretory proteins have been identified that lack the characteristic hydrophobic signal sequence *(see ref. 7 for a recent review)*. The mechanism(s) involved in alternative routing of secretory proteins are presently not known. The mechanism by which hydrophobic routing signals mediate targeting to the secretory route is in important aspects understood and will be discussed below.

2.1.2. Entry in the Secretory Route

The presence of a secretory route signal in a polypeptide ensures that these proteins enter the secretory route at specific membranes. Generally, this is accomplished by the fact that translation of these proteins is completed on membrane-bound ribosomes. In eukaryotic cells, the ribosomes are associated to parts of the endoplasmic reticulum (the RER, rough endoplasmic reticulum), that are studded with ribosomes, giving it its characteristic rough morphology as seen by electron microscopy. In bacteria, specific regions of the cytoplasmic membrane are the functional equivalent of the RER.

2.1.3. Secretory Route Signals

Most secretory route proteins have cleavable signal peptides that share a number of characteristics. These are in the case of eukaryotic cells as follows:

1. An N-terminal location;
2. A length of 5–35 residues;
3. A net positively charged region within the first 2–10 residues;
4. A central core region of at least 9 neutral or hydrophobic residues capable of forming an alpha-helix;
5. A turn-inducing amino acid next to the hydrophobic core; and
6. A specific cleavage site for a signal peptidase (6).

Bacterial signal peptides differ only in some minor aspects. For example, signal peptides of gram-positive bacteria mostly have a longer and more basic N-terminus (8). Signal peptides from one secretory protein are able to replace those of other proteins targeted to a comparable site. For example, the signal peptides of *E. coli* periplasmic and outer membrane proteins (9), of vacuolar and secreted proteins (10) of *S. cerevisiae* and of constitutively secreted and storage granule proteins of multicellular eukaryotes (11) can be exchanged without affecting the targeting of the mature protein. In heterologous environments, e.g., expression of eukaryote genes in bacterial cells, signal peptides of some proteins have been shown to mediate efficient secretion. For example, insulin and ovalbumin are excreted by *E. coli* upon expression of these genes (*12*), and yeast is able to excrete efficiently various secretory proteins from multicellular eukaryotes. However, exceptions in this respect have also been described, suggesting that to some extent, differences in the translocation mechanism and/or related machinery exist between different types of organisms.

2.1.4. The Cotranslational Translocation Mechanism in Eukaryotes

The mechanism by which signal sequences direct polypeptides to the ER in eukaryotes is still incompletely understood (*see* e.g., ref. *13* for a recent review on this topic). After initial synthesis of a polypeptide on cytoplasmic ribosomes, the N-terminal sequence emerges and is recognized by the signal recognition particle (SRP). Binding of SRP probably slows down rather than completely arrests further translation. Subsequently, the SRP-ribosome-nascent polypeptide complex binds to the ER via interaction of SRP with its receptor (the docking protein). The latter then binds GTP, replacing GDP, resulting in the release of SRP from the signal sequence and ribosome. Next upon GTP hydrolysis the cycle is completed by dissociation of SRP from its receptor. Subsequently, the actual transfer of the polypeptide chain through the membrane occurs. It has to be noted that SRP is thus not involved in the translocation event itself, but in the initiation of the process. It is believed that SRP functions as an antifolding device keeping the polypeptide chain in a translocation competent state (*13*).

The mechanism by which a polypeptide is transferred across the membrane is still not completely resolved. It has been proposed that transport in some cases may occur directly through the lipid bilayer without participation of proteins. However, there is compelling evidence that the majority of polypeptides are transported through a hydrophilic translocation tunnel assembled from membrane proteins (*14*). The so-called signal sequence receptor, a major integral ER protein, has been proposed to be a component of such a translocation complex at least in multicellular eukaryotic cells (*13,15*). Furthermore, a role in the translocation event has been postulated for ribosome binding receptor(s) that are probably identical to the so-called ribophorins, being abundant integral membrane proteins confined to the RER. The signal sequence is believed to insert into the membrane as a helical hairpin, the N-terminus of the signal peptide normally remaining at the cytoplasmic side of the membrane. The mature part of the polypeptide is pulled across the bilayer and the signal peptide remains in the membrane owing to strong hydrophobic interactions (*13*). This type of transfer is referred to as "tail first." Signal peptides with no positive charge at the N-terminus are able to insert into the membrane differently, i.e., the N-terminus is at the trans side of the membrane ("head first insertion").

2.1.5. Other Mechanisms Mediating Translocation of Secretory Route Signal-Containing Polypeptides

Recent literature suggests that different mechanisms are used for translocation of secretory route signal-containing polypeptides (*16*). In eukaryotic cells, the translocation of some secretory route proteins appears not tightly coupled to translation.

This is thought to be due to the fact that these polypeptides adopt no strong secondary structure because they are relatively short (*17*), the ribosome remains attached to the nascent polypeptide (*18*), or because cytosolic components (heat shock proteins/ unfoldases) prevent formation of secondary structure (*see*, e.g., refs. *19,20*). In prokaryotes, the majority of secretory route polypeptides may in fact be posttranslationally translocated, at least as is suggested by in vitro studies (*see* refs. *21,22*). For a recent review regarding similarities and dissimilarities in translocation of secretory proteins in eukaryotes and prokaryotes, *see* ref. *23*.

2.1.6. Cleavage of the N-Terminal Secretory Route Signal (Signal Peptide)

In the endoplasmic reticulum of eukaryotes and the cytoplasmic membrane of bacteria special endoproteases occur, the so-called signal peptidases, responsible for removal of the N-terminal signal sequence (*24*). There is evidence that signal peptides are removed cotranslationally in the case of polypeptides that are cotranslationally translocated. The specificity of signal peptidases is highly conserved: Prokaryotic signal peptides are released by eukaryotic signal peptidases and vice versa. Bacterial and eukaryotic signal peptides are rapidly further degraded after cleavage from the precursor polypeptide. Information on proteases involved in the latter process is still very limited (*see* ref 25).

2.1.7. Noncleavable Internal Secretory Signals: An Anchoring of Membrane Proteins

Intergral membrane proteins of the vacuolar system and bacterial cytoplasmic membranes show the presence of one or more transmembrane domains existing of about 20 hydrophobic amino acids that resemble strongly the hydrophobic core of signal peptides. The cleavable signal peptide and noncleavable signal sequences are functionally related. When cleavage of a N-terminal signal peptide from a soluble secretory protein is prevented, anchoring via this segment in the membrane occurs. Furthermore, noncleavable internal signal sequences, at least when located near the N-terminus, can initiate cotranslational translocation across membranes similar to signal peptides. An additional important feature of internal signal sequences is that they stop further extrusion of the polypeptide across the membrane in a direction, (therefore also called "stop transfer sequences") (26). A polypeptide with a cleavable signal pepdtide and one stop transfer sequence will have a (N-terminal) domain on the trans side of the membrane and a (C-terminal) domain on the cis, cytoplasmic side, i.e., it will be bitopic. When a polypeptide contains a second stop transfer sequence the transfer of the protein will be polytopic, exposing both termini at the trans side and an internal domain at the cytoplasmic side of the membrane. (For more detailed information about membrane protein topogenesis, *see*, e.g., refs. 26–28.)

2.2. Early Modifications of Proteins in the Secretory Route

Secretory route proteins undergo a number of modifications very rapidly. In eukaryotes, some of these modifications occur truly cotranslationally.

2.2.1. The Cleavage of Signal Peptide

As discussed above, the proteolytic removal of the N-terminal signal peptide by the signal peptidase is a very early event that occurs often cotranslationally in eukaryotes (see, e.g., ref. 3).

2.2.2. Protein Folding

There is compelling evidence that folding of proteins may start already cotranslationally in eukaryotes (29–30). A major rate limiting step in protein folding is the formation of disulfide bonds. This process is catalyzed by the enzyme PDI (protein disulfide isomerase) that is found in the endoplasmic reticulum of all cells examined (31). Intrachain or interchain disulfide bonds have been shown to be cotranslationally formed in several proteins, e.g., in nascent rat albumin 11 of its 17 disulfide bonds are generated by the time that translation is completed (32). A second rate limiting step in protein folding is the cis-trans isomerization of prolylpeptide bonds (33). The existence of an enzyme catalyzing proline isomerization, the prolyl-peptidyl prolyl cis trans isomerase, has been described (34). Proline isomerization is believed to occur already to some extent cotranslationally.

Further folding of polypeptides takes place posttranslationally and is mostly within 15 minutes completed (30).

2.2.3. N-Linked Glycosylation

Most secretory route proteins are cotranslationally glycosylated by transfer of a complex oligosaccharide (GlcNAc2Man9Glc3) en bloc from a dolichol phosphate lipid carrier to asparagine residues occurring in the tripeptide sequence Asn-X-Thr/Ser in which X may not be proline or aspartate (see ref. 35 for a review on this topic). It has become clear that not all potential asparagine residues are actually glycosylated. N-linked glycans are particularly found in polypeptide segments that favor a beta-turn conformation.

N-glycosylation is generally thought to occur as the polypeptide is being transferred into the ER, while it is still in an unfolded state (29,35,36).

An established function of N-linked glycans is their role in promoting folding of proteins. Elegant studies on vesicular stomatitis virus (VSV) G protein have revealed that N-linked glycans are essential for correct folding of the protein and subsequent transport to the cell surface (29).

2.2.4. Initial Processing of N-Linked Glycans

The N-linked oligosaccharide chains are immediately processed following their attachment to the polypeptide (see ref. 35). The three terminal glucose moieties are removed by alpha-glucosidases I and II. Glucose trimming has been shown to occur cotranslationally in the case of VSV G protein.

Further processing of N-linked glycans is different in yeast and higher eukarayotes. In *S. cerevisiae*, a single alpha 1-2 linked mannose is substituted by an alpha 1-3 linked mannose (37). This process is essential for the subsequent oligosaccharide chain elongation of glycoproteins in the Golgi apparatus of yeast (see below). In higher eukaryotes, one terminal mannose residue is removed by the action of the endoplasmic reticulum alpha-mannosidase yielding (GlcNAc2Man8) N-linked glycans (35).

Subsequent processing of N-linked glycans occurs when the glycoproteins have been transferred to the Golgi apparatus. Very recently it has been shown that the initial modification required for the formation of mannose-6-phosphate moieties in N-linked glycans of soluble lysosomal proteins is probably initiated earlier (38).

2.2.5. Oligomerization

Many secretory route proteins consist of multiple subunits, either covalently or noncovalently linked homo- or hetero-oligomers (30). In principle, assembly of oligomers cannot be cotranslationally completed.

However, it has been shown that nascent heavy chains of immunoglobulins can be already disulfide bonded to completed light chains after about 70% of the heavy chain has been translated (39). The efficiency of oligomerization is variable and the half-maximal assembly of different oligomers ranges between five minutes to two hours after synthesis of the components (30). It is generally thought that proteins can proceed further into the secretory pathway only after they have acquired a completely folded three-dimensional structure and, if required, assembly in oligomers has been completed. Specific proteins such as BiP (Binding Protein) appear to play a role in the retention of incorrectly folded proteins or incompletely assembled oligomers in the endoplasmic reticulum (40). BiP is closely related to the cytoplasmic heat shock protein hsp 70. It is assumed that these proteins act as detergents; they bind to hydrophobic surfaces but at intervals use the energy of ATP hydrolysis to change their conformation to a nonbinding state. This would lead to (temporary) solubilization of proteins in an unfolded state. For recent reviews regarding heat shock (related) proteins that are thought to function in a variety of processes such as protein translocation, folding, assembly, and retention (*see* refs. *41–45*).

2.2.6. Formation of Lipid Anchors

Some secretory route proteins are modified by covalent binding of lipids or fatty acids. This group of modifications occurs posttranslationally and the function is believed to be anchoring of proteins to the membrane.

Bacterial lipoproteins. These proteins show a thio-ester ester linked diglyceride to N-terminal cysteine of which the amino group is also acyled (46). This characteristic lipid moiety is synthesized at the cell surface as follows. To a cysteine residue that is close to the N-terminus of the newly formed polypeptide glycerol is linked via a thio-ester bond. Next, fatty acylation of glycerol takes place. Subsequently, by the action of a specific lipoprotein signal peptidase the cysteine-residue becomes N-terminal and its free amino acid is fatty acyled.

Complex glycosylated phosphatidylinositol anchors. In eukaryotes, several membrane proteins have recently been found to be anchored in the membrane via glycosylphosphatidyl inositol attached to their carboxy termini (47–50). These lipoproteins are released from membranes upon incubation with PI-specific phospholipase C. The lipid anchors are attached very shortly after synthesis of the polypeptides. Little is known about the synthesis of the glycosyl PI-anchor. Most likely the PI anchor is first completely assembled and subsequently attached to the polypeptide. No consensus sequence for this type of lipid modification has so far been identified.

Palmitoylation. Some secretory route membrane proteins are palmitoylated via ester linkages to internal serine (hydroxylamine resistent oxyesters) or internal cysteine residues (hydroxylamine sensitive thioesters) (47,48,52,53). In some cases, palmitoylation occurs very rapidly while proteins are still in the ER, in other cases while they reside in membranes of the Golgi apparatus. The palmitoyl moieties in proteins identified so far are never the sole membrane anchor. Indications exist that palmitoylation is also a cytosolic process and consequently that palmitoyl chains may reside in the cytoplasmic leaflet of membranes (48).

Myristoylation. A membrane anchor can also be created by myristoylation of N-terminal glycines (47,48,52,53). The bond between the myristate carbonyl and the amino group is an amide. Myristoylation occurs in the cytoplasma and myristoylated proteins, although often anchored to membranes of the vacuolar system, are not true secretory route proteins.

2.2.7. Proline and Lysine Hydroxylation

In special cases, such as procollagen synthesis, amino acid side chain modifications by enzymes such as proline hydroxylase and lysine hydroxylase occur in the ER during or shortly after synthesis (54).

2.3. Further Routing in the Vacuolar System

After entering the vacuolar system at the ER and initial modification in this compartment, some proteins are retained (resident ER proteins) but the majority are further transported to final destinations within and outside eukaryote cells (*see* ref. 40 for a recent review).

It has been shown for some resident ER proteins that their signal for retention in the ER resides in a C-terminal sequence consisting of four amino acids: Lys-Asp-Glu-Leu (KDEL). There is evidence that KDEL-containing proteins are not simply retained by permanent anchoring to an immobilized membrane receptor. The KDEL-receptor appears to be involved in a mechanism that retrieves proteins that have left the ER (40). The retrieval is thought to occur from some compartment intermediate between ER and Golgi apparatus, called the salvage compartment (55). In yeast, a similar mechanism is operative and it appears that the C-terminal sequence HDEL is preferably used (56).

Proteins that do not reside in the ER exit from the compartment at specialized regions adjacent to Golgi apparatus that lack bound ribosomes. After leaving the ER proteins enter the Golgi apparatus. This organelle is composed of a stack of about six or more discrete membrane-bound cisternae from which small vesicles bud or fuse, especially at the rims (57). After entering at the cislateral face (the side adjacent to the ER) most glycoproteins may transverse the complete Golgi apparatus to leave at the opposite translateral face. The accompanying intercompartmental transport occurs in vesicles, and is a rapid unidirectional process with high specificity.

Within the Golgi apparatus, different subcompartments are distinguishable (cis, medial, and trans Golgi) on the basis of a distinct content on oligosaccharide processing enzymes (35). How proteins are specifically retained in individual Golgi cisternae is presently not known.

Adjacent to the translateral Golgi cisternae is a network of tubules located with coated buds and vesicles that is called the trans Golgi reticulum or network (TGN) (57–60). It is generally thought that in the TGN, sorting occurs of constitutively secreted proteins from proteins destined to specific intracellular compartments such as lysosomes (vacuoles) and storage granules or to specific cell surface domains in polarized cells (61–63).

Sorting of proteins to the lysosomes or storage granules seems to involve specific clathrin coated domains of the TGN that gradually bud, condense, and acidify. Sorting to the apical membrane also appears clathrin mediated (61–63). The mechanisms underlying sorting in the TGN are still largely unknown. There is some consensus that constitutively secreted proteins lack a specific sorting signal and are secreted by a default pathway (61–64).

It is not precisely known how sorting of soluble proteins to storage granules takes place. There are indications that sorting signals represent specific protein domains that depend on the tertiary structure rather than the primary structure of proteins.

Specific pH-sensitive receptors are thought to be involved in the sorting process (64). In polarized cells, e.g., epithelial cells or hepatocytes, the plasma membrane is divided by tight junctions into two different domains, the apical membrane (oriented toward the outside and containing microvilli) and the basolateral membrane (oriented toward the inside). The two membrane domains have a unique protein and lipid composition (65). It is presently not completely clear what the nature of sorting signals is. Very recently, it has been shown that modification of proteins with a glycosyl-phosphatidylinositol anchor mediates sorting to the apical membrane (50).

The best understood sorting mechanism so far is the mannose-6-phosphate mediated transport of solubilized lysosomal enzymes in mammalian cells (62,66). As described in more detail below, N-linked oligosaccharides in soluble lysosomal proteins undergo a specific modification resulting in the formation of mannose 6-phosphate moieties. These phosphomannosyl-bearing proteins bind in the Golgi apparatus to specific receptors (the 275 and 46 kDa mannose-6-phosphate specific receptors, M6PR) and receptor-ligand complexes are segregated in the TGN from the constitutive secretory pathway. Subsequently, dissociation of receptors and ligand occurs in highly acidic compartments, followed by transfer of discharged ligand (i.e. lysosomal proteins) to lysosomes and recycling of free receptor to the Golgi apparatus. The 275kD M6PR is also located on the cell surface and known to be involved in uptake of phosphomannosyl-bearing proteins from the surrounding medium. It is clear that integral lysosomal membrane proteins (62) and the membrane-associated lysosomal hydrolase glucocerebrosidase (67) are transported independently of M6P residues to lysosomes. The sorting signal in integral lysosomal membrane proteins appears to be located in their C-terminal cytoplasmic domains (62). How sorting of lysosomal proteins that lack a cytoplasmic domain and mannose-6-phosphate moieties, such as glucocerebrosidase, is mediated is presently not known. The existence of alternative mechanisms operative in sorting of soluble lysosomal proteins in mammalian cells is indicated by I-cell disease (Mucolipidosis II) (68). In this inherited disorder in humans, the formation of phosphomannosyl residues is defective. Nevertheless, soluble lysosomal enzymes appear to be normally transported to lysosomes in various cell types and tissues (see, e.g., ref. 68).

Furthermore it is clear that sorting of soluble proteins to vacuoles in plants and yeast (the equivalents of the mammalian lysosome) is not dependent on phosphomannosyl or any other sorting signal in N-linked glycans (69,70). It has been shown for some vacuolar hydrolases in yeast that the essential information for targeting is located at the N-terminus (71). Again, it appears that no simple consensus sequence is involved in sorting of vacuolar proteins. Less is known for the sorting of plant vacuolar proteins (72,73). Expression of the plant vacuolar storage protein phytohemagglutinin in yeast results in efficient transport to vacuoles, indicating that the plant vacuolar sorting signal is recognized by the yeast sorting pathway (72). In some vacuolar proteins in yeast and plants a segment of about 20 amino acids that is N-terminal after removal of the signal peptide appears essential for efficient sorting, and delivery to vacuoles is found to be accompanied by further proteolytic processing at the N-terminus (cleavage of the so-called pro-segment).

2.4. Further Modification of Proteins En Route in the Vacuolar System

As proteins traverse through the Golgi apparatus to their final locations, further modifications may occur.

2.4.1. Processing of N-Linked Glycans

Since the modification of N-linked oligosaccharides is different in mammalian and plant cells (35,74) compared to that in yeast (37) these processes are separately discussed.

In mammalian and plant cells, the GlcNac2Man8 ("high mannose type") oligosaccharides in proteins entering the Golgi apparatus may be modified to a variety of structures including "hybrid-type" or "complex-type" oligosacccharides. Individual glycosylation sites on proteins tend to have characteristic oligosaccharides and when heterogeneity is encountered, it usually consists of closely related oligosaccharide structures (35). It seems that the polypeptide structure strongly influences the type of modification that oligosaccharide chains undergo. Oligosaccharides committed to becoming "complex-type" structures appear relatively more exposed compared to those committed to remaining "high mannose type" (75). The final oligosaccharide structure assembled on a specific protein glycosylation site is further determined by the time and order in which that glycoprotein encounters the processing glycosidases and glycosyltransferases and their specificity. The various enzymes involved in processing of oligosaccharides are located in specific regions of the Golgi apparatus (35,76). The high degree of compartmentalization of these processing enzymes has been used to analyze passage of glycoproteins through different Golgi subcompartments. The formation of a biantennary sialylated (complex-type) oligosaccharide occurs as follows. In the cis-most cistermae alpha-mannosidase I removes residual alpha 1-2 mannose residues to yield GlcNAc2Man5 oligosaccharides. In medial Golgi cisternae, GlcNAc-transferase I catalyzes the addition of a GlcNAc residue to one branch of the structure. Subsequently, alpha mannosidase II catalyzes the release of alpha 1-3 and alpha 1-6 linked mannose residues to yield GlcNAc2Man3Gi-NAc oligosaccharides. Next, one GlcNAc is added by GlcNAc transferase II. In the case of N-linked glycans in mammalian cells, only occasionally, but in plant cells generally fucose residue(s) are added by fucosyltransferase to GlcNAc residues. Action of GlcNAc transferase IV instead of GlcNAc transferase II results in addition of two GlcNAc residues to a single mannose residue creating a tri-antennary instead of bi-antennary structure. A so-called "hybrid-type" oligosaccharide is assembled if alpha 1-3 and alpha 1-6 mannose residues are not released by the action of mannosidase II. Finally, in translateral Golgi cisternae and probably even in the TGN, galactosyltransferase and sialyltransferase may catalyze the addition of galactose and sialic acid residues to give as a final product a GlcNAc2Man3GlcNAc2Gal2SA2 bi-antennary sialylated complex type oligosaccharide.

N-linked glycans are rarely further modified by sulfation of mannose and GlcNAc residues and O-acetylation of sialic acid residues (reviewed in ref. 35).

In yeast, the modification of N-linked glycans is markedly different. The GlcNAc2Man6 oligosaccharides are processed to large mannan structures that may contain as many as 150 mannose residues (37).

N-linked oligosaccharides in soluble lysosomal enzymes and integral lysosomal membrane proteins in mammalian cells appear to undergo rather unique modifications. As mentioned before, soluble lysosomal proteins acquire mannose-6-phosphate moieties in (some of) their oligosaccharides that play an important role in routing of the glycoprotein to lysosomes (35,66). First, a N-acetyl-glucosaminyl-1-phospho-6-mannose diester is generated in high mannose type oligosaccharides by the action of "phosphotransferase." In mid-Golgi cisternae, the covering GlcNac residue is removed by "phosphodiesterase" and a mannose-6-phosphate moiety is generated. Phosphates may be linked to five different mannose residues of the high mannose type oligosaccharide but individual oligosaccharide chains contain maximally two phosphates. It is presently believed that the addition of the first covered phosphate occurs in the endoplasmic reticulum whereas the second is added in cis-Golgi cisternae. The specificity of the phosphotransferase for soluble lysosomal proteins is based on protein–protein interactions. The transferase recognizes a protein domain that depends on the tertiary structure rather than on the primary structure.

N-linked glycans in integral lysosomal membrane proteins are often sialylated polylactosamine-type structures (62). They are composed of repeating units of the sequence Gal,1-4,GlNAc,'-3 attached to the GlcNAc2Man3 core.

Polylactosamine synthesis is initiated by the so-called extension enzyme, a specific GlcNAc-transferase (78). The integral lysosomal membrane proteins identified so far are particularly rich in N-linked glycans that are thought to serve as a protection against degradation in the lysosome (62).

2.4.2. O-Linked Glycosylation

In eukaryotes, hydroxyl groups of serine or threonine residues of some secretory proteins and membrane proteins are O-glycosylated in the Golgi apparatus (79,80). The peripheral structure of O-linked glycans is quite variable, but the internal structure is more uniform and consists of N-acetylgalactosamine (GalNAc) often followed by galactose. Further fucose, N-acetylglucosamine, N-acetylgalactosamine, galactose, and sialic acid residues can be incorporated in O-linked glycans. In contrast to biosynthesis of N-linked glycans, O-linked oligosaccharides involve no core glycosylation. In O-linked glycan synthesis, individual glycosyl-moieties are sequentially added.

In yeast, O-linked glycosylation is already initiated in the ER. Mannose residues are transfered from dolichol-1-mannose to serine and/or threonine residues. In the Golgi apparatus, additional mannose residues may be added after transfer from GDP-mannose (81).

2.4.3. Proteoglycan Oligosaccharide Processing

Serine residues in proteoglycans such as chondroitin sulfate, heparan sulfate, and dermatan sulfate are linked to a tetrasaccharide composed of xylose-galactose-galactose-glucuronic acid (79). Further elongation occurs by addition of disaccharides that differ among proteoglycan classes. TGN oligosaccharide chains are terminally O-sulfated. After addition of diasaccharide, extensive further modification may occur. This processing is the most spectacular in the case of heparin and involves N-deacetylation, N-sulfation, epimerization to iduronic acid, and finally O-sulfation (82).

2.4.4. Tyrosine Sulfation

Various secretory proteins are known to undergo tyrosine sulfation. This is a late process occurring in the lateral Golgi cisternae or the TGN. Presently, a consensus sequence for sulfotransferase action has not been identified. Evidence has been presented that sulfation of proteins is of importance for their transport from the trans Golgi region (*see* ref. *83* for a recent review).

2.4.5. Proteolytic Processing

Many proteins in the vacuolar system undergo some type of proteolytic modification beyond the stage of the endoplasmic reticulum. To differentiate between the different forms of proteolytically modified proteins, the following terminology is used: A precursor is a form of a protein that still has to be proteolytically modified to yield the final ("mature") polypeptide chain; a pre-pro-peptide is a polypeptide containing a cleavable signal sequence (pre-region), one (or more) sequences that are ultimately removed (pro-segment(s)), and the mature polypeptide moiety. Some examples of secondary processing of propeptides to mature proteins are discussed.

In yeast one of the best understood forms of secondary proteolytic modification is that of the pro-alpha factor protein (84). Four copies of the alpha-factor peptide are present in the propeptide.

In the Golgi apparatus, it is initially processed by the product of the KEX 2 gene, a calcium dependent, neutral serine protease that cleaves between basic residues. A KEX1-encoded carboxypeptidase and an aminopeptidase catalyse further processing to mature alpha-factor peptides.

In mammalians, a number of secretory proteins are endoproteolytically processed, such as proalbumin and the family of hormone precursors (e.g., proinsulin, proglucagon). Endoproteolytic cleaving between basic residues and carboxypeptidase and aminopeptidase activities are generally involved. The identity of the proteases that are responsible for proteolytic modification is not completely clear. Very recently, a human structural homolog for the KEX2 endoprotease (fur gene product) has been identified (85). It is a calcium dependent, serine thiolase cleaving between basic residues. The enzyme appears to be located in the Golgi apparatus. It is known for some time that proteolytic processing of prohormones appears to be associated with incorporation of precursors in secretory granules budding from the trans Golgi region (63,64). Soluble lysosomal proteins are nearly all proteolytically modified in one or more steps to their mature forms (62). The precise site of initial posttranslational proteolytic modification is not exactly known. It is believed to occur in highly acidic structures that are no longer part of the constitutive secretory route. Further proteolytic modification steps most likely take place in the lysosomes and can be very slow. Very similar observations have also been made for vacuolar proteins in plants and yeast (see, e.g., ref. 73). Since it has been found that some hydrolases are inactive as precursors, it has been proposed that late proteolytic modification would be an elegant mechanism to transport these proteins as inactive forms (62). Although this may be true for some hydrolases, a growing number of hydrolases are found to be active as precursors as well (86,87).

2.4.6. Peptide α-Amidation

About half of the bioactive peptides found in the nervous and endocrine systems possess a C-terminal α-amide group (88). In most cases, amidation is of importance for biological activity of the peptides. Amidation occurs probably in secretory granules by enzyme catalyzed modification of peptides with C-terminal glycine. The reaction requires oxygen and ascorbate as substrates and renders glyoxylate, dehydroascorbate, and amidated peptide as products.

2.5. Continued Protein Modification in the Vacuolar System?

In recent years, it has become clear that many membrane proteins recycle repeatedly between the cell surface and the trans Golgi region. It has been shown for some proteins that during renewed passage through the trans Golgi region, (re)sialylation of N-linked glycans can occur (89). It has been proposed that recycling may even involve earlier Golgi compartments, although this is still controversial (89,90). It is conceivable that renewed passage of proteins through the trans Golgi region may allow other forms of modification than that of N-linked glycans as well. However, no evidence for this assumption has so far been presented.

3. Proteins of Chloroplasts, Mitochondria, Peroxisomes, and Nuclei and Their Modification

3.1. Chloroplasts

3.1.1. Posttranslational Import

Most proteins localized in chloroplasts are encoded by the nuclear genome and are posttranslationally imported into the organelle. Polysomes are not found in association with chloroplast envelope. Within the chloroplast, distinct compartments are distinguishable with a specific protein composition: the envelope outer membrane, the intermembrane space, the envelope inner membrane, the stroma, thylakoid membranes, and thylakoid lumen.

3.1.2. N-Terminal Transit Peptide as Import Signal

The majority of chloroplast proteins studied are synthesized as larger precursors with an N-terminal "transit peptide" (reviewed in refs. 91,92). The transit peptide targets the protein to the organelle and ensures import into the stroma where it is proteolytically cleaved. Transit peptides are positively charged and rich in serine and alanine but show little homology in primary sequence (93). It is thought that specific conserved structural features in transit peptides interact with the import system (91).

3.1.3. Import Apparatus

Little is known so far about the components of the import apparatus in the envelope membrane. A 66 kDa protein is believed to be involved (94) and a role for a 30 kDa protein is suggested by studies using anti-idiotypic antibodies (95,96). The 30 kDa protein is present at contact sites between outer and inner membrane of the chloroplast envelope. Import of precursors into chloroplasts occurs at these contact sites.

3.1.4. Energy Dependence

Import of proteins in chloroplasts requires ATP. It is currently thought that ATP is needed in the cytoplasm for binding of precursors to the chloroplast import apparatus and for making precursors competent for membrane translocation: ATP is needed in the stroma for the actual translocation across the envelope (97).

3.1.5. Competence Factors and Import

Cytosolic factors are required for efficient import of proteins into chloroplasts. These (competence) factors are thought to bind to precursors to create a conformational state competent with translocation (91). The role of competence factors in chloroplast biogenesis may be similar to that of comparable factors in import in mitochondria as is discussed below.

3.1.6. Initial Processing: Removal of Transit Peptide

A 180 kDa stromal processing peptidase removes transit peptides from imported chloroplast proteins during or immediately after translocation (98).

3.1.7. Intra-Chloroplast
Routing and Further Processing

There is relatively little information concerning the mechanisms involved in targeting of proteins to the outer and inner membranes of the chloroplast envelope. The import of nuclear encoded proteins into the thylakoid lumen has been more intensively studied. These proteins have been found to contain two independent routing signals: the N-terminal transit peptide ensuring transport to the stroma, followed by a hydrophobic domain ("thylakoid transfer domain") ensuring import into the thylakoid lumen (99). After import into stroma, the N-terminal transit peptide is removed and after import into the thylakoid lumen, membrane-bound thylakoid peptidase removes the thykaloid transfer sequence. The thylakoid-transfer domains resemble closely the secretory route signal sequences in prokaryotes and eukaryotes (91, see Section 2.). Insertion of proteins into the thylakoid membrane instead of complete translocation is accomplished by the presence of stop transfer sequences, analogous to the mechanism found for secretory membrane proteins.

In the thylakoid lamella, further proteolytic processing of proteins may occur as well as palmitoylation (100,101).

3.1.8. Evolutionary Aspects

It is generally thought that cyanobacteria represent the free progenitors of chloroplasts. It has indeed been found that these bacteria import proteins into their thylakoid lumen posttranslationally by means of N-terminal signal sequences that resemble the thylakoid transfer sequences. An attractive model has been proposed that states that only chloroplast import signals have to be added to prokaryotic precursor proteins in order to target them from the plant cell cytoplasm to the thylakoid lumen (91). The function of the chloroplast targeting signals is the transport of proteins to their original site of synthesis in the ancestral organism, i.e., the stroma, from where the ancestral transport system still functions and mediates targeting to the thylakoid lumen (91).

3.2. Mitochondria

The vast majority of mitochondrial proteins are coded for by the nucleus and are therefore imported (see refs. 102–105 for recent reviews).

3.2.1. Post- or Cotranslational Import

Although in vitro experiments suggest that import of precursors into mitochondria occurs efficiently posttranslationally it is not completely clear whether this reflects the in vivo situation.

Mitochondria from growing but not from resting cells show surface bound polysosomes (*106*). It is recently thought that most precursors may be imported cotranslationally, although posttranslational import may also occur. Within mitochondria, four different compartments are distinguishable: the outer membrane, the intermembrane space, the innermembrane, and the matrix.

3.2.2. N-Terminal Prepeptide as Import Signal

The import of nuclear encoded mitochondrial proteins has been intensively and successfully studied (for recent reviews *see* refs. *102–105*). The vast majority of nuclear encoded mitochondrial proteins have been found to be synthesized as larger precursors. The routing signal for import is located in an N-terminal extension, the "prepeptide." Again, there appears to be little homology in primary structure. A characteristic feature of prepeptides is that they form amphiphilic alpha helices with periodically arranged basic amino acids that are exposed at one face (*107*).

3.2.3. Import Apparatus

Competition experiments have indicated that mitochondria possess at least four distinct import receptors (for prepeptide containing precursors, the ATP/ADP carrier (AAC), porin, and apocytochrome c) (*108*). The exact nature of these receptors is still not established. Import most likely occurs at contact sites of the inner and outer membrane (*109*).

3.2.4. Energy Requirements

Import in mitochondria requires ATP as well as the electrochemical gradient, contrary to the situation in chloroplast biogenesis. ATP is most likely needed at several stages as described in the model discussed below (*104*).

3.2.5. Competence Factors and Import: a Model

Similar to import of nuclear encoded proteins in chloroplasts, that of precursors in mitochondria requires cytosolic factors. It has been proposed that these (competence) factors are bound to both the mature part and the prepeptide of the precursor (*105*). The latter competence factor is thought to optimize the presentation of the routing signal and subsequent interaction with the import receptor. After binding to the receptor, the factor is released (this probably requires ATP hydrolysis). Subsequently, insertion of the prepeptide occurs across contact sites. Next, the mature part is unfolded with concomitant release of the second cytosolic competence factor. Finally, translocation occurs, followed by proteolytic removal of the prepeptide and folding. In the latter process, intraorganellar competence factors (chaperonins) may be involved (reviewed in ref 110).

3.2.6. Initial Processing: Removal of the Prepeptide

After import of the precursor, the prepeptide is removed by a specific membrane-bound protease, the prepeptidase. A number of mitochondrial proteins are not proteolytically modified, some of them having internal routing signals (AAC, porin, apocytochrome c).

3.2.7. Intramitochondrial Transport and Further Processing

Proteins destined for the inner membrane or the intermembrane space enter first the matrix where the prepeptide is removed. Next they are re-exported to the inner membrane or intermembrane space using the ancestral protein secretory pathway (see Section 2.).

In the intermembrane space, some proteins undergo further proteolytic modification (111) and specific enzymes catalyze the addition of heme groups to cytochromes (112).

3.3. Peroxisomes

Peroxisomes (or their equivalents, glyoxysomes and microbodies) are present in almost all eukaryote cells. Peroxisomes arise from preexisting ones by a process of fission. The organelles have no endogenous DNA and are thus entirely dependent on import of proteins (113,114).

3.3.1. Posttranslational Import

Peroxisomal proteins appear to be posttranslationally imported as mature size polypeptides (113,114). Exceptions regarding the absence of proteolytic modification are acylCoA oxidase and 3-oxoacylCoA thiolase (115). Their proteolytic processing appears, however, not to be linked to import in peroxisomes.

3.3.2. The Peroxisomal Routing Signal(s)

A number of peroxisomal proteins contain an SKL (serine-lysine-leucine) sequence at their C-termini or located internally. There is growing evidence that this tripeptide can act as a routing signal for peroxisomes (116). For example, when the SKL tripeptide is fused to bacterial chloramphenicol, acyltransferase, or to mouse dihydrofolate reductase, these proteins are directed to peroxisomes. Other, still putative, routing signals have been reported: a routing signal contained in the N-terminal part of thiolase (117), and two signals in acylCoA oxidase of *Candida tropicalis*, one close to the N-terminus and one near the center of the protein (118).

Intriguing is the finding that some peroxisomal proteins are also found in other subcellular compartments and that some proteins show interspecies variability in the subcellular localization, e.g., alanine: glyoxylate aminotransferase (AGT1), being peroxisomal in humans, mitochondrial in dog, and showing a dual localization in rat, reviewed in ref. 119. The cDNA coding for the human hepatic AGT1 and that of rat suggest that the human AGT1 has lost its mitochondrial targeting signal owing to a point mutation. Some patients with primary Hyperoxaluria type 1 direct AGT1 to mitochondria instead of peroxisomes. This is owing to a mutation resulting in restoration of the mitochondrial targeting signal (120).

3.3.3. Energy Requirement and Competence Factors

Import of proteins into peroxisomes requires ATP but no electrochemical gradient (121). It is thought that in analogy to biogenesis of mitochondria and chloroplasts cytosolic factors are required to render proteins competent for transport into peroxisomes. At present, no clear information about import receptors is available. The cDNA encoding the abundant 69–70 kDa integral peroxisomal membrane protein has recently been cloned (122). The nucleotide sequence predicts that this protein belongs to a super-family of ATP-binding proteins involved in various biological processes including membrane transport. It is still entirely speculative that this protein is involved in protein translocation across the peroxisomal membrane.

3.3.4. Defects in Peroxisomal Import?

In humans, a number of peroxisomal disorders occur (123,124). Some of these diseases are characterized by a general loss of peroxisomal functions: Cerebro-hepato-renal (Zellweger) syndrome, neonatal Adrenoleukodystrophy (NALD), Hyperpipecolic acidaemia (HPA) and the infantile form of Refsum disease (IRD). Since in materials from patients with Zellweger syndrome and related disorders instead of normal peroxisomes peroxisomal ghosts are detected, i.e., membrane-bound peroxisomes lacking the normal matrix proteins (125,126), it is thought that these diseases are due to deficiencies in the import of proteins in peroxisomes. Elucidation of the molecular basis of these disorders may render crucial insight in the mechanism involved in peroxisome biogenesis.

3.3.5. Modification of Proteins

With the exception of the proteolytic processing of some peroxisomal proteins, little is known about other protein modifications in peroxisomes. It is clear that the vast majority of peroxisomal proteins are not glycoproteins (113). However, it has been claimed that catalase, an abundant matrix protein, may be a glycoprotein (127). Some peroxisomal enzymes require addition of a prosthetic group to obtain full biological activity. For example, heme in catalase and FAD in acyl CoA oxidase and urate oxidase. It is not known whether prosthetic groups are added before or after import of these proteins in peroxisomes. A similar lack of knowledge represents the presence of tetrameric catalase in peroxisomes. Either tetrameres of catalase are imported, or alternatively, oligomerization of catalase polypeptides may occur in the peroxisome. In the light of findings made concerning translocation competence of polypeptides, the latter mechanism appears more attractive. It was recently observed that binding of the irreversible inhibitor aminotriazole to catalase inhibits import into peroxisomes (128). This effect is reminiscent of observations made with respect to import into mitochondria of a fusion protein consisting of a prepeptide and dihydrofolate reductase. Methotrexate is an irreversible inhibitor of dihydrofolate reductase and inhibits mitochondrial import of the fusion protein. It is thought that covalent binding of methotrexate interferes with unfolding of the fusion protein that is required for import (129).

3.4. Nucleus

A double lipid bilayer contiguous with the endoplasmic reticulum (called the nuclear envelope) surrounds the nucleus. The inner membrane of the envelope is associated with a network of intermediate-type microfilaments, the karyoskeleton, composed of proteins called laminins. The nuclear envelope contains pores with a defined ultrastructure. Transport of macromolecules in and out of the nucleus occurs through the nuclear pores.

3.4.1. Posttranslational Import and Nuclear Routing Signal

The nucleus contains a wide variety of proteins that appear to be imported from the cytosol posttranslationally. For recent reviews *see* refs. *130–132*. In vitro studies have revealed that nuclear proteins are selectively taken up and retained by the nucleus. Signals mediating routing of nuclear proteins to the nucleus, called karyophilic signals, have been identified. Again, no similarity in sequence has been found, except for some more general structural features; karyophilic signals show an overall basic charge due to presence of numerous lysine or arginine residues and a turn-inducing proline or glycine residue. There is evidence that some nuclear proteins contain more than one karyophilic signal.

3.4.2. Import into the Nucleus

The protein import into the nucleus is restricted to passage through the nuclear pore. The import mechanism is still largely unknown (*130–132*). At present, it is clear that import is dependent on ATP, and there is evidence for interaction of nuclear proteins (prior to import) with some receptor in the nuclear pore.

3.4.3. Processing

Karyophilic signals are permanent features of nuclear proteins. In contrast to other routing signals, they are not removed by proteolytic modification. This probably ensures efficient rerouting of nuclear proteins after dissociation of the nuclear envelope during mitosis or meiosis. Some posttranslational modifications have been reported to occur in the nucleus. For example, proteolytic processing and myristoylation of specific proteins (*133,134*). Of interest in this connection is the recently discovered existence of glycosylated proteins as constituents of the nuclear pore (*135,136*), *see also* Section 4. The N-acetylglucosamine modified nuclear pore proteins play an essential role in the nuclear import event. Wheat germ agglutinin, a lectin that reacts with N-acetylglucosamine-moieties, has been found to inhibit potently the import of proteins into the nucleus (*137*).

3.4.4. Posttranslational Uptake of Cytosolic Proteins by Lysosomes

Intracellular proteins are rapidly degraded in lysosomes in response to serum deprivation in the case of cultured cells or in response to starvation in the case of some tissues. Evidence has been presented that a category of cytosolic proteins sharing peptide sequences that are related to the sequence Lysine-Phenylalanine-Glutamate-Arginine-Glutamine (KFERQ) are rapidly imported into lysosomes upon serum deprivation (*see* ref. *138* for a recent review). A role for a heat shock 70 kDa protein is indicated by the finding that proteins sharing the KFERQ-like sequences are specifically recognized by this protein. These findings resemble in some aspects the observations made for translocation of proteins across membranes of the endoplasmic reticulum, mitochondria, or chloroplasts. However, it should be stressed that direct translocation of proteins containing KFERQ-like sequences across the lysosomal or endosomal membrane has yet to be demonstrated.

4. Other Protein Modifications

4.1. Glycosylation

There is growing evidence for the occurrence of protein glycosylation outside the vacuolar system (for a recent review *see* ref. *139*). Several sugartransferases have been reported to be present in nuclei (mannosyl-, galactosyl-, N-acetylglucosaminyl-, N-acetylgalactosaminyl, and sialyltransferase) and the cytosol (galactosyl-transferase). Furthermore, a number of glycoproteins have been detected outside the vacuolar system. For example,

1. Proteins in the cytoplasma and nucleus with single N-acetylglucosaminyl residues glycosidically attached to serine or threonine residues;

2. Unusual types of O-linked mannose-containing proteoglycans in the cytoplasma of rat brain;
3. Unique types of heparin sulfate in rat liver nuclei;
4. Glycosyl residues bound to the hydroxyl moiety of tyrosine residues of the cytoplasmic primer of glycogen synthesis.

The precise sites of most of these modifications are not known and the enzymes involved are only poorly characterized. However, it appears that glycosylation of proteins is not a modification that is completely restricted to the vacuolar system.

4.2. Lipid Modification

Palmitoylation and myristoylation (*see* Section 2) both occur in the cytosol and are thought to mediate anchoring of proteins to membranes. More recently in the cytosol of eukaryotic cells, another form of covalent lipid modification of proteins has been detected, i.e., the attachment of a prenyl-moiety via a thioester linkage to cysteine moieties (for a review *see* ref. 140). The modification occurs on a cysteine residue near the carboxy terminus and is part of the consensus sequence CAAX (where A is an aliphatic amino acid and X any amino acid). Two different forms of prenoylation have been identified so far. In most cases, the prenyl group appears to be farnesyl but also geranyl have been identified. Prenoylation is inhibited by compactin or mevolin that block the formation of mevalonate, an intermediate in isoprenoid synthesis. Well known examples of prenoylated proteins are yeast mating factors and eukaryotic ras proteins as well as nuclear laminins. It has been noted that prenoylation is often accompanied by another set of modifications. In the mature forms of some proteins, the prenoylated cysteine is carboxyterminal and methylated.

4.3. Ubiquitin

In eukaryotic cells, ubiquitin, an abundant cellular protein, can be covalently attached to a variety of proteins. For recent reviews *see* refs. *141–145*. Ubiquitin is a globular protein with a basic, an acidic and hydrophobic face. The carboxyterminus of ubiquitin can be conjugated to amino-groups of target proteins.

Ubiquitin can be conjugated to ε-amino groups of one or more lysines, to already conjugated ubiquitin (polyubiquitination) or to the α-amino group at the N-terminus. Ubiquitin conjugation is a multi-step process. In an initial activating step, a thioester is formed between the C-terminus of ubiquitin and an internal cysteine-residue in an ubiquitin-activating enzyme. The activated ubiquitin is next transferred to cysteine residues of ubiquitin-carrier proteins that donate the ubiquitin to the target proteins. Various functions have been proposed for ubiquitin conjugation. Ubiquitin-conjugation has been shown to "tag" protein for subsequent degradation. There is growing evidence that ubiquitination is also involved in DNA-repair, cell cycle control, and stress response (reviewed in ref. *145*). It has been proposed that ubiquitin may function in a similar manner to some heat-shock proteins by (in this case covalent) interaction of its possible contact faces with the target protein (*145*).

4.4. Phosphorylation and Dephosphorylation

Proteins in various intracellular locations can be reversibly phosphorylated at specific serine, threonine, or tyrosine residues. Phosphorylation (catalyzed by protein kinases) and dephosphorylation (catalyzed by protein phosphatases) play a crucial role at various steps in regulation of cellular processes. Because of the complexity of processes involved in protein (de)phosphorylation the reader is referred to refs. *146,147* for recent reviews.

5. Identification and Manipulation of Protein Modification: Practical Aspects

5.1. Identification of Frequently Encountered Modifications in Eukaryotic Proteins

In the preceding sections, a variety of protein modifications have been described. In this section, some commonly used approaches to identify the more frequent forms of protein modification are discussed.

5.2. Strategies to Identify the Occurrence of Proteolytic Processing

5.2.1. When the cDNA Encoding a Particular Protein Has Been Cloned and Purified Protein Is Available

Comparison of the amino acid sequence predicted by the nucleotide sequence of cDNA with the amino sequence of (C- and N-termini) of purified protein will reveal any proteolytic modification.

5.2.2. When a Suitable Antibody Against a Protein Is Available

After metabolic labeling of cells with radioactive amino acid, the protein of interest is selectively purified by binding to antibodies. Using a brief period of labeling with radioactive amino acid, followed by a chase in the presence of nonlabeled amino acid labeled proteins of different age can be obtained. These proteins can be subjected to separation, for example, using polyacrylamide gel electrophoresis in the presence of sodium dodecylsulfate under reducing conditions. After visualization by autoradiography, major changes in molecular mass of a protein during its life span will be detected. Sequencing of C- and N-termini of different molecular mass forms will reveal whether these changes are owing to proteolytic modification (Table 1).

5.2.3. When No Antibody Is Available and Different Proteins Are Suspected To Be Biosynthetically Related

The most direct approach is analysis of the amino acid sequence of the proteins. Another possibility is to digest the proteins similarly with appropriate endoproteases to obtain peptide fragments. Analysis of the properties of the fragments will reveal whether the proteins yield some identical fragments, as is expected for proteins that are derived from the same precursor.

Table 1
Major Protein Modifications in Higher Eukaryotes

I. *Vacuolar system*

   a. Endoplasmic reticulum
     Removal of leader peptide/signal peptide
     Formation of inter- and intrachain sulfide bonds
     N-linked glycosylation of asparagine residues
     Initial trimming of N-linked glycans (glucosidases,
       ER-mannosidase)

   b. Golgi apparatus/trans Golgi network
     Further trimming of N-linked glycans (mannosidases I/II)
     Formation of complex-type N-linked glycans
     (oligosaccharyltransferases)
     Formation of O-linked glycans
     Sulfation
     Secondary proteolytic modification

II. *Mitochondria. chloroplasts*
    Removal of "prepeptide"/"transitpeptide"
    Acylation, secondary proteolytic modification

III. *Nucleus*
    O-linked glycosylation
    Linkage of prenyl-moieties

5.3. Strategies to Identify the Presence of (Differently Processed Forms) of N-Linked Glycans

5.3.1. When a Large Amount of Pure Protein Is Available

After hydrolysis, the presence of sugar moieties can be established by various chemical and physical methods. N-linked glycans can be released (upon protease digestion) with glycanase N/glycopeptidase F and their structure determined using chromatographic techniques or NMR analysis.

Table 2
Glycosidases Commonly Used in N-Glycan Analysis

Glycosidase	Specificity	Does not cleave
Endo-β-galactosidase	Internal Gal-β1-4GlcNAc or Gal β 1-4 Glc in poly N-acetyllactosamine glycans	Terminal β-1-4Gal bonds
Endoglycosidase D	(Man 5) high mannose-type glycans	(Man 6-9) glycans
Endoglycosidase F	High mannose, biantennary hybrid, or complex-type glycans	Tri/tetra-antennary structures
Endoglycosidase H	High mannose or hybrid glycans	Complex-type structures
Glycopeptidase F	All N-linked glycans	O-linked glycans
Neuraminidase	Terminal sialic acid	
β-Galactosidase	Terminal galactose	
β-N-acetyl-D-glucosaminidase	Terminal N-acetylglucos amine	

5.3.2. When Only Limited Amounts of Protein Can Be Purified

The incorporation of radioactive label in oligosaccharide chains may help to increase the sensitivity of detection. For example, cells can be grown in the presence of radioactive mannose.

The protein may be radioactively labeled by growing cells in the presence of radioactive amino acid. Subsequently, pure protein or crude preparations can be incubated with a variety of endo- or exo-glycosidases that specifically hydrolyze certain linkages (*see* Table 2). The effect on apparent molecular mass of the protein of interest can be tested.

The affinity of lectins for binding to the protein can be tested. Lectins are proteins that bind to specific oligosaccharide-structures (*see*, e.g., ref. *156*). Lectins covalently immobilized to sedimentable beads can be conveniently used to establish binding of a particular protein. When binding to lectin-beads is oligosaccharide specific, it is sensitive to competition with the appropriate sugar (analog). Some lectins are also available as conjugates with peroxidase. They

Table 3
In Vivo Inhibitors of Posttranslational Processes

Inhibitor	Reaction inhibited	Ref.
Compactin and other HMG–CoA reductase inhibitors	Dolichol synthesis	148
2-Dexoxy-fluoro D-glucose	Dol-Man; Dol-Glc synthesis	148
Tunicamycin	Dol-PP-GlcNAc synthesis	148
Deoxymannojirimycin	Mannosidase I	149
Swainsonine	Mannosidase II	150
Bromoconduritol A/B	α-Glucosidase II	150
Castanospermine	α-Glucosidase I	150
Deoxynojirimycin	α-Glucosidase I and II	150
Bestatin	Aminopeptidases	151
E-64	Thiolproteases	152
Leupeptin	Serine and thiol proteases	151
Pepstatin	Acid proteases	151
Okadaic acid	Protein-serine/threonine phosphatases	153
Chlorate	Protein sulfation	154

Known side effects. Swainsonine is an inhibitor of lysosomal α-mannosidase; castanospermine and deoxynojirimycin of lysosomal β-glucosidase. Inhibition of HMG-CoA reductase affects the total isoprenoid metabolism (140). 2-Deoxy-fluoro-D-glucose will affect energy status of cells. Tunicamycin is a mixture of homologs of which some inhibit protein synthesis. Separation of different homologs is possible by reversed phase HPLC (155).

can be used for detection of oligosaccharide structures in proteins immobilized to nitrocellulose or wells of microtiter plates. Furthermore, some lectins can be purchased as biotinylated compounds. Very sensitive detection of bound biotinylated lectins is possible using streptavidine conjugated with some probe.

Another possibility to identify N-linked oligosaccharides is to expose cells to compounds that interfere with the formation and processing of these oligosaccharides. If a protein contains N-linked glycans, inhibition of glycosylation or glycan processing will affect the apparent properties, e.g., molecular mass with gel electrophoresis in the presence of sodium dodecylsulphate. Table 3 shows a number of the compounds that can be used to interfere with N-linked glycan formation and processing.

5.4. Strategies to Identify
the Presence of Lipid Modifications

A first indication for the presence of a lipid modification can be obtained by determining membrane-association of the protein of interest. In most cases, lipid modification is sufficient to ensure membrane anchoring. The presence of covalent lipid also increases the apparent hydrophobicity of the protein as can be tested with hydrophobic affinity chromatography or Triton X-114 phase partitioning. The demonstration of marked hydrophobicity of a protein does not imply that covalent lipid is present since it may be caused by hydrophobic amino acid sequences as well.

5.4.1. Identification of the Presence
of Complex Glycosylated
Phosphatidylinositol Anchors

This lipid anchor can be cleaved upon incubation with PI-specific phospholipase C. Removal of the lipid anchor will affect the apparent molecular mass and hydrophobicity of the protein. The exact composition of the lipid anchor can be established by a combination of chemical and physical techniques.

5.4.2. Identification of the Presence
of Palmitoyl- or Myristoyl Residues

Cells can be metabolically labeled with radioactive (precursors) of the fatty acids. Palmitate is generally released by hydroxylamine treatment, whereas myristate is resistant.

5.5. Manipulation of Protein Modification

5.5.1. "Specific" Inhibitors

Some protein modifications are known to be inhibited in vivo by specific compounds. Table 3 gives an overview of some "specific inhibitors." Employing the agents listed in Table 3, it should be taken in mind that the reported specificity of inhibitors is not always established in great detail. Moreover, some compounds are not commercially available in pure form and can be contaminated with other potential effectors. Furthermore, it should be realized that inhibition of one particular cellular process often results in secondary effects. This may seriously complicate the interpretation of findings made.

5.5.2. "Aspecific" Manipulation

Protein modification processes can be manipulated less selectively by interfering with intracellular ion-gradients, ATP concentrations, transport pathways and organization of the cytoskeleton, variation of temperature to which cells are exposed or the induction of changes in differentiation state of cells. Some examples of such manipulations are now discussed.

5.5.3. Manipulation of Ion-Gradients, in Specific Proton Gradients

Intracellular ion-gradients can be effected by a variety of ionophores or permeant weak bases. Particularly often employed is monensin, a carboxylic ionophore that exchanges protons for sodium ions (157). It is well documented that monensin causes major changes in the morphology of the (*trans* elements) of the Golgi apparatus (158). In the presence of low concentrations of monensin, processing of N-linked glycans to complex-type structures is inhibited (159). At high concentrations, other processes, such as receptor-mediated endocytosis and lysosomal functions, are impaired (*see*, e.g., ref. 159).

Permeant weak bases, such as ammonia, methylamine, and chloroquine, are able to pass membranes in unprotonated form, but in acidic compartments, these compounds accumulate in charged protonated form, resulting in an increase of pH. Permeant weak bases are sometimes referred to as acidotropic or lysosomotropic amines. The latter term is misleading, since all acidic intracellular compartments are affected by these amines, including trans Golgi structures and endosomes, and even cytosolic pH may be increased. A striking effect of permeant weak bases is the induction of swelling and alkalinization of lysosomes, thus inhibiting low pH-dependent lysosomal processes such as proteolysis and deglycosylation. However, effects on sorting in the trans Golgi region and general inhibition of intracellular membrane flow have also been observed. For a review on the effects of permeant weak bases *see* ref. *160*.

Very recently, the existence of a more specific effector of intraorganellar pH, bafilomycin, has been reported (161). Bafilomycin inhibits the activity of V-type protontranslocating ATPases that mediate acidification of endosomes, trans Golgi compartments, and lysosomes (162).

In cultured cells, the transport of newly formed lysosomal enzymes to lysosomes and their concomitant modification appears sensitive to extracellular conditions (163). It was noted that small differences in culture medium pH, at least in the presence of bicarbonate/carbondioxide, influenced the rate and efficiency of transport of newly synthesized hydrolases to lysosomes. The overall composition of N-linked glycans on these proteins seems to depend on their rate of transit through the trans Golgi region. Reminiscent effects of extracellular pH were recently reported for glycosphingolipid synthesis in the Golgi apparatus (164). More recently, it has been observed that osmolarity and amino acid composition of culture medium also exert effects on rate and efficiency of transport of hydrolases to lysosomes. It has been found that the transport of newly formed hydrolases, and concomitantly their N-linked glycan processing, correlates with cytosolic pH (Aerts, J. M., unpublished observations).

5.5.4. Inhibition of Transport
in the Vacuolar System

There has been and still is an intense search for conditions that specifically inhibit transport between successive compartments of the vacuolar system of higher eukaryotes. Such conditions would be of enormous help in elucidation of the mechanisms underlying sorting processes. At present, only one such condition has been detected. The fungal drug Brefeldin A is reported to inhibit specific transport from the ER to the Golgi apparatus by blocking membrane flow from but not back to the ER. For some recent applications of Brefeldin A see refs. 165–167. Drugs that inhibit microtubule organization such as colchicine, nocodazole, and taxol result in fragmentation and dispersion of the Golgi apparatus (168). However, interpretation of the effect of these inhibitors is complicated since the identity and functioning of the Golgi apparatus appear not to be completely lost (169,170). Transport in the vacuolar system has also been perturbed by inhibition of ATP production (anerobic conditions or by uncouplers of oxidative phosphorylation) and by variation of temperature. It has been observed in this way that at 10°C transport from the ER stops, and that at 20°C transport from the trans Golgi region is inhibited as well as transport between early and late endosomes (reviewed in ref. *171*).

5.5.5. Differentiation State

The modification of proteins may be affected by the differentiation state of cells. A remarkable example in this respect are some lysosomal membrane proteins (lamps) in human HL60 cells. Upon differentiation to granulocyte-like cells, lamps are more expressed at the cell surface and their N-glycans are predominantly poly-N-acetyllactosamine complex-type owing to increased activity of the extension enzyme (*172*).

6. Modification and Recombinant Products

Economic production of eukaryotic proteins in heterologous systems has become increasingly more important because of medical and veterinary interest. As described that many proteins have to undergo a variety of modifications to obtain their final structural features. In prokaryotes and simple eukaryotes, a number of the modification processes occurring in higher eukaryotes are absent. The advantages and disadvantages of producing heterologous eukaryotic proteins in different cell types are discussed.

6.1. Production in Bacteria

Major advantages of bacteria are the availability of simple expression vectors and ease at which they can be grown in high quantities. Their major disadvantage is the almost complete absence of modification machinery. It is generally preferred that heterologous proteins are secreted. In principle, this can be ensured for nonsecretory proteins by creating a construct with cleavable secretory signal sequence. Accumulation in the cytoplasm is disadvantageous because of degradation by proteases, of aggregation and inactivation of the heterologous protein, and difficulties in purification of the protein of interest. A practical problem encountered with bacteria is that the export machinery is often inefficient and easily saturated. The absence of the glycosylation machinery in bacteria also forms a serious restriction in application as expression system since many proteins of interest are glycosylated proteins.

6.2. Production in Yeast

Major advantages of the use of yeast as an expression system are the relative low costs of culture and the presence of a secretory pathway similar to that in higher eukaryotes. A clear disadvantage is that some modification processes of higher eukaryotes are absent. For example, N-linked oligosaccharides are processed to highly mannosylated structures in yeast and not to complex type structures found in higher eukaryotes. In principle, mannosylation defective mutant strains can be used to prevent formation of highly mannosylated glycans. A large variety of heterologous proteins of commercial interest is produced in S. cerevisiae, including insulin, somatostatin, interferon and amylase (see, e.g., ref. 173).

6.3. Production in Insect Cells

The production of heterologous proteins in insect cells using baculovirus vectors is of increasing interest. N-linked glycan modification in insect cells closely resembles that in mammalian cells (174).

6.4. Production in Mammalian Cells

The major advantage of using mammalian cells to produce heterologous eukaryotic proteins forms the likelihood that the produced proteins are correctly folded, modified, and secreted (175). The high costs of mammalian cell culture are a serious economic restriction.

7. Clinical Implications

The molecular basis of inherited defects in humans is established for a rapidly increasing number of disorders. A considerable number of defects have been found to be owing to a specific change in a particular protein leading to defective modification.

For example, one of the most frequent inherited disorders in humans is cystic fibrosis. The major mutation underlying this disorder is thought to lead to incomplete N-linked glycosylation of a protein that is essential in the regulation of chloride channels (*176*). So far, only few examples are known of primary defects in enzymes involved in protein modification processes. Examples are the defective formation of mannose-6-phosphate moieties with the lysosomal storage disorders Mucolipidosis II and III (*see* Section 2.), a deficiency of GlcNAc transferase II activity resulting in congenital dyserythropoietic anemia type II (*177*), and defects in enzymes involved in collagen synthesis (*178*). Specific abnormalities in N-linked glycan composition have been observed in relation to rheumatic arthritis, liver cirrhosis, and other diseases (*179–182*), however, it is questionable whether these are primary defects. The still limited number of disorders that is ascribed so far to primary defects in processing enzymes may be explained by the fact that such defects are often incompatible with life, since the properties of categories of proteins will be affected. In this connection, it has to be realized that some protein modifications are highly sensitive to interference (*see* Section 5.). Exposure to specific inhibitors of modification processes may therefore also lead to clinical abnormalities.

Finally, a recent breakthrough in therapy of Gaucher disease, the most frequent lysosomal storage disorder in humans, should be mentioned since it illustrates nicely the importance of insight in protein modification. In patients with Gaucher disease, the activity of a lysosomal hydrolase, glucocerebrosidase, is deficient (*183*). This results in massive lysosomal accumulation of the lipid glucocerebroside in lysosomes of macrophages. In the past, attempts were made to correct the defect by administering intravenously glucocerebrosidase isolated from human placenta.

The efficacy of this treatment was not high until it was realized that modification of the N-linked oligosaccharide composition of placental glucocerebrosidase would improve targeting of the enzyme to the desired cells (*184*). Recently, patients have been treated with glucocerebrosidase of which the oligosaccharides are modified in such a way that it contains only mannose residues as terminal sugar moieties. This ensures selective uptake by macrophages and delivery to their lysosomes since macrophages contain a specific mannose-binding receptor involved in endocytosis. Treatment with this modified enzyme preparation has been reported to result in spectacular improvement of clinical complications in Gaucher patients (*185*).

Bibliography

1. Milstein C., Brownlee G. G., Harrison T. M., and Matthews M. B. (1972), *Nature* **239**, 117–120.
2. Blobel G. and Sabatini D. (1970), *J. Cell Biol.* **45**, 130–145.
3. Blobel G. and Dobberstein B. (1975), *J. Cell Biol.* **67**, 835–851.
4. Walter P., Gilmore R. and Blobel G. (1984), *Cell* **38**, 5–8.
5. Meyer D. and Dobberstein B. (1982), *Nature* **297**, 647–650.
6. von Heijne G. (1986), *Nucleic Acid Res.* **14**, 4683–4690.
7. Muesch A., Hartmann E., Rohde K., Rubartelli A., Sitia R., and Rapoport T. A. (1990), *Trends Biochem. Sci.* **15**, 86–89.
8. Gascuel O. and Danchin A. (1986), *J. Mol. Evol.* **24**, 130–142.
9. Tommassen J., van Tol H., and Ligtenberg B. (1983), *EMBO J.* **2**, 1275–1279.
10. Johnson L. M., Bankaitis V. A., and Emr S. D. (1987), *Cell* **48**, 875–885.
11. Cornet P. (1983), *Bio/Technol.* **1**, 589–594.
12. Fraser J. H. and Bruce B. J. (1987), *Proc. Natl. Acad. Sci. USA* **75**, 3936–3940.
13. Rapoport T. A. (1990), *Trends Biochem. Sci.* **15**, 355–358.
14. Gilmore R. and Blobel G. (1985), *Cell* **42**, 497–505.
15. Gorlich H., Prehn S., Hartmann E., Herz J., Otto A., Kraft R., Wiedmann M., Knespel S. M., Dobberstein B., and Rapoport T. A. (1990), *J. Cell Biol.* **111**, 2283–2294.
16. Wickner W. T. and Lodish H. F. (1985), *Science* **230**, 400–407.
17. Muller, G. and Zimmermann, R. (1988) *EMBO J.* **7**, 639–648.
18. Garcia P. D., Hsiung-Du J. H., Rutter W. J., and Walter P. (1988), *J. Cell. Biol.* **106**, 1093–1104.
19. Deshaies R. J., Koch B. D., Werner-Washburne M., Craig E. A., and Schekman R. (1988), *Nature* **322**, 800–805.
20. Chirico W., Walter M. G., and Blobel G. (1988), *Nature* **322**, 805–810.
21. Freudl R. (1988), *J. Biol. Chem.* **263**, 344–349.
22. Josephson L. and Randall L. L. (1981), *Cell* **25**, 191–195.
23. Pugsley A. P. (1989), *Protein Targeting*, Acad. Press, San Diego
24. Jackson R. C. and Blobel G. (1977), *Proc. Natl. Acad. Sci. USA* **74**, 5598–5602.
25. Simon K., Perara E., and Lingappa V. R. (1887), *J. Cell Biol.* **104**, 1165–1172.
26. Yost C., Hedgpeth J., and Lingappa V. R. (1983), *Cell* **34**, 759–766.
27. Szczesna-Skorupa E., Browne N., Mead D., and Kemper B. (1988), *Proc. Natl. Acad. Sci USA* **85**, 738–742.
28. Jennings M. L. (1989), *Ann. Rev. Biochem.* **58**, 43–68.
29. Rose J. K. and Doms R. W. (1988), *Ann. Rev. Cell Biol.* **4**, 257–288.
30. Hurtley S. M. and Helenius, A. (1989), *Annu. Rev. Cell. Biol.* **5**, 277–308.

31. Freedman R. B. (1984), *Trends Biol. Sci.* **9**, 438–441.
32. Peters T. and Davidson K. L. (1982), *J. Biol. Chem.* **257**, 8847–8853.
33. Freedman R. B. (1987), *Nature* **329**, 196–197.
34. Lang K., Schmid F. X., and Fisher G. (1987), *Nature* **329**, 268–270.
35. Kornfeld R. and Kornfeld S. (1985), *Ann. Rev. Biochem.* **54**, 631–664.
36. Rothman J. E. and Lodish H. F. (1977), *Nature* **269**, 775–780.
37. Kukurazinska M. A., Bergh M. L. E., and Jackson M. (1987), *Ann. Rev. Biochem.* **56**, 915–944.
38. Lazzarino D. and Gabel G. A. (1988), *J. Biol. Chem.* **263**, 10,118–10,126.
39. Bergman L. W. and Kuehl W. M. (1979), *J. Biol. Chem.* **254**, 8869–8876.
40. Pelham H. R. B. (1989), *Ann. Rev. Cell Biol.* **5**, 1–24.
41. Pelham H. R. B. (1986), *Cell* **46**, 959–961.
42. Gething M. J. and Sambrook J. (1990), *Sem. Cell Biol.* **1**, 65–72.
43. Ellis R. J. and Hemmingsen S. M. (1989), *Trends Biol. Sci.* **14**, 339–342.
44. Schlesinger M. J. (1990), *J. Biol. Chem.* **265**, 12,111–12,114.
45. Wiech H., Stuart R., and Zimmermann R. (1990), *Sem. Cell Biol.* **1**, 55–63.
46. Wu. H. C. and Tokunaga M. (1986), *Curr. Top. Microbiol. Immunol.* **125**, 127–158.
47. Cross G. A. (1987), *Cell* **48**, 179–181.
48. Sefton B. M. and Buss J. E. (1987), *J. Cell Biol.* **104**, 1449–1453.
49. Ferguson M. A., Homans S. W., Dwek R. A., and Rademacher T. W. (1988), *Science* **239**, 753–759.
50. Lisanti M. P. and Rodriguez-Boulan E. (1990), *Trends Biol. Sci.* **15**, 113–118.
51. Mc Ilhinney R. A. J. (1990), *Trends Biol. Sci.* **15**, 387–390.
52. Towler D. A., Gordon J. I., Adams S. P., and Glaser L. (1988), *Ann. Rev. Biochem.* **57**, 69–100.
53. Schultz A. M., Hendersson L. E., and Oroszlan (1988), *Ann. Rev. Cell Biol.* **4**, 612–648.
54. Kivirikko K. I. and Myllyla R. (1986), *Ann. NY Acad. Sci.* **460**, 187–210.
55. Warren G. (1987), *Nature* **327**, 17–18.
56. Dean N. and Pelham H. R. B. (1990), *J. Cell Biol.* **111**, 369–377.
57. Rambourg A. and Clermont Y. (1990), *Eur. J. Cell Biol.* **51**, 189–200.
58. Geuze H. J., Slot J. W., Strous G., Hasilik A., and von Figura K. (1984), *J. Cell Biol.* **98**, 2047–2054.
59. Roth J., Taatjes D. J., Lucocq J. M., Weinstein J., and Paulson J. C. (1985), *Cell* **43**, 287–295.
60. Cooper M. S., Cornell Bell, A. Chernjavsky, A., Dani J. W., and Smith S. J. (1990), *Cell* **61**, 135–145.
61. Griffiths G. and Simons K. (1986), *Ann. Rev. Cell Biol.* **5**, 483–526.
62. Komfeld S. and Mellman I. (1989), *Ann. Rev. Cell Biol.* **5**, 483–526.
63. Pfeffer S. and Rothman J. E. (1987), *Ann. Rev. Biochem.* **56**, 829–852.
64. Kelly R. B. (1985), *Science* **230**, 25–32.
65. Simons K. and Fuller S. D. (1985), *Ann. Rev. Cell. Biol.* **1**, 243–288.
66. von Figura K. and Hasilik A. (1986), *Annu. Rev. Biochem.* **55**, 167–193.
67. Aerts J. M. F. G., Schram A. W., Strijland A., van Weely S., Jonsson L. M. V., Tager J. M., Sorrel S. H., Ginns E. I., Barranger J. A., and Murray G. J. (1988), *Biochem. Biophys. Acta* **964**, 303–308.

68. Nolan C. M. and Sly W. S. (1989), *The Metabolic Basis of Inherited Disease* (Scriver C. R., Beaudet A. L., Sly W. S., Valle D., eds.), pp. 1589–1601.
69. Klionsky D. J., Banta L. M., and Emr S. D. (1988), *Moll. Cell. Biol.* **8**, 2105–2116.
70. Schwaiger H., Hasilik A., von Figura K., Wiemken A., and Tanner W. (1982), *Biochem. Biophys. Res. Commun.* **104**, 950–956.
71. Valls L. A., Winther J. R., and Stevens T. H. (1990), *J. Cell Biol.* **111**, 361–368.
72. Tague B. W. and Chrispeels M. J. (1987), *J. Cell. Biol.* **105**, 1971–1979.
73. Matsuoka K., Matsumoto S., Hattori T., Machida Y., and Nakamura K. (1990), *J. Biol. Chem.* **265**, 19,750–19,755.
74. Kaushal G. P. and Elbein A. D. (1989), *Methods Enzymol.* **179**, 452–457.
75. Dahms N. M. and Hart G. W. (1986), *J. Biol. Chem.* **261**, 13,186–13,196.
76. Farquhar M. G. (1986), *Annu. Rev. Cell Biol.* **1**, 447–488.
77. Lang T., Reitman M., Tang J., Roberts R. M., and Kornfeld S. (1984), *J. Biol. Chem.* **259**, 14,663–14,671.
78. Carlsson S. R. and Fukuda M. (1990), *J. Biol. Chem.* **265**, 20,488–20,495.
79. Kornfeld S.and Kornfeld R. (1980), Plenum, New York, pp. 1–34.
80. Niemann H., Boschek B., Evans D., Rosing M., Tamura T., and Klenk H-D. (1982), *EMBO J.* **1**, 1499–1504.
81. Hasselbeck A. and Tanner W. (1983), *FEBS* **158**, 335–338.
82. Lindahl U., Feingold D., and Roden L. (1986), *Trends Biochem. Sci.* **11**, 221–225.
83. Huttner W. B. (1988), *Ann. Rev. Physiol.* **50**, 363–376.
84. Fuller R. S., Sterne R. E., and Thorner J. (1988), *Ann. Rev. Physiol.* **50**, 345–362.
85. Bresnahan P. A., Leduc R., Thomas L., Torner J., Gibson H. L., Brake A. J., Barr P. J., and Thomas G. (1990), *J. Cell Biol.* **111**, 2851–2859.
86. Hasilik A., von Figura K., Conzelmann E., and Sandhoff K. (1982), *Eur. J. Biochem.* **125**, 317–321.
87. Oude Elferink R. P. J., Brouwer-Kelder E. M., Surya I., Strijland A., Kroos M., Reuser A. J. J., and Tager J. M. (1984), *Eur. J. Biochem.* **139**, 489–495.
88. Eipper B. A. and Mains R. E. (1986), *J. Cell Biol.* **103**, 265–273.
89. Snider M. D. and Rogers O. C. (1986), *J. Cell Biol.* **103**, 265–273.
90. Neefjes J. J., Verkerk J., Broxterman H. J. G., van der Marel G. A., van Boom J. H., and Ploegh H. L. (1988), *J. Cell Biol.* **107**, 79–87.
91. Smeekens S., Weisbeek P., and Robinson C. (1990), *Trends Biochem. Sci.* **15**, 73–76.
92. Weisbeek P., Hagemann J., de Boer D., Pilon R., and Smeekens S. (1989), *J. Cell Sci. Suppl.* **11**, 199–223.
93. von Heijne G., Stepphun J., and Herrman R. (1989), *Eur. J. Biochem.* **180**, 535–545.
94. Cornwell K. and Keegstra K. (1987), *Plant Physiol.* **85**, 780–785.
95. Pain D. and Blobel G. (1990), *J. Cell Biol.* **111**, 1825–1838.
96. Schnell D., Blobel G., and Pain D. (1990), *J. Cell Biol.* **111**, 1825–1838.
97. Robinson C., Ellis R. J. (1984), *Eur. J. Biochem.* **142**, 337–342.
98. Olson L., Theg S., Selman B., and Keegstra K. (1989), *J. Biol. Chem.* **264**, 6724–6729.

99. Smeekens S., Bauerle C., Hageman J., Keegstra K., and Weisbeek P. (1986), *Cell* **46,** 365–375.
100. Herrin D. and Michaels A. (1985), *FEBS* **184,** 90–95.
101. Matoo A. K. and Edelman M. (1987), *Proc. Natl. Acad. Sci. USA* **84,** 1497–1501.
102. Attardi G. and Schatz G. (1988), *Annual Review of Cell Biology* **4,** 290–335.
103. Hartl F. U., Pfanner N., Nicholson D. W., and Neupert W. (1989), *Biochem. Biophys. Acta* **983,** 1–45.
104. Pfanner N. and Neupert W. (1990), *Annu. Rev. Biochem.* **59,** 331–353.
105. Neupert W., Hartl F-U., Craig E. W. and Pfanner N. (1990), *Cell* **63,** 447–450.
106. Suissa M. S. and Schatz G. (1982), *J. Biol. Chem.* **257,** 13,048–13,055.
107. von Heijne G. (1986), *EMBO J.* **5,** 1335–1342.
108. Pfaller R., Steger H. F., Rassow, Pfanner N., and Neupert W. (1988), *J. Cell Biol.* **107,** 2483–2490.
109. Schleyer M. and Neupert W. (1985) *Cell* **43,** 339–350.
110. Halberg R. L. (1990), *Sem. Cell Biol.* **1,** 37–45.
111. Hartl F-U., Ostermann J., Guiard B., and Neupert W. (1987), *Cell* **51,** 1027–1073.
112. Nicholson D. W., Kohler H., and Neupert W. (1987), *Eur. J. Biochem.* **164,** 147–157.
113. Lazarow P. B. and Fujiki Y. (1985), *Annu. Rev. Cell Biol.* **1,** 489–530.
114. Borst P. (1989), *Biochim. Biophys. Acta* **1008,** 1–13.
115. Miura S., Mori M., Takiguchi M., Tatibana M., Furuta S., Miyazawa S., and Hashimoto T. (1984), *J. Biol. Chem.* **259,** 6397–6402.
116. Gould S. J., Keller G. A., Hoskin N., Wilkinson J., and Subramani S. (1989), *J. Cell Biol.* **108,** 1657–1664.
117. Swinkels B. W., Gould S. J., Bodnar A. G., Rachubinski R. A., and Subramani S. (1990), *J. Cell Biol.* **111,** 386a
118. Small G. M., Szabo L. J., Lazarow P. B. (1988), *EMBO J.* **7,** 1167–1173.
119. Wiemer E. (1990), Biogenesis of Peroxisomes in relation to Disorders of Peroxisome Assembly, Thesis University of Amsterdam, The Netherlands.
120. Purdue P. E., Takada Y., and Danpure C. J. (1990), *J. Cell Biol.* **111,** 2341–2351.
121. Imanaka T., Small G. M., and Lazarow P. B. (1987), *J. Cell Biol.* **105,** 2915–2922.
122. Kamijo K., Taketani S., Yokota S., Osumi T., and Hashimoto T. (1990), *J. Biol. Chem.* **265,** 4534–4540.
123. Wanders R. J. A., Heijmans H. S. A., Schutgens R. B. H., Barth P. G., van den Bosch H., and Tager J. M. (1988), *J. Neurol. Sci.* **88,**1–33.
124. Lazarow P. B. and Moser H. W. (1989), McGraw-Hill, New York, pp. 1479–1509.
125. Santos M. J., Imanaka T., Shio H., Small G. M., and Lazarow, P. B. (1988), *Science* **239,** l536–1538.
126. Wiemer E. A. C., Brul S., Just W. W., van Driel R., Brouwer E. M., van den Berg M., Weijers P., Schutgens R., van den Bosch H., Schram A. W., Wanders R., and Tager J. M. (1990), *Eur. J. Cell Biol.* **50,** 407–417.
127. Pegg M., Crane D., and Masters C. (1986), *Biochem. Int.* **12,** 831–838.

128. Middelkoop E., Strijland A., and Tager J. M. (1991) *FEBS Lett.*, **279**, 79–82.
129. Eiler M. and Schatz G. (1986), *Nature* **322**, 228–232.
130. Dingwall C. and Laskey R. A. (1986), *Annu. Rev. Cell Biol.* **2**, 367–390.
131. Newport J. and Forbes D. J. (1987), *Annu. Rev. Biochem.* **56**, 535–565.
132. Finlay D. R., Newmeyer D. D., Hartl P. M., Horecka J., and Forbes D. J. (1989), *J. Cell Sci. Suppl.* **11**, 225–242.
133. Lehner C. F. et al. (1986), *Proc. Natl. Acad. Sci. USA* **83**, 2096–2099.
134. Streuli C. H. and Griffin B. E. (1987), *Nature* **326**, 619–622.
135. Holt G. D., Snow C. M., Senior A., Haltiwanger R. S., Gerace L., and Hart G. W. (1987), *J. Cell Biol.* **104**, 1157–1164.
136. Davis L. I. and Blobel G. (1987), *Proc. Natl. Acad. Sci. USA* **84**, 7552–7556.
137. Yoneda Y., Imamoto-Sonobe N., Yamaizumi M., and Uchida T. (1987), *Exp. Cell Res.* **173**, 586–595.
138. Dice J. F. (1990), *Trends Biochem. Sci.* **15**, 305–309.
139. Hart G. W., Haltiwanger R. S., Holt G. D., and Kelly W. G. (1989), *Annu. Rev. Biochem.* **58**, 841–874.
140. Glomset J. A., Gelb M. H., and Farnsworth C. C. (1990), *Trends Biochem. Sci.* **15**, 139–142.
141. Finley D. and Varshavky A. (1985), *Trends Biochem. Sci.* **10**, 343–346.
142. Herschko A. (1988), *J. Biol. Chem.* **263**, 15,237–15,240.
143. Rechtersteiner M. (1987), *Annu. Rev. Cell Biol.* **3**, 1–30.
144. Ciechanover A. and Schwartz A. L. (1989), *Trends Biochem. Sci.* **14**, 483–488.
145. Jentsch S., Seufert W., Sommer T., and Reins H. A. (1990), *Trends. Biochem. Sci.* **15**, 1851–1858.
146. Cohen P. (1989), *Annu. Rev. Biochem.* **58**, 453–458.
147. Kemp B. E. and Pearson R. B. (1990), *Trends Biochem. Sci.* **15**, 342–346.
148. Schwartz R. T. and Datema R. (1983), Methods Enzymol, **83**, 432–443.
149. Fuhrmann U., Bause E., and Ploegh H. (1985), *Biochim. Biophys. Acta* **825**, 95–110.
150. Schwarz R. T. and Datema R. (1984), *Trends Biochem. Sci.* **9**, 32–34.
151. Umezawa H. (1982), *Annu. Rev. Microbiol.* **36**, 75–79.
152. Hanada K. (1987), *Agric. Biol. Chem.* **42**, 523–536.
153. Cohen P., Holmes C. F. B., and Tsukitani Y. (1990), *Trends Biochem. Sci.* **15**, 98–102.
154. Bauerle P. A. and Huttner W. (1986), *Biochem. Biophys. Res. Commun.* **141**, 870–877.
155. Mahoney W. C. and Duksin D. (1980), *J. Chromatgr.* **198**, 506–510.
156. Osawa T. and Tsuji T. (1987), *Ann. Rev. Biochem.* **56**, 21–42.
157. Sandaux R., Sandaux J., Gavach C., and Brun C. (1983), *Biochim. Biophys. Acta* **684**, 127–132.
158. Tartakoff A. M. (1983), *Cell* **32**, 1026–1028.
159. Ledger P. W. and Tanzer M. L. (1984), *Trends Biochem. Sci.* **9**, 317–324.
160. Dean R. T., Jessup W., and Roberts C. R. (1984), *Biochem. J.* **217**, 27–40.
161. Bowman E. J., Siebers A., and Altendorf K. (1988), *Proc. Natl. Acad. Sci. USA* **85**, 7972–7976.
162. Nelson N. and Taiz L. (1989), *Biochem. Sci.* **14**, 113–116.

163. Aerts J. M. (1988), Biochemical studies on glucocerebrosidase in reladon to Gaucher disease, Thesis, University of Amsterdam, the Netherlands.

164. Iber H., van Echten G., Klein R. A., and Sandhoff K. (1990), *Eur. J. Cell Biol.* **52**, 236–240.

165. Mitsumi Y., Misumi Y., Miki K., Takatsuki A., Tamura G., and Ikehara Y. (1986), *J. Biol. Chem.* **261**, 11,398–11,403.

166. van Echten G., Iber H., Takatsuki A., and Sandhoff K. (1990), *Eur. J. Cell Biol.* **51**,135–139.

167. Lippincott-Schwartz J., Yuan L. C., Bonifacino J. S., and Klausner R. (1989), *Cell* **56**, 801–813.

168. Turner J. R. and Tartakoff A. M. (1989), *J. Cell Biol.* **109**, 2081–2088.

169. Salas P. J., Misek E., Vega-Salas D., Gunderson D., Cereijido M., and Rodrigues-Bolan E. (1986), *J. Cell Biol.* **102**, 1853–1867.

170. Ho W. C., Allan V. J., van Meer G., Berger E. G., and Kreis T. E. (1989), *Eur. J. Cell Biol.* **48**, 250–263.

171. Tartakoff A. M. (1987), The Secretory and Endocytic Paths, Wiley, New York.

172. Lee N., Wang W.-C., and Fukuda M. (1990), *J. Biol. Chem.* **265**, 20,476–20,487.

173. Gwynne, D. I. (1987), *Bio/technol.* **5**, 713–719.

174. Luckow V. A. and Summers M. D. (1988), *Bio/technol.* **6**, 47–55.

175. Zettmdssl G. (1987), *Bio/technol.* **5**, 720–725.

176. Cheng S. H., Gregory R. J., Marshall J., Paul S., Souza D. W., White G. A., O'Riordan C., and Smith A. E. (1990), *Cell* **63**, 827–834.

177. Fukuda M. M., Dell A., and Scartezzini P. (1987), *J. Biol. Chem.* **262**, 7195–7206.

178. Byers P. H. (1989), *The Metabolic Basis of Inherited Disease* (Scriver C. R., Beaudet A. L., Sly W. S., Valle D. M eds.), McGraw-Hill, New York, pp. 2805–2842.

179. Rademacher T. W., Parekh R. B., and Dwek R. A. (1988), *Glycobiology Ann. Rev. Biochem.* **57**, 785–838.

180. Axford J. S., Lydyard P. M., Isenberg D. A., Mackenzie L., Hay F. C., and Roitt I. M. (1987), *Lancet* 1486–1489.

181. Pekeiharing J. M., Hepp E., Kamerling J. P., and Leijnse B. (1988), *Ann. Rheum. Dis.* **47**, 91–95.

182. Jaeken J., van Eijk H. G., van der Heul C., Corbeel L., Eeckels L., and Eggermont E. (1984), *Clin. Chem. Acta* **144**, 245–247.

183. Barranger J. A. and Ginns E. I. (1989) *The Metabolic Basis of Inherited Disease* (Scriver C. R., Beaudet A. L., Sly W. S., Valle D., eds.), McGraw-Hill, New York, pp. 1677–1698.

184. Furbish F. S., Steer C. J., Krett N. L., and Barranger J. A. (1981), *Biochim. Biophys. Acta* **673**, 425–435.

185. Barton N. W., Furbish F. S., Murray G. J., Garfield M., and Brady R. O. (1990), *Proc. Natl. Acad. Sci. USA* **87**, 1913–1916.

Seven

Protein Fragmentation

Linda A. Fothergill-Gilmore

1. Why Fragmentation of Proteins?

Proteins are large molecules. A great many studies to investigate protein structure and function require that a protein be fragmented in order to produce pieces that are suitable for further detailed investigation. Some of the main types of experiments that require protein fragmentation are summarized in Section 1. of this chapter. The next two sections are concerned with the methods used to generate fragments of a protein; these can be grouped conveniently depending on whether the fragmentation causes denaturation or not. The final section of this chapter considers the use of recombinant protein technology to produce protein fragments.

1.1. Sequence Determination

Peptide sequence information may be required for a host of different reasons. These may be concerned with the investigation of enzyme mechanism, the characterization of posttranslational processing, the design of oligonucleotide probes, and many others. Please refer to Section 1.1. of the chapter entitled "Peptide Sequence Determination" for a fuller description of the reasons why protein sequence information might be needed.

There are two main strategies to obtain peptide sequences (*see* Chapter 8) that depend on whether the requirement is for a complete sequence, or for only a limited amount of sequence information.

From: *Protein Biotechnology* • F. Franks, ed.   ©1993 The Humana Press Inc.

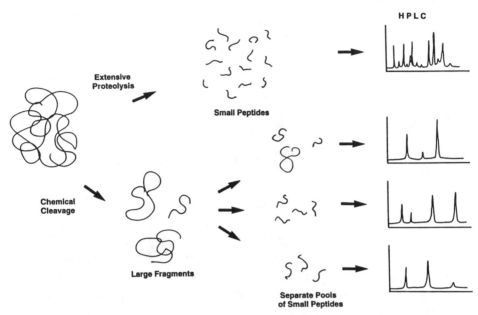

Fig. 1. Sequence determination.

For partial sequences, it may be enough to subject a protein to Edman degradation directly. However, many proteins have a blocked N-terminus, and are thus not susceptible to Edman degradation. If this is the situation, then it is necessary to cleave the protein to generate fragments with free N-terminal residues (Fig. 1). It is ideal to choose a method that will cleave the protein at only a single peptide bond, and thus avoid the need for separation and purification of the fragments. If this is not possible, then it is usually most convenient to use a cleavage procedure that gives only a small number of fragments. The strategy to obtain a complete protein sequence involves the production of complete sets of relatively easily purified fragments. These fragments should on average be 20–40 residues long, and the different sets of fragments should provide overlap information.

Both chemical and enzymic methods are commonly used in peptide sequencing (*see* Chapter 8). The most useful methods are usually those that are highly specific so that the potential cleavage sites are well understood. It is entirely acceptable to use methods that cause complete denaturation of the fragments because the Edman degradation procedure itself will soon denature the fragments.

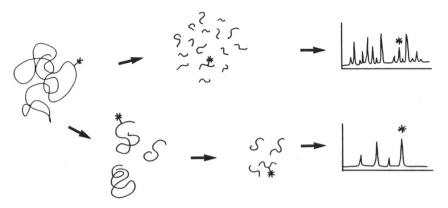

Fig. 2. Isolation of modified residues.

1.2. Isolation of Modified Residues

Proteins frequently contain chemical moieties not derived from amino acids. These may be added to the protein in vivo once it has been synthesized on the ribosome, or in vitro as part of an investigation to study protein structure and function. In vivo posttranslational modifications may include phosphorylation, glycosylation, myristylation, and ADP-ribosylation. Experiments in vitro may be concerned with chemical modification to identify important residues, or may involve the characterization of sites of crosslinking.

The methods to generate and purify peptides containing modified residues are usually very similar to those used for peptide sequence determination (Fig. 2). In addition, the procedures all require a strategy to identify the peptide containing the modified residue, and these frequently exploit radiolabeling or spectrophotometry. In most cases, the peptide is then subjected to sequence determination, and it is quite acceptable to use fragmentation procedures that cause denaturation of the fragments.

1.3. Release of Domains from Membranes

Integral membrane proteins, such as receptors or transport proteins, possess a stretch of hydrophobic amino acid residues that are buried within the lipid bilayer of membranes. The intra- and extracellular domains provide the ligand-binding and catalytic

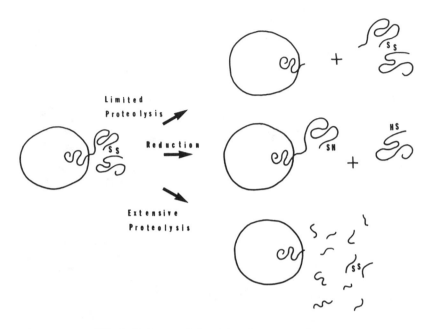

Fig. 3. Release of domains from membranes.

regions of the molecule, and experiments to investigate the structure and function of membrane proteins often have a requirement for the specific isolation of the different domains. There are three main experimental approaches, all of which use the protein inserted in the membrane of a cell (microsome) or a liposome as starting material. Limited proteolysis can be used in an attempt to release the entire extracellular domain of the protein in a more-or-less native state (Fig. 3). In this case, the procedures described in Section 3. would be used in order to minimize denaturation. A substantial number of membrane proteins, such as immunoglobulin-type receptors, contain extracellular domains with more than one subunit linked by disulfide bridges. Reducing agents can be used to release these subunits (*see* Section 2.3.), probably in an unfolded denatured state. Alternatively, extracellular domains can be digested completely to small peptides (*see* Section 2.2.), leaving the transmembrane and intracellular domains intact for further study.

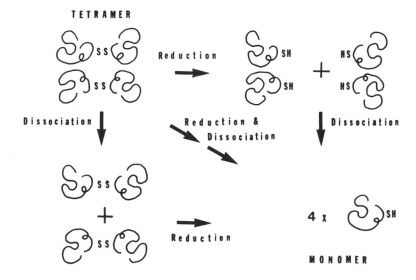

Fig. 4. Separation of subunits.

In many studies of membrane proteins, it is possible to generate microsomes or liposomes that are inside-out with respect to the physiologically normal orientation. In this way, the intracellular domains can become accessible to the procedures described in the preceding paragraph.

1.4. Separation of Subunits

It is probably true to say that most proteins are oligomeric. There may be multiple copies of the same subunit, or there may be assemblies of several different types of subunit. Intracellular proteins are held together by noncovalent forces (electrostatic, hydrogen bonding, hydrophobic interactions, and so on), whereas extracellular proteins may have disulfide bridges or (more rarely) other types of covalent link, as in extracellular matrix proteins.

Fragmentation of a protein with covalent links requires the use of reducing agents or of other types of chemical cleavage (Fig. 4). All of these procedures are likely to cause extensive denaturation of the protein, and it is thus probably not possible to obtain the individual subunits in a native state. By contrast, it is some-

Fig. 5. Separation of domains.

times feasible to separate noncovalent oligomers into separate subunits with conformations approaching the native state. A simple procedure is just to make a dilute solution of the protein. Proteins that are not dissociated by this method will probably require treatment with denaturing agents such as guanidine or urea.

1.5. Separation of Domains

Domains of proteins can be considered to be compact folded units within the proteins. Domains are stable assemblies of elements of regular secondary structure (reviewed by Richardson [1]), such as the nucleotide-binding Rossmann fold found in dehydrogenases, or the alpha/beta barrel originally described in triose phosphate isomerase, and now observed in many other enzymes. The separation of a protein into domains requires the use of methods that are likely to cleave only those peptide bonds occurring in exposed loops between the domains (Fig. 5). Thus, the most suitable procedures involve limited proteolysis as described in Section 3.

1.6. Removal of Tails or Loops

A comparison of different isoenzymes often reveals that variations in control properties reside in the terminal regions. For example, phosphorylation sites frequently occur near the N- or C-terminus of a protein, as in pyruvate kinase (2). It appears from an evolutionary point of view that it is relatively easy for mutations near to the ends of proteins to be fixed by selection, presumably because the termini tend to be flexible surface regions, and changes can be tolerated without major disruption of the native structure of the protein. These observations are also true for other types of protein that are present in an organism in multiple forms. Therefore, it can often be of considerable interest to treat a protein in such a way

Fig. 6. Removal of tails.

that it has lost its flexible tail (Fig. 6). It may be desirable to study the peptide released in order to characterize its structure and function, or the aim may be to investigate the properties of the tailless form of the protein. In both cases, the methods of limited proteolysis described in Section 3. will probably be most appropriate.

2. Fragmentation with Denaturation

The procedures described in this section will most often be used in conjunction with peptide sequence studies. Additional relevant information and comments can be found in Chapter 8, and cross references will be made as appropriate. Further useful information and protocols can be found in Allen (3).

2.1. Chemical

Chemical methods to cleave proteins have two main advantages: (a) specificity and (b) infrequency of susceptible peptide bonds. These procedures are thus suitable for the generation of relatively simple mixtures of large fragments. Four methods are in common use:

1. Cyanogen bromide cleavage at the C-terminal side of methionine residues (4);
2. Hydroxylamine cleavage of asparagine–glycine peptide bonds (5);
3. o-Iodosobenzoic acid cleavage on the C-terminal side of tryptophan residues (6); and
4. Mild acid cleavage of aspartate-proline peptide bonds (7).

Examples of strategies to use these methods are described in Chapter 8 (*see* Fig. 17 for a summary flow diagram).

2.2. Enzymic

Proteolytic enzymes serve a great many functions in biological systems, including digestion of nutrient proteins, processing of blood clotting and fibrinolysis components, recycling of denatured proteins, and so on. Many of these naturally occurring proteolytic enzymes are of great use to the protein chemist. A list of the enzymes most commonly used to produce peptides for sequence studies is given in Table 1, together with an indication of their specificity. (Proteolytic enzymes that have very high substrate specificity, and thus cleave proteins at only a few peptide bonds, are mentioned in Section 3.2.) Strategies and some experimental protocols relevant to the use of proteolytic enzymes in sequence studies are given in Chapter 8.

2.3. Reducing Agents

Oligomeric proteins that are held together by covalent bonds usually have disulfide bridges providing the links between the subunits. Disruption of these bonds is readily achieved by treatment with reducing agents, such as dithiothreitol or mercaptoethanol. However, the bonds can easily reoxidize, and precautions should be taken to prevent disulfide bonds from reforming. A convenient approach is to alkylate the sulfhydryl groups, and the use of 4-vinylpyridine to yield the pyridylethyl derivative of cysteine (8,9) can be recommended. This particular derivative is preferred if the protein is to be subjected to peptide sequence analysis, because of its ease of separation in the HPLC systems of automated sequencers and amino acid analyzers. It should also be mentioned that dithiothreitol is the preferred reducing agent, not only because of its less objectionable smell, but also because it can be used in lower concentrations. In addition, the oxidized form of dithiothreitol does not interchange with protein sulfhydryls and thus does not give rise to undesirable mixed disulfides.

3. Fragmentation Avoiding Denaturation

The experimental approaches described in this section will most often be used in attempts to generate large fragments of proteins that have retained some measure of their native structure.

The conditions are chosen to be as mild as possible, and to cleave as few peptide bonds as possible.

Native proteins are generally folded into compact structures that are inherently resistant to proteolysis. For this reason, proteins that are to be subjected to extensive proteolysis (*see* Sections 1.1. and 1.2.) are usually denatured first. Many native proteins are not entirely resistant to proteolytic attack, however, because they possess one or more peptide bonds that are unusually susceptible to particular proteases. Such susceptible bonds may be in exposed loops linking globular domains, or may just be on the surface of the tightly folded protein. Ends of protein chains are frequently quite flexible and may also be accessible to proteolytic enzymes. There are two main approaches to achieve specific cleavage of unusually sensitive peptide bonds: (a) limitation of the activity of broad specificity proteases, or (b) the use of proteases with high substrate specificity.

3.1. Partial Proteolysis

In theory, any of the proteolytic enzymes listed in Table 1 could be used for the partial cleavage of proteins. In practice, however, only a few of the enzymes have been in routine use. Trypsin, pepsin, subtilisin, and papain are among those that have been used successfully *(10–13)*. The susceptibility to proteolysis varies from protein to protein and from peptide bond to peptide bond, and it is thus not possible to give a detailed protocol that is generally applicable. Relatively small changes in pH, ionic strength, temperature, the presence of substrates or other ligands, and stabilizing agents such as glycerol may have a large effect on the rates of peptide bond cleavage.

Several generalizations can be made about the strategies for achieving partial proteolysis.

1. It is sensible to try several different proteolytic enzymes with a range of different specificities.
2. It is usually necessary to limit the activity of the proteolytic enzymes in some way. This may be done either by the use of very low enzyme to substrate ratios (say, 1–1000 by weight), or by very short times of digestion (say, ~1 min).

Table 1
Proteolytic Enzymes Used in Peptide Sequence Studies

Enzyme	Usual source	Specificity	Comments
1. Trypsin	bovine pancreas	C-terminal side of lysine and arginine	Has low level of inherent "pseudotrypsin" activity, that is chymotrypsin-like. Lys-Pro and Arg-Pro bonds resistant
2. Clostripain	*Clostridium histolyticum*	C-terminal side of arginine	May also cleave at occasional lysines
3. Arginine-specific protease	mouse submaxillary gland	C-terminal side of some arginines	Also known as Endoproteinase Arg-C
4. Lysine-specific protease	*Lysobacter enzymogenes*	C-terminal side of lysine	Also known as Endoproteinase Lys-C
5. Lysine-specific protease	*Armillaria mellea*	N-terminal side of lysine	May also cleave at C-terminal side of some arginines
6. Protease II	*Myxobacter*, strain AL-1	N-terminal side of lysine	
7. Staphylococcal protease	*Staphylococcus aureus* (strain V8)	C-terminal side of glutamate or aspartate	Conditions may be adjusted to give enhanced specificity for glutamate—Glu-Pro bonds resistant—Also known as V8-protease; Endoproteinase Glu-C

8. Endoproteinase Asp-N	*Pseudomonas fragi*	N-terminal side of aspartate and cysteic acid	
9. Chymotrypsin	bovine pancreas	C-terminal side of aromatic and some hydrophobic residues	May cleave occasionally at other residues—X-Pro bonds resistant
10. Thermolysin	*Bacillus thermoproteolyticus*	N-terminal side of hydrophobic residues	Useful for subfractionation of purified peptides
11. Protease	*Astacus*	N-terminal side of Ala, Thr, Ser, Gly	
12. Pepsin	porcine gastric mucosa	many peptide bonds, especially between aromatic and hydrophobic residues	Low pH optimum, so prior denaturation of substrate not necessary, and disulfide interchange low
13. Subtilisin	*Bacillus subtilis*	broad specificity	Specificity too broad to be generally useful
14. Papain	*Carica papaya*	broad specificity	Sometimes useful in partial proteolysis
15. Elastase	porcine pancreas	broad specificity	

3. It is usually advantageous to be able to stop the digestion in a controlled manner. The use of one of the trypsin inhibitors can be a particularly convenient way to stop trypsin digestion, and has the advantage that it should not cause any denaturation of the target protein. Alternatively, immobilized proteolytic enzymes can conveniently be separated from a protein and its fragments.

4. Pilot experiments should be done with a range of conditions, and with the use of SDS-polyacrylamide gel electrophoresis or HPLC to follow the course of digestion.

3.2. Proteolytic Enzymes with High Substrate Specificity

Many cellular processes are controlled by proteolytic activation of inactive precursors. Thus, for example, zymogens are converted to active enzymes, prohormones are cleaved to hormones, and the blood clotting, fibrinolysis, and complement cascades all involve successive proteolytic steps. All of the proteolytic enzymes involved in these and other processes are characterized by high substrate specificity to ensure that appropriate control is maintained. Potentially all of these enzymes could be used to achieve fragmentation of a protein of interest into a small number of fragments. Examples of enzymes that have been used in this way include plasmin, thrombin, and rennin (14–16). The experimental approach is to set up pilot digestions with as many enzymes as can be obtained, and then to examine the products by electrophoresis or HPLC.

4. Recombinant Protein Technology to Express Fragments

The techniques of genetic engineering can be very suitable for the production of fragments of proteins. However, it is of course necessary to have isolated and cloned the DNA encoding the protein, and to have determined the nucleotide sequence of the gene or cDNA. The methods of oligonucleotide-directed mutagenesis can then be used to introduce deletions or stop codons as appropriate in order to specify the production of the desired fragment. Once the altered form of the gene has been obtained, it is necessary to devise a suitable expression system for the recombinant protein. Potentially, this experimental approach can lead to the convenient production of milligram to gram quantities of recombinant proteins or their fragments. The same system can also be used for site-directed mutagenesis studies.

Bibliography

1. Richardson J. S. (1981), *Adv. Prot. Chem.* **34**, 147–339.
2. Humble E. (1980), *Biochim. Biophys. Acta* **626**, 179–187.
3. Allen G. (1989), *Sequencing of Proteins and Peptides*, 2nd ed., Elsevier, Amsterdam.
4. Gross E. and Witkop B. (1962), *J. Biol. Chem.* **237**, 1856–1860.
5. Bornstein P. and Balian G. (1977), *Methods Enzymol.* **47**, 132–145.
6. Mahoney W. C. and Hermodson M. A. (1979), *Biochemistry* **18**, 3810–3814.
7. Landon M. (1977), *Methods Enzymol.* **47**, 145–149.
8. Tarr G. E. (1986), *Methods of Protein Microcharacterization*, (Shively J. E., ed.), Humana, Clifton, NJ, pp. 162–163.
9. Hayes J. D., Kerr L. A., and Cronshaw A. D. (1989), *Biochem. J.* **264**, 437–445.
10. Thorley-Lawson D. A. and Green N. M. (1975), *Eur. J. Biochem.* **59**, 193–200.
11. Geisow M. J. and Beaven G. H. (1977), *Biochem. J.* **161**, 619–625.
12. Richards F. M. and Vithayathil P. J. (1959), *J. Biol. Chem.* **234**, 1459–1465.
13. Porter R. R. (1959), *Biochem. J.* **73**, 119–126.
14. Pohl G., Källström M., Bergsdorf N., Wallén P., and Jörnvall H. (1984), *Biochemistry* **23**, 3701–3707.
15. Leavis P. C., Rosenfeld S., and Lu R. C. (1978), *Biochim. Biophys. Acta* **535**, 281–286.
16. Jollés J., Fiat A. M., Alais C., and Jollés P. (1973), *FEBS Lett.* **30**, 173–176.

Eight

Peptide Sequence Determination

Linda A. Fothergill-Gilmore

1. Reasons for Needing Protein Sequence Information

The reasons for requiring amino acid sequence information are many. A scientific project might need to elucidate an enzyme mechanism, or design a probe to isolate a gene, or find a good antigen for vaccine development. Two experimental approaches are available for the determination of amino acid sequences. One strategy is to purify the particular protein or peptide and to determine its sequence by direct chemical procedures, usually with the assistance of automated instruments. Alternatively, the sequence can be deduced from the nucleotide sequence of the corresponding genes or cDNA. Recent technological developments have enabled direct protein sequencing to be done conveniently and at very high levels of sensitivity.

1.1. Determination of Protein Structure and Function

The properties of a protein (or peptide) depend fundamentally on the order of amino acids that are polymerized to form its primary structure. It is the primary structure that determines the folding of a protein to attain its characteristic secondary and tertiary structures. Moreover, the chemical properties of the amino acid side chains form the basis for substrate specificity, catalysis,

From: *Protein Biotechnology* • F. Franks, ed. ©1993 The Humana Press Inc.

antigenicity, thermal stability, and a host of other properties. A knowledge of the amino acid sequence is therefore crucial to an understanding of a protein's function.

The next few paragraphs will be devoted to a brief review of the chemical basis of protein structure and function. This topic is considered in more detail in Chapters 1 and 2. Proteins are linear polymers of amino acids linked together by amide bonds. The polypeptide backbone is unbranched, but may occasionally be crosslinked through certain of the amino acid side chains. The peptide bond is a special example of an amide linkage, and has its own characteristic properties that confer many of the distinctive features of the polypeptide chain. Three properties are of particular importance:

1. The electrons associated with the peptide bond are delocalized such that the carbon–oxygen bond is less than double, and the carbon–nitrogen bond has partial double bond character;
2. The peptide bond is planar; and
3. The two α-carbon atoms are almost always at opposite corners of the amide plane in the *trans* configuration.

These features are illustrated in Fig. 1.

The chemical nature of an amino acid residue is conferred by its side chain. There are 20 commonly occurring amino acid side chains, and it is critically important for a scientist doing protein sequencing to understand the properties of these side chains. Of course it is also essential to have a good appreciation of the chemistry of amino acid side chains in order to understand protein function. The structures of the amino acid side chains are given in Fig. 2. A summary of the names of the amino acids and their abbreviations can be found in Table 1. If you are not already familiar with the one letter code for the amino acids, then it is important that you learn it because this is the *lingua franca* of protein sequencing and protein structure in general.

Amino acids can be divided into two major sets: *polar/hydrophilic* and *nonpolar/hydrophobic*. Within the polar set are found amino acids that can have either charged or uncharged side chains. The nonpolar set includes both aromatic and aliphatic amino acids.

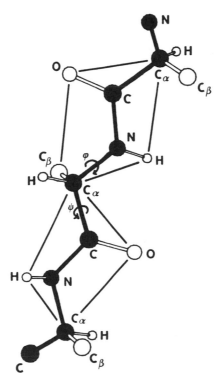

Fig. 1. The planes of two peptide bonds are joined by the tetrahedral bonds of an α-carbon. The two planes can rotate about the α-carbon bonds: rotation about the Cα–C bond is defined by an angle *psi*, and rotation about the Cα–N bond is defined by an angle *phi*. Rotation is positive when viewed from the α-carbon.

These two major sets of amino acids overlap, and those found in the overlap are considered to be ambivalent to water. A third category of amino acids includes those with *small* side chains. Glycine is an extreme example, with its side chain consisting of only a single hydrogen atom. The Venn diagram in Figure 3 shows the relationships of the 20 amino acids according to the properties just mentioned.

Proteins have evolved to function in aqueous environments and it is thus not surprising that the hydrophilic amino acids tend to be located on the surface, with the nonpolar hydrophobic resi-

Amino acid	Structure of amino acid residue in proteins	Residue weight	pKa range
Alanine	$-NH-CH-CO-$ $\quad\quad\ \|$ $\quad\quad CH_3$	71.1	NA
Arginine	$-NH-CH-CO-$ $\quad\quad\ \|$ $\quad\quad (CH_2)_3$ $\quad\quad\ \|$ $\quad\quad NH$ $\quad\quad\ \|$ $\quad\quad C=NH_2$ $\quad\quad\ \|$ $\quad\quad NH_2\ +$	156.2	11.6-12.6
Asparagine	$-NH-CH-CO-$ $\quad\quad\ \|$ $\quad\quad CH_2$ $\quad\quad\ \|$ $\quad\quad CONH_2$	114.1	NA
Aspartic acid	$-NH-CH-CO-$ $\quad\quad\ \|$ $\quad\quad CH_2$ $\quad\quad\ \|$ $\quad\quad COO^-$	115.1	3.0-4.7
Cysteine	$-NH-CH-CO-$ $\quad\quad\ \|$ $\quad\quad CH_2$ $\quad\quad\ \|$ $\quad\quad SH$	103.1	NA
Glutamic acid	$-NH-CH-CO-$ $\quad\quad\ \|$ $\quad\quad (CH_2)_2$ $\quad\quad\ \|$ $\quad\quad COO^-$	129.1	3.0-4.7
Glutamine	$-NH-CH-CO-$ $\quad\quad\ \|$ $\quad\quad (CH_2)_2$ $\quad\quad\ \|$ $\quad\quad CONH_2$	128.1	NA

Glycine	$-NH-CH-CO-$ $\quad\quad\quad\ \|$ $\quad\quad\quad\ H$	57.1	NA
Histidine	$-NH-CH-CO-$ $\quad\quad\quad\ \|$ $\quad\quad\quad\ CH_2$	137.2	5.6-7.0
Isoleucine	$-NH-CH-CO-$ $\quad\quad\quad\ \|$ $\quad\quad\quad\ CH-CH_3$ $\quad\quad\quad\ \|$ $\quad\quad\quad\ CH_2$ $\quad\quad\quad\ \|$ $\quad\quad\quad\ CH_3$	113.2	NA
Leucine	$-NH-CH-CO-$ $\quad\quad\quad\ \|$ $\quad\quad\quad\ CH_2$ $\quad\quad\quad\ \|$ $\quad\quad\quad\ CH-CH_3$ $\quad\quad\quad\ \|$ $\quad\quad\quad\ CH_3$	113.2	NA
Lysine	$-NH-CH-CO-$ $\quad\quad\quad\ \|$ $\quad\quad\quad\ (CH_2)_4$ $\quad\quad\quad\ \|$ $\quad\quad\quad\ NH_2^+$	128.2	9.4-10.6
Methionine	$-NH-CH-CO-$ $\quad\quad\quad\ \|$ $\quad\quad\quad\ (CH_2)_2$ $\quad\quad\quad\ \|$ $\quad\quad\quad\ S$ $\quad\quad\quad\ \|$ $\quad\quad\quad\ CH_3$	131.2	NA
Phenylalanine	$-NH-CH-CO-$ $\quad\quad\quad\ \|$ $\quad\quad\quad\ CH_2$	147.2	NA

Proline	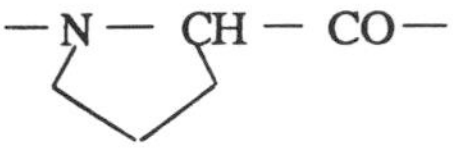 $-N-CH-CO-$ (cyclic ring)	97.1	NA
Serine	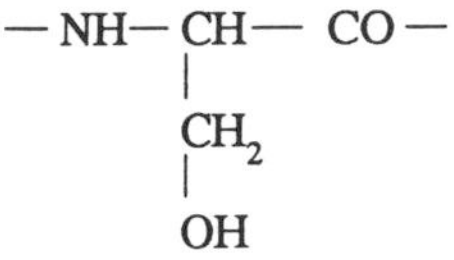 $-NH-CH-CO-$ $\quad\ \ CH_2$ $\quad\ \ OH$	87.1	NA
Threonine	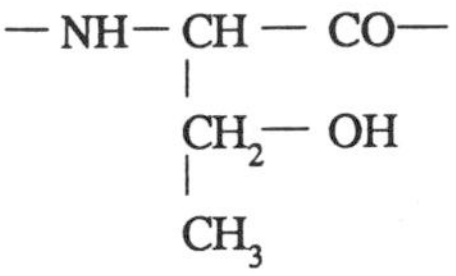 $-NH-CH-CO-$ $\quad\ \ CH_2-OH$ $\quad\ \ CH_3$	101.1	NA
Tryptophan	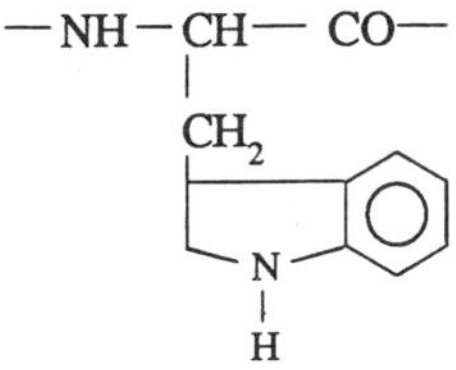 $-NH-CH-CO-$ $\quad\ \ CH_2$ (indole ring, N–H)	186.2	NA
Tyrosine	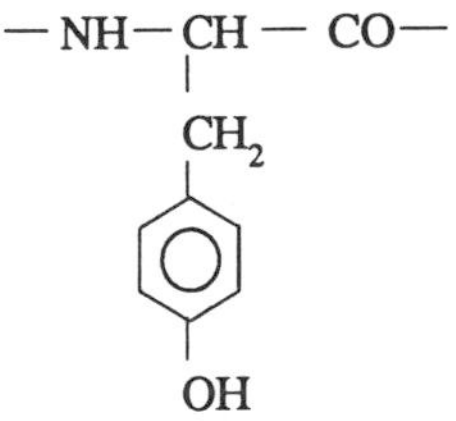 $-NH-CH-CO-$ $\quad\ \ CH_2$ (benzene ring) $\quad\ \ OH$	163.2	9.8-10.4
Valine	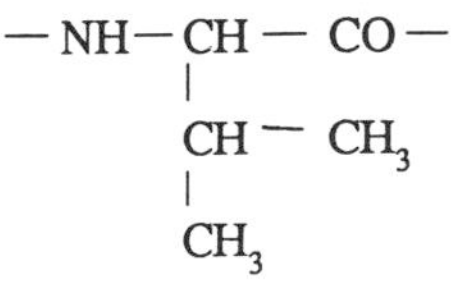 $-NH-CH-CO-$ $\quad\ \ CH-CH_3$ $\quad\ \ CH_3$	99.1	NA

Fig. 2. The twenty commonly occurring amino acids. The structures of the ionisable side chains are given in the form that will be most abundant at pH 7. The residue weights are for the uncharged forms of the side chains. It is relevant to note that the pKa of the amino group at the N-terminus of a peptide or protein is in the range 7.9–10.0, and is thus lower than that of the amino group in the lysine side chain. Similarly, the pKa of the carboxyl group at the C-terminus of a peptide or protein is in the range 1.8–2.5 and is consequently lower than those of the aspartic and glutamic acid side chains. NA means not applicable.

Table 1
Abbreviations for Amino Acids

Amino acid	Three-letter abbreviation	One-letter symbol	Amino acid	Three-letter abbreviation	One-letter symbol
Alanine	Ala	A	Histidine	His	H
Arginine	Arg	R	Isoleucine	Ile	I
Asparagine	Asn	N	Leucine	Leu	L
Aspartic acid	Asp	D	Lysine	Lys	K
Asparagine or			Methionine	Met	M
aspartic acid	Asx	B	Phenylalanine	Phe	F
Cysteine	Cys	C	Proline	Pro	P
Glutamine	Gln	Q	Serine	Ser	S
Glutamic acid	Glu	E	Threonine	Thr	T
Glutamine or			Tryptophan	Trp	W
glutamic acid	Glx	Z	Tyrosine	Tyr	Y
Glycine	Gly	G	Valine	Val	V

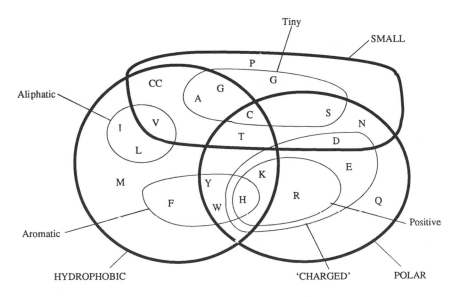

Fig. 3. The 20 amino acids grouped according to a selection of physico-chemical properties that are important for protein function and for protein sequencing. The abbreviation C indicates the reduced thiol form of cysteine, whereas CC indicates the oxidized disulfide form.

dues buried in the interior. In the latter category, the amino acids valine, leucine, isoleucine, methionine, and phenylalanine are almost invariably on the inside of a protein. Glycine and alanine are so small that they can apparently be accommodated in the interior or on the surface with equal facility. The neutral polar residues serine, threonine, asparagine, and glutamine are usually found on the outside, but can be inside if their polar groups are "neutralized" by hydrogen bonding to other like residues or to the carbonyl C=O group of the polypeptide backbone. These residues are often involved in crosslinking of two peptide chains by means of hydrogen bonds. Tyrosine and tryptophan are located inside and outside, but when tyrosine is inside, its hydroxyl group is usually hydrogen-bonded.

The remaining polar amino acids (aspartic acid, glutamic acid, lysine, arginine, and histidine) are commonly on the outside of proteins, and can exist in either charged or uncharged form, depending on the pH of the surroundings. At acidic pH, aspartic acid and glutamic acid have uncharged carboxyl groups, whereas lysine, arginine, and histidine each is protonated and carries a positive charge. Under basic conditions, the carboxyl groups of aspartic acid and glutamic acid will be negatively charged, and lysine, arginine, and histidine will be uncharged. The actual ratio of the acidic to the basic form of a given residue depends on its strength as an acid or base. The pH at which the two forms are present in equal amounts is the pK, and the usual ranges of these are indicated in Fig. 2. These values are of crucial importance to many of the chemical modification and cleavage procedures described elsewhere in the course.

Two of the twenty amino acids, cysteine and proline, have not yet been mentioned because they do not fit neatly into the above classification system, and because they have rather special properties of their own. The cysteine side chain has a thiol group under reducing conditions, but can be oxidized to form a disulfide, even under relatively mild conditions such as atmospheric oxygen. The properties of the reduced and oxidized forms are quite different, and this can have a profound influence on the structure and properties of a protein. The differences in properties can also lead to considerable problems for the protein chemist. It is thus usual practice to modify the cysteine side chains before undertaking protein sequencing studies.

Proline is the only amino acid in which the side chain loops back to reattach to the main polypeptide chain. This has two main consequences: 1. Proline has a secondary amino group instead of the primary amino group present in all the other amino acids. It thus behaves differently in many of the reactions used in protein chemistry as will be seen later. 2. Proline forces a bend in the polypeptide backbone and can thus disrupt an α-helix and exert other constraints on the folding of a protein.

The detailed crystal structures of perhaps 200 proteins are now known, and to a first approximation it is apparent that the polypeptide backbone is found in four types of structure: α-helices, β-strands, turns, and random conformations. These structures frequently occur in multiple assemblies such as the helix-turn-strand-turn-helix motifs found in many enzymes (*see* Richardson *(7)* for an excellent review of the different types of structural motifs found in proteins). An example of this type of folding pattern is found in the structure of the glycolytic enzyme phosphoglycerate mutase, as shown in Fig. 4. As might be expected, the nature of the side chains of amino acid residues has a major role to play in determining the conformation of the polypeptide backbone. Thus, as mentioned above, proline frequently serves to disrupt α-helices by forcing a bend in the polypeptide backbone. At the other end of the spectrum, the tiny side chain of glycine introduces few constraints and this residue can be found in many types of structure. In some cases where the polypeptide backbone has a tight bend, glycine is the only residue that could be accommodated. It is outside the scope of this introduction to discuss all the possible influences of primary structure in determining the overall conformations of proteins. However, it is relevant to mention that it is possible to predict with a reasonable degree of success (about 65% accuracy) the secondary structure of a protein (α-helices, β-strands, and so on) from a knowledge of the primary structure (amino acid sequence).

1.2. DNA Probes
and Isolation of Gene or cDNA

The advent in the 1970s of convenient and efficient methods for DNA sequencing has meant that a very large proportion of sequence determination is done at the DNA level. This approach is

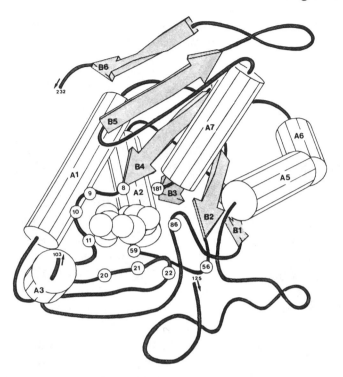

Fig. 4. The conformation of the polypeptide chain in yeast phosphogly-cerate mutase. The elements of regular secondary structure are labeled A1–A3 and A5–A7 for the α-helices (drawn as cylinders), and B1–B6 for the β-strands (drawn as large arrows). The general conformation of the protein backbone is indicated in dark "ribbon" shapes. A portion of the polypep-tide chain including helix 4 has been removed to reveal the α-carbon posi-tions of residues at the active site, as well as the position occupied by bound ligands as indicated by the space-filling drawing of 3-phosphoglycerate located by crystal soaking experiments (from L. A. Fothergill-Gilmore and H. C. Watson (1989) *Adv. Enzymol.* **62,** 227–313).

particularly advantageous for very large proteins or for proteins that are present in very small quantities or are difficult to isolate. Of course, the DNA sequencing approach is essential for charac-terizing those portions of a genome that are not expressed into protein.

An early step in a DNA sequencing project is to isolate the gene or cDNA of interest. Conceptually, this is a straightforward process usually involving the construction of a DNA library, and then screening the library for a clone containing the desired gene or cDNA. In practice, the correct clone is not always easily found, and some of the difficulties can be attributed to the procedures for screening the DNA library. One of the main approaches for screening is to use a radioactively labeled oligonucleotide of about 15–20 bases that is complementary to the DNA to be isolated. The more perfectly matched the oligonucleotide is to its complementary sequence, the more specific the hybridization and screening.

Frequently at the beginning of a project, there is no knowledge of the DNA sequence, and thus, the oligonucleotide probe must be designed on the basis of knowing a suitable short region of protein sequence. The degeneracy of the genetic code means that it is not possible to specify a single perfectly complementary oligonucleotide by this approach. However, judicious choice of the amino acid sequence for amino acids with low degeneracy will ensure that well matched oligonucleotides will be present as a relatively high proportion of the mixture synthesized to correspond to the amino acid sequence. Thus, sequences with methionine and tryptophan (single codons) and aspartic acid, asparagine, glutamic acid, glutamine, histidine, lysine, phenylalanine, tyrosine, and cysteine (two codons) are most suitable. Sequences with leucine, serine, and arginine (six codons) are particularly unfortunate.

How much protein sequence information is required to design a suitable oligonucleotide probe? Clearly, if it were possible to ensure that the sequence contained only amino acids of low degeneracy, then 5–7 residues would be sufficient. However, most sequences contain a mixture of suitable and unsuitable amino acids, and in practice, a sequence of at least 30–40 residues is frequently required. This length of sequence can be obtained quite readily from a single protein or peptide sample with the use of an automated protein sequencer. The usual strategy is to attempt initially to determine the N-terminal sequence of the whole protein. Unfortunately, a large proportion of all proteins (about 50%) have chemically blocked N-termini, and it is thus often necessary to resort to methods to cleave the protein, and to sequence one of the resulting fragments. The strategies and methods for obtaining limited sequence information are considered in more detail later in this Chapter (*see* Section 2.1.).

1.3. Posttranscriptional
and Posttranslational Processing

A knowledge of the nucleotide sequence of a gene reveals a great deal concerning the protein encoded by the gene and the control of its expression, but tells nothing of the processing events that may occur after the DNA has been transcribed into RNA. Basically, processing may take place at three levels:

1. Alternative splicing of an RNA primary transcript to form different mature mRNAs;
2. Maturation of a precursor form of a protein; and
3. Chemical modification of a mature protein.

In all these cases, the molecular details of the processing are probably best established by protein sequencing.

Many examples are now known in which a single RNA primary transcript can give rise to more than one protein product. The isoenzymes of pyruvate kinase provide a good example. There are four pyruvate kinase isoenzymes, but it is now quite clear that there are only two genes. The M1 (muscle) and M2 (kidney) isoenzymes arise by differential RNA splicing of the same RNA primary transcript in different tissues (8). The two isoenzymes differ by only 20 amino acid residues clustered in one region of the protein molecule that is involved in intersubunit contacts, and that corresponds to a single exon (9). The M2 isoenzyme displays cooperativity, whereas the M1 isoenzyme does not, and the exon shuffling is thus crucial for conferring different enzymic properties. The L (liver) and R (erythrocyte) isoenzymes are also encoded by a single gene, but in this case, the two different enzymes result from the use of different promoters (10). The R isoenzyme is 31 amino acid residues longer at the N-terminus.

There are numerous examples of the processing of an inactive precursor protein to give rise to active products. Probably some of the best characterized systems involve the polypeptide hormones, with insulin being a prime example. Digestive enzymes are also synthesized as inactive precursors, and there are a great many other examples where there is a need to hold a protein in an inactive form prior to its release in an active form in response to a specific stimulus. Other examples of protein processing include the targeting of proteins for secretion, where the signal sequence is cleaved off during secretion to yield the mature protein. The details of the mechanisms of all of these types of posttranslational processing can profitably be studied by protein sequencing.

Protein sequencing in conjunction with mass spectrometry is the method of choice for characterizing the posttranslational chemical modification of proteins. Such modifications can occur both in vivo and in vitro and, for example, can take the form of phosphorylation, glycosylation, ADP-ribosylation, myristylation, or the formation of Schiff bases. The purification of a modified peptide with subsequent sequencing and characterization of the modifying group enables these important processes to be studied in detail.

1.4. Immunological Studies

There are many aspects of immunology that can benefit from a protein sequencing approach. In the area of designing novel vaccines, for example, a comparison of the sequences of a likely protein antigen isolated from different sources can identify potential epitopes or functionally conserved regions. Peptides corresponding to these regions can then be isolated or synthesized, and tested against a panel of antibodies. In this way, it can be possible to map the epitopes of an antigen, and to use this information for vaccine development. Antibodies can also be used to purify antigens from cell extracts or from other materials. The antigen can then be characterized by protein sequencing as a prelude to a wide range of research and development projects.

1.5. Identification of Isoenzymes

Many organisms such as mammals and higher plants have several different genes that encode enzymes catalyzing the same reaction. These isoenzymes frequently have somewhat different properties, and are often expressed in a tissue-, organelle-, or developmental-specific manner. It is fundamental to an understanding of the biology of isoenzymes to be able to characterize their expression. It may also be of considerable clinical relevance to be able to identify the presence of a particular isoenzyme because of changes in expression that occur after malignant transformation. Thus, for example, the presence of the muscle-type of phosphoglycerate mutase in the adult brain is indicative of neoplastic tissue, and the level of expression appears to correlate with the degree of malignancy of the tumor (11).

The purification of a small quantity of an isoenzyme and the subsequent determination of a limited amount of protein sequence is a relatively convenient and completely unequivocal way of establishing the identity of an isoenzyme.

1.6. Quality Control

Peptides can be made in the laboratory by synthetic chemical procedures and proteins can be made by recombinant DNA techniques. In both these cases, there is a need to ascertain how well the product conforms to that expected, especially if the peptide or protein is to be used as a drug. In the case of peptide synthesis, incomplete unblocking prior to each coupling step can lead to the production of complicated mixtures of peptides that differ in length. The less efficient the unblocking and coupling, the more heterogeneous the product. Probably the best method for measuring the degree of heterogeneity is to subject the product to Edman degradation in an automated sequencer. The quantitative estimation of the amount of the different amino acids at each position gives a good idea of how successful the peptide synthesis has been. Sequence information of this type is now usually required before a peptide preparation is approved for drug use.

The problems that can beset the production of a recombinant protein are quite different, and can be a major concern, especially in high expression systems. Thus, there are numerous examples of vector instability, incorrect transcription, and improper protein processing. In many cases it is considered appropriate to express a recombinant protein in a system where it eventually is found as an insoluble intracellular inclusion body. The recovery of the protein requires procedures that are potentially damaging to it, and it is essential that the final product be fully characterized, including partial or even complete protein sequence determination. The requirement for protein sequencing procedures for quality control means that protein chemists in the biotechnology industry now comprise a major proportion of all such scientists.

2. Strategies for Sequence Determination

This section of the chapter will be devoted to a discussion of the strategies that can be adopted to obtain protein sequence information. Frequently, as in the case of the design of oligonucleotide probes, the requirement is for the determination of only a limited stretch of amino acid sequence. The approaches to obtain a partial sequence are somewhat different from those adopted to determine a complete sequence, and the two topics will thus be considered in separate sections. The presentation of methods and strategies is easier (and, we hope, more interesting) when based on an actual example. We have chosen the recently discovered peptide hormone, hair growth stimulating hormone (HGSH) as a suitable example that will enable us to illustrate most of the points about sample preparation and technique.

Baldness is a condition that affects a great many people, especially men. It is often regarded as a sign of aging, and there would be considerable demand for a treatment that could halt or even reverse the balding process. A few years ago, a senior protein chemist at a leading pharmaceutical firm was evaluating the efficacy of as many existing treatments for baldness as he could track down. He happened to learn of a folk remedy in use in Alaska that appeared to be extraordinarily effective in restoring hair growth. Further enquiry revealed that the treatment consisted of the topical application of an extract of the skin of a large animal that had been dug from the ground. The details of the species of animal or the formulation of the extraction method were not available, but seemed sufficiently promising to be worth investigating further. Several bits of information indicated that the animal was a fossil mammoth (*Mammuthus*), and samples of skin from a frozen specimen were obtained. These samples were subjected to a variety of extraction procedures and a bioassay involving bald mice was developed. A procedure using urea extraction and gel filtration proved to be particularly successful, and eventually a small protein of M_r 11,000 was isolated. This protein was extremely effective in the mouse assay, and an immediate decision was made to determine the primary structure of the protein as a prelude to developing it as a pharmaceutical product. The protein was named hair growth stimulating hormone or HGSH.

Because of the need to obtain results as quickly as possible, it was decided to pursue two different approaches simultaneously. One research group had the goal of obtaining a limited amount of amino acid sequence to enable the gene encoding HGSH to be isolated and sequenced. The other group aimed to determine the sequence of HGSH directly. A spirit of friendly competition soon arose between the groups, and the race was on.

2.1. Limited Sequence Information

2.1.1. N-Terminal Sequence

The usual approach to obtain sufficient protein sequence information for the design of oligonucleotide probes is initially to attempt automated Edman degradation of the intact protein. For this purpose, it is necessary to have a minimum of about 100 pmol of protein, with 500–1000 pmol as optimum quantities. In the case of HGSH, this corresponds to about 1.1 µg minimum, and 5.5–11 µg optimally. Fortunately, the results from the automated sequencing give a quantitative measure of the amount of each amino acid derivative released at each cycle of Edman degradation. This means that it is quite acceptable for the protein sample to be only 80–90% pure. With this level of purity, the amounts of contaminating proteins would be so small proportionally that it is not difficult to interpret the main sequence. Please see Sections 4. and 5. for descriptions of the chemistry of Edman degradation and the identification of the amino acid phenylthiohydantoin (PIH) derivatives released at each cycle.

A sample of about 5 µg of purified HGSH was taken for automated sequencing and the results awaited with eagerness. Inspection of the first chromatograms from the sequencer showed that a low yield sequence of about 20 pmol was obtained. The most likely interpretation of these results is that the observed sequence corresponded to a minor contamination of the HGSH sample, and that the HGSH protein itself had a blocked N-terminus. Unfortunately, about half of all proteins are N-terminally modified by, for example, acetyl groups such that the α-amino group is not free to react with the Edman reagent. Occasionally it is possible to use enzymic methods to remove acyl groups attached to the α-amino group or to remove a pyrrolidone carboxylic acid residue (a cyclized and N-terminally blocked form of glutamine). However, the enzymic methods are not reliably useful, and if the amount of protein sample is limited, it is probably more productive to cleave the protein and sequence one or more fragments.

2.1.2. Cleavage and Sequencing of Fragments

The ideal method for cleaving a blocked protein to release a free N-terminus for subsequent sequencing is one that will very specifically hydrolyze only one peptide bond. If the cleavage is limited to one peptide bond, then it is not necessary to separate the fragments prior to sequencing because the fragment corresponding to the N-terminal portion of the protein will still be blocked. The avoidance of separation procedures saves both time and sample. It is perhaps relevant to point out that there can be substantial problems of recovery of material if one is working in the low pmol range. A rough rule of thumb is that about half a sample will be lost (primarily by adsorption to surfaces) for every procedure used. The recent availability of high-sensitivity instruments (such as those in use in this course) has enabled a great many more proteins to be successfully sequenced than would previously have been possible. But it is important to adopt a strategy that involves as few steps as possible.

There are several potentially useful procedures for limited cleavage of peptide bonds, and in order to be able to choose rationally which one to use, it is advantageous to have a reasonably accurate amino acid composition of the protein. Table 2 shows the results of an amino acid analysis of HGSH after hydrolysis of the peptide bonds in 6N HCl at 110°C for 20 h. Before the HGSH was hydrolyzed, it was treated with 4-vinylpyridine to form pyridylethyl derivatives of any cysteine residues that might be present. This modification reaction should *always* be done before amino acid analysis and before sequencing on any samples that might have cysteine residues. As mentioned in Section 1.1., cysteine side chains can readily exist in different forms depending on the reducing or oxidizing conditions of their surroundings. The differences in the properties of the various forms of cysteine mean that it is very difficult to quantitate cysteine unless a stable deriva-

Table 2
Amino Acid Composition of HGSH[a]

Amino acid	pmol	Residues/mol	Number of residues
Asx	347	9.9	10
Glx	434	12.4	12
Ser	224	6.4	6
Gly	371	10.6	11
His	94.5	2.7	3
Arg	137	3.9	4
Thr	130	3.7	4
Ala	175	5.0	5
Pro	214	6.1	6
Tyr	63.0	1.8	2
Val	207	5.9	6
Met	133	3.8	4
PE-Cys[b]	109	3.1	3
Ile	168	4.8	5
Leu	238	6.8	8
Phe	102	2.9	3
Lys	217	6.2	6
Trp	—	1.1	1
Total			99

[a]The results are the averages of three analyses, and are uncorrected for destruction during hydrolysis, or incomplete hydrolysis of peptide bonds (*see* Section 6.). Tryptophan content was determined on a separate sample after hydrolysis with mercaptoethanesulfonic acid. The amino acids are given in the order of their elution from the HPLC column in the amino acid analyzer. The "number of residues" correspond to the numbers established from the sequence determinations.

[b]Abbreviation: PE-Cys, pyridylethyl-cysteine.

tive is formed. The pyridylethylation reaction is simple and can be done in either the vapor phase or in solution (*12*). The pyridylethyl derivative of cysteine is stable and also elutes in a convenient position in the HPLC system used by both the amino acid analyzer and the sequencer. Other types of cysteine modification reaction can be readily done, such as carboxymethylation (*12*), but the derivatives do not separate well on the HPLC systems.

The amino acids released by acid hydrolysis of HGSH were derivatized with phenylisothiocyanate prior to separation by reverse-phase HPLC as is described in more detail in Section 6. The determination of an accurate amino acid composition is one of the most difficult procedures routinely done in a biochemistry laboratory, and many of the potential problems are also discussed in Section 6.

Of the various methods for cleaving peptide bonds, five are probably most useful for obtaining a small number of fragments:

1. Cyanogen bromide cleavage at the C-terminal side of methionine residues;
2. o-Iodosobenzoic acid cleavage at tryptophan residues;
3. Cleavage at arginine residues by clostripain after blocking of lysine residues by succinylation;
4. Mild acid cleavage at aspartyl–proline peptide bonds; and
5. Hydroxylamine cleavage at asparaginyl–glycine bonds.

Which methods did the HGSH team choose to use? The initial strategy was to search for a method that would cleave only one peptide bond.

2.1.2.1. CNBr Cleavage

CNBr cleavage is a particularly convenient method because it can be done in the vapor phase. This means that a sample can be spotted onto a glass fiber disk from the sequencer, treated with CNBr, dried, and placed in the sequencer. There is no need to transfer the sample from tube to tube, or to separate reaction byproducts prior to sequencing. However, inspection of the amino acid composition of HGSH reveals that there are four methionine residues, and cleavage with CNBr would thus be expected to produce five fragments. This method proved to be of some use for the determination of the complete sequence of HGSH (*see* Section 2.2.), but was not considered to be very suitable for obtaining limited sequence information.

2.1.2.2. Tryptophan

Tryptophan is one of the most difficult amino acids to determine in an amino acid composition because it is destroyed by hydrochloric acid hydrolysis. Alternative methods of analysis are available, but they are both less reliable and less sensitive (*see* Section 6.). Nevertheless, the HGSH protein chemists deemed it worthwhile to commit the material and effort to measuring the tryptophan content, and they were rewarded because the protein has only one

```
        5        10       15       20
(W)  L A G D S P I G G Y Q L R N D S K P G D E
```

Fig. 5. N-terminal amino acid sequence of the fragment resulting from cleavage at the tryptophan residue of HGSH.

tryptophan. A sample of HGSH (about 500 pmol or 5.5 µg) was treated with *o*-iodosobenzoic acid and then desalted on a microbore HPLC system. A rather poorly resolved double peak was obtained that was pooled, concentrated, and loaded into the sequencer. There was no need to try to ensure that the cleavage procedure was completely efficient or to devise a method to separate the expected two fragments, because of the blocked N-terminus of one of the fragments. An easily interpreted sequence of 22 residues was obtained as shown in Fig. 5. The yield of the initial PTH-leucine at 75 pmol was lower than would normally be expected from a 500 pmol sample, but the *o*-iodosobenzoic acid cleavage procedure is known to be inefficient and the HGSH team was not surprised. The inefficiency of the method means that it is not very suitable for cleaving proteins where there are several tryptophan residues. In these cases, it is not uncommon to obtain complicated peptide mixtures resulting from partial cleavage at all the possible tryptophans.

There was initial elation at obtaining the sequence, but this soon diminished when it was realized that the sequence was really not very good for specifying an oligonucleotide probe. The few amino acids with two codons were flanked with those with four or six codons. It was decided that it was necessary to try alternative cleavage methods to obtain more sequence.

2.1.2.3. EXAMINATION OF THE AMINO ACID COMPOSITION

Examination of the amino acid composition showed that HGSH has four arginine residues, and thus cleavage with clostripain would probably not be very useful to obtain limited sequence information. Reference to the next section, however, will show that it was the best method for helping to establish the entire sequence of HGSH.

It is obviously not possible to deduce from the amino acid composition whether aspartyl–proline or asparaginyl–glycine bonds are present in HSGH. The most reasonable approach for

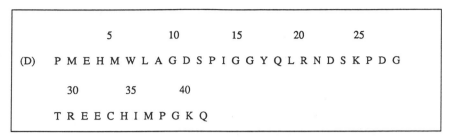

		5		10		15		20		25	

(D)   P M E H M W L A G D S P I G G Y Q L R N D S K P D G

       30   35   40

       T R E E C H I M P G K Q

Fig. 6. N-terminal sequence of the fragment of HGSH released by mild acid cleavage. The cysteine residue was identified as the pyridylethyl derivative.

these methods is to take a small amount of the protein (say, about 100 pmol) and test whether it is cleaved by examining the treated sample on an analytical reverse-phase HPLC system. Alternatively, the treated protein can be examined by SDS-polyacrylamide gel electrophoresis. Of the two cleavage methods, the more convenient is the mild acid cleavage because it can be done in the vapor phase (as described for the CNBr cleavage). Both methods were tried on small samples of HGSH, and the HPLC separations showed that the protein was susceptible to cleavage with mild acid, whereas hydroxylamine had no effect. Accordingly, about 500 pmol of HSGH was spotted onto a sequencer disk, treated with acid vapor, dried, and placed in the sequencer. A high yield (420 pmol) and trouble-free sequence of 41 residues was obtained as shown in Fig. 6.

A comparison of this sequence with that obtained earlier clearly shows that the two peptides cover the same region of the protein, with residue 7 corresponding to the N-terminus of the o-iodosobenzoic acid cleaved peptide. The N-terminal portion of the mild acid cleaved peptide contains two methionines as well as the tryptophan and is thus particularly suitable for specifying an oligonucleotide probe. This was made and hybridized well in a Southern blot to genomic DNA isolated from a *Mammuthus* tissue sample. The HGSH team with the goal of isolating the gene was now well satisfied that they had a suitable probe for screening the DNA libraries they were in the process of constructing.

2.1.3. Detection and Sequencing of Labeled Peptides

The chemical versatility of amino acid side chains means that the different residues can participate in a wide variety of *intra-* and *inter-*molecular reactions. These reactions are often of considerable interest because they may involve active site residues or may in some other way be indicative of an essential function. A useful way of studying these reactions is to make use of labeled ligands that then enable a modified peptide to be readily detected even from a complex mixture. Some examples of this type of approach would include the use of radioactive phosphorus to characterize a phosphorylation site, tritiated sodium borohydride to reduce a Schiff base, or trinitrobenzene sulfonate to form a colored derivative with reactive lysine residues. It is not feasible to discuss here the wide range of modification reactions that can be useful (that would be a book in itself), but one type of labeling experiment done on HGSH can be mentioned as an illustration of the general approach.

Disulfide bridges can play a major role in determining the three-dimensional structure of a protein, and it was thus of considerable interest to ascertain whether two of the three cysteine residues of HGSH (*see* Table 2) might be involved in a disulfide. The HGSH scientists adopted the straightforward experimental approach of carboxymethylating the protein with ^{14}C-iodoacetic acid in the presence and absence of the reducing agent, dithiothreitol (DTT) (the method is described in [12]). They found by radioactivity measurements, the incorporation of one mole of reagent per mole of HGSH in the absence of DTT, and three moles of reagent per mole of HGSH in its presence. It was thus clear that HGSH has one disulfide and one free thiol. Which of the three cysteine residues is the free thiol? This was established by digesting a sample of HGSH (about 500 pmol) that had been carboxymethylated in the absence of DTT, with the proteolytic enzyme *Staphylococcal* protease (*see* next section) and separating the peptides by reverse-phase HPLC. Twelve peaks were observed that were collected into fractions and examined for radioactivity. Only one peak was labeled, and this material was put into the sequencer. It corresponded to a small peptide with the sequence: T R A C E. A portion of the PTH-amino acid at each cycle was taken for counting, and it was found that most of the radioactivity was in residue 4. These results therefore show unequivocally that the free thiol involves this cysteine residue. As a corollary, the other two cysteines must be involved in a disulfide bridge.

2.2. Complete Sequences

The initial strategy to be adopted for the determination of a complete sequence will depend on the size of the protein to be sequenced. For small proteins such as HGSH, the usual first step is to use an efficient cleavage procedure—usually with proteolytic enzymes—that will yield peptides of about 20–30 residues in length. These peptides are purified, usually by reverse phase HPLC, and then sequenced. It is subsequently necessary to obtain evidence for the order in which the first collection of peptides occur in the protein. This is commonly done by the use of a second complementary cleavage procedure, followed by purifying and partially sequencing the second collection of peptides. These results are usually sufficient to establish the overlapping sequences, although sometimes a third cleavage procedure must be used to finish off the last bit of sequence determination.

For proteins larger than about 200 residues, the results of limited cleavage procedures—usually chemical—can be of considerable use to divide the protein up into pieces of tractable size (of about 100 residues). Several of these methods have already been encountered in the discussion in Section 2.1. on how to obtain limited sequence information. We will consider the applications of these procedures to complete sequence determination in the following section.

We now turn to the HGSH research group that had embarked on the direct determination of the complete amino acid sequence of HGSH. The strategies and procedures they adopted will serve as examples of a general approach that could be successfully applied to most proteins. It is important to stress here that the purity of a protein for complete sequencing must be much greater than that for limited sequencing. The reason for the higher standard is that peptides produced by cleavage of the parent protein may often be recovered in variable yield. Indeed some peptides will be only sparingly soluble, and may not even be recovered at all. It therefore becomes very difficult to distinguish between a con-

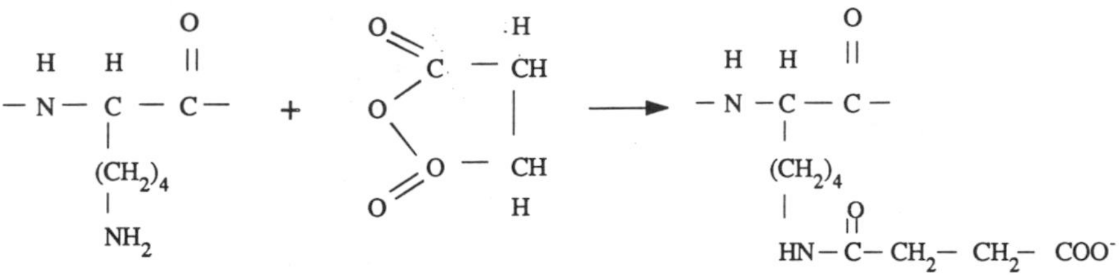

Fig. 7. Modification of lysine residues by succinic anhydride. Anhydrides of other dicarboxylic acids can also be used, commonly maleic anhydride and citraconic anhydride. In these cases, the modified lysines can be unblocked by treatment with dilute acid. The reversibility of the lysine modification is potentially useful for experiments to first block and then reveal proteolytic cleavage sites. However, in practice, the unblocking rarely attains 100%, and this in turn can lead to remarkably complex peptide mixtures.

taminating sequence and one that has been recovered in low yield. The goal should be for the protein to be at least 95% pure, and preferably rather better than that.

2.2.1. Cleavage Procedures (Part 1)

2.2.1.1. PROTEOLYTIC ENZYMES—CLOSTRIPAIN

A particularly useful first step in the determination of the sequence of a protein is to cleave the polypeptide chain at arginine residues. This is usually a nonabundant amino acid, with the consequence that the peptide mixture is relatively simple, and also the peptides are often of a suitable size for automated sequencing. Two different commonly available proteolytic enzymes will cleave at the C-terminal side of arginine: trypsin and clostripain (13). The latter enzyme is to be preferred because it has a narrower specificity and will cleave preferentially at arginine residues, although it often cleaves at lysines as well. (The pseudotrypsin activity present in many preparations of trypsin means that peptides resulting from cleavage at chymotrypsin-like sites are quite often found.) A successful and convenient way to ensure that cleavage is limited to arginines, is to modify the lysine side chains such that they are no longer susceptible to cleavage. Succinylation with succinic anhydride (see Fig. 7) is an excellent method that has the added important feature of improving the solubility of the resulting peptides.

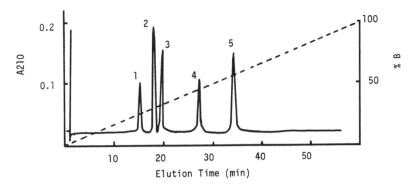

Fig. 8. Separation of succinylated clostripain peptides from HSGH by reverse phase HPLC. The peptides were eluted with a linear gradient of Solvent A (0.1% trifluoroacetic acid) and Solvent B (80% acetonitrile).

Poor peptide solubility can be an intractable problem, especially after chemical cleavage procedures. Another advantage of the succinylation procedure is that PTH-succinyl lysine elutes in a convenient position in the HPLC system of the sequencer.

With this background in mind, the HGSH team decided to use succinylation followed by clostripain digestion as their first approach. A sample of 500 pmol of HGSH was succinylated, desalted, and then treated with clostripain. The peptides were separated by reverse phase HPLC as shown in Fig. 8. Please see Chapter 2 for a discussion of the theory and applications of liquid chromatography. Fractions corresponding to the peaks were collected in microcentrifuge tubes and concentrated ready for amino acid analysis and sequencing. It is important to stress that the fractions were not taken to dryness. Once again, peptide solubility can be a major headache, and some peptides are very difficult to recover if they have been dried onto the surface of a tube. The results of amino acid analysis are given in Table 3, and the results of the sequencing are summarized in Fig. 9.

A number of points can be made from these results:

1. A total of 59 residues of clearly identified sequence was obtained by this method.
2. Another method (at least) will be required to complete the sequence determination.

Table 3
Amino Acid Compositions of Clostripain Peptides[a]

Amino acid	Peptide									
	R1		R2		R3		R4		R5	
Asx	3.8	(4)	1.9	(2)	2.9	(3)	0.8	(1)	—	—
Glx	1.2	(1)	2.1	(2)	6.2	(6)	2.0	(2)	1.1	(1)
Ser	0.8	(1)	1.8	(2)	2.7	(3)	—	—	—	—
Gly	0.9	(1)	3.1	(3)	1.9	(2)	1.0	(1)	3.8	(4)
His	—	—	0.8	(1)	0.9	(1)	0.8	(1)	—	—
Arg	0.9	(1)	0.9	(1)	0.7	(1)	—	—	0.8	(1)
Thr	0.8	(1)	—	—	1.7	(2)	0.7	(1)	—	—
Ala	—	—	2.0	(2)	1.0	(1)	1.0	(1)	1.0	(1)
Pro	1.1	(1)	2.1	(2)	1.2	(1)	—	—	2.1	(2)
Tyr	—	—	0.9	(1)	—	—	—	—	0.8	(1)
Val	—	—	—	—	2.9	(3)	—	—	3.1	(3)
Met	—	—	1.8	(2)	0.9	(1)	—	—	0.8	(1)
PE-Cys[b]	—	—	—	—	1.0	(1)	0.9	(1)	0.9	(1)
Ile	—	—	0.9	(1)	1.9	(2)	—	—	1.7	(2)
Leu	—	—	1.9	(2)	0.9	(1)	1.8	(2)	1.9	(3)
Phe	—	—	—	—	1.0	(1)	—	—	1.9	(2)
Suc-Lys[b]	1.1	(1)	—	—	3.2	(3)	2.1	(2)	—	—
Trp	ND	—	ND	(1)	ND	—	ND	—	ND	—
Total		(11)		(22)		(32)		(12)		(22)

[a]The results are uncorrected for destruction during hydrolysis, or incomplete hydrolysis of peptide bonds. Tryptophan was not determined. The results are expressed as residues/mole of peptide, with the values in brackets corresponding to those established by the sequence determination. The dash indicates less than 0.3 residues/mole in the composition, or not present in the sequence.

[b]Abbreviations: PE-Cys, pyridylethyl-cysteine. Suc-Lys, succinyl-lysine.

3. Peptide R2 is probably derived from the N-terminus because it is blocked and gave no sequence.
4. Peptide R4 is probably derived from the C-terminus because it has no arginine residue.
5. The relatively disappointing result from the sequencing of R5 was because of the prolyl-proline sequence which did not degrade well in the sequencer (*see* Section 5. for a discussion of this problem).

R1	N D S K P D G A E T (R)
R2	no sequence
R3	E E C H I M P G K G Q K V S S E N L I N D A (F) ? ? (V)...
R4	T H E K Q K L C G L A D
R5	A C E V G G M F L I P P V Y L ? ? (G)...

Fig. 9. The N-terminal amino acid sequences of the clostripain peptides. The amino acids in brackets were assigned with some uncertainty; the symbol '?' means that no assignment was unequivocal in that cycle.

2.2.1.2. Proteolytic Enzymes—Staphylococcal Protease

The cleavage procedure chosen for the second stage of the complete sequence determination of HGSH was digestion with the extracellular protease from *Staphylococcus aureus* (strain V8). This enzyme has reasonable specificity for cleavage on the C-terminal side of glutamic acid residues, although it will sometimes attack aspartic acids as well *(14)*. This enzyme thus provides a very satisfactorily complementary cleavage to that of clostripain. A sample of 300 pmol of HGSH was digested with *Staphylococcal* protease and the peptides separated by reverse phase HPLC (Fig. 10). Fractions corresponding to the peaks were collected in microcentrifuge tubes and concentrated ready for amino acid analysis and sequencing. The initial sequence results were watched closely so that the minimum of redundant sequencing was done in order to save sequencer time and running costs. The results of amino acid analysis are given in Table 4, and the results of the sequencing are summarized in Figure 11.

Several points can be made from these results:

1. The peptides generally gave good sequence results. It was decided in the cases of E1 and E7 to stop the sequencer after 5 residues because the remainder of the sequence was already known.
2. Peptide E5 is probably derived from the N-terminus because it is blocked and gave no sequence.

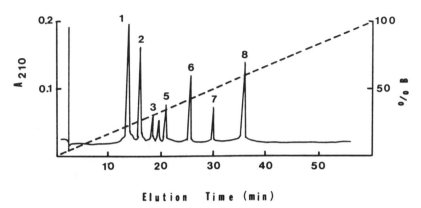

Fig. 10. Separation of *Staphylococcal* protease peptides from HGSH by reverse phase HPLC. A 300 pmol sample of the protein was dissolved in 200 µL of 0.1*M* NH$_4$HCO$_3$, and 10 µL of a solution of *Staphylococcal* protease (10 µg/mL) was added. Digestion was for 2 h at 37°C. The peptides were eluted with a linear gradient of Solvent A (0.1% trifluoroacetic acid) and solvent B (80% acetonitrile).

3. Peptides E2, E3, E4, E5, and E6 all apparently resulted from cleavage at glutamic acid residues. Peptides E1 and E7 have aspartic acid residues at their C-termini. From this evidence it is not possible to decide which peptide is from the C-terminus of HGSH, but a comparison with the results from the clostripain peptides shows that it must be peptide E7. Peptide E2 rather surprisingly has two glutamic acid residues at its N-terminus. It is very occasionally observed that double acid residues are resistant to cleavage by *Staphylococcal* protease.

4. The initial sequence results from peptide E8 indicated that it would contain the prolyl–proline sequence that had been troublesome for the clostripain peptide R5. The sequencer conditions were accordingly changed for residues 8 and 9 to give increased cleavage times (*see* Section 5.). This enabled rather more sequence past the prolines to be determined than in the case of peptide R5.

Table 4A
Amino Acid Compositions of *Staphylococcal* Protease Peptides[a]

	Peptide							
Amino acid	E1		E2		E3		E4	
Asx	2.9	(3)	—	—	—	—	1.8	(2)
Glx	1.1	(1)	4.2	(4)	1.0	(1)	1.1	(1)
Ser	0.8	(1)	1.8	(2)	—	—	1.0	(1)
Gly	2.8	(3)	1.9	(2)	—	—	1.0	(1)
His	0.9	(1)	0.9	(1)	—	—	—	—
Arg	0.8	(1)	0.9	(1)	1.0	(1)	—	—
Thr	—	—	0.9	(1)	0.8	(1)	—	—
Ala	1.0	(1)	—	—	1.0	(1)	—	—
Pro	1.1	(1)	1.2	(1)	—	—	—	—
Tyr	0.9	(1)	—	—	—	—	—	—
Val	—	—	0.9	(1)	—	—	—	—
Met	0.8	(1)	0.9	(1)	—	—	—	—
PE-Cys[b]	—	—	1.0	(1)	0.9	(1)	—	—
Ile	0.9	(1)	0.8	(1)	—	—	—	—
Leu	1.8	(2)	—	—	—	—	—	—
Phe	—	—	—	—	—	—	—	—
Lys	—	—	2.1	(2)	—	—	1.1	(1)
Trp	ND	(1)	ND	—	ND	—	ND	—
Total		(18)		(18)		(5)		(7)

[a]The results are uncorrected for destruction during hydrolysis, or incomplete hydrolysis of peptide bonds. Tryptophan was not determined. The results are expressed as residues/mole of peptide, with the values in brackets corresponding to those established by the sequence determination. The dash indicates less than 0.3 residues/mole in the composition, or not present in the sequence.

[b]Abbreviation: PE-Cys, pyridylethyl-cysteine.

2.2.2. Assembly of Complete Sequence from Overlapping Sequences (Part 1)

The combination of the results from the two enzymic digests enabled a large portion of the sequence to be assembled, as can be seen in Fig. 12. The overlapped peptides clearly fall into three stretches of sequence. It is likely that the peptides in sequence I overlap those of sequence II with the residues T R. Similarly,

Table 4B
Amino Acid Compositions of *Staphylococcal* Protease Peptides[a]

Amino acid	E5		E6		E7		E8	
Asx	1.0	(1)	2.8	(3)	0.9	(1)	—	—
Glx	0.9	(1)	2.1	(2)	1.0	(1)	1.1	(1)
Ser	0.9	(1)	0.8	(1)	—	—	—	—
Gly	—	—	—	—	0.9	(1)	3.9	(4)
His	—	—	—	—	—	—	0.9	(1)
Arg	—	—	—	—	—	—	0.8	(1)
Thr	—	—	0.9	(1)	—	—	0.8	(1)
Ala	1.0	(1)	1.1	(1)	1.0	(1)	—	—
Pro	1.0	(1)	1.1	(1)	—	—	2.3	(2)
Tyr	—	—	—	—	—	—	0.8	(1)
Val	—	—	1.8	(2)	—	—	3.0	(3)
Met	0.9	(1)	—	—	—	—	0.9	(1)
PE-Cys[b]	—	—	—	—	0.9	(1)	—	—
Ile	—	—	0.9	(1)	—	—	1.7	(2)
Leu	—	—	0.9	(1)	1.7	(2)	1.8	(3)
Phe	—	—	0.9	(1)	—	—	1.9	(2)
Lys	—	—	1.2	(1)	2.1	(2)	—	—
Trp	ND	—	ND	—	ND	—	ND	—
Total		(6)		(14)		(9)		(22)

[a]The results are uncorrected for destruction during hydrolysis, or incomplete hydrolysis of peptide bonds. Tryptophan was not determined. The results are expressed as residues/mole of peptide, with the values in brackets corresponding to those established by the sequence determination. The dash indicates less than 0.3 residues/mole in the composition, or not present in the sequence.
[b]Abbreviation: PE-Cys, pyridylethyl-cysteine.

sequences II and III probably overlap with the sequence T H E. Peptide R4 in sequence III must be located at the C-terminus of HGSH because it does not have an arginine residue at its C-terminus. However, the evidence is not very strong for these overlaps, and additional information should be obtained.

The reason why the sequencer run of peptide E1 was stopped after only 5 cycles is that it was obvious that this peptide corresponds to a portion of the sequence established by the other HGSH

E1   H M W L A ... (the sequence was stopped after 5 cycles)

E2   T R E E C H I M P G K G Q K V S (S) (E)

E3   T R A C E

E4   S K P D G D E

E5   No sequence

E6   N L I N D A F T S V K V Q E

E7   K Q K L C ... (the sequence was stopped after 5 cycles)

E8   V G G M F L I P P V Y L L V G F (G) ...

Fig. 11. The N-terminal sequences of the *Staphylococcal* peptides. The amino acids in brackets were assigned with some uncertainty.

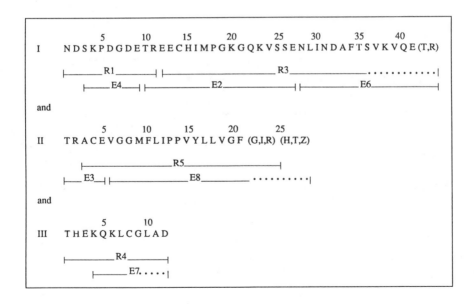

Fig. 12. Assembly of clostripain and *Staphylococcal* protease peptides from HGSH. The sequenced portions of the peptides are indicated by the continuous lines, and the unsequenced portions by the dotted lines.

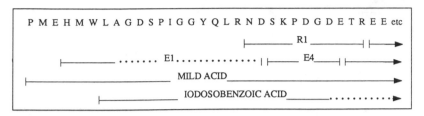

Fig. 13. Assembly of sequence I from Fig. 12 and the sequences determined from the mild acid cleavage and *o*-iodosobenzoic acid cleavage fragments. The sequenced portions of the peptides are indicated by the continuous lines, and the unsequenced portions by the dotted lines.

research group, which had used two different chemical cleavage
procedures chosen to yield limited sequence information. If their
sequence results are taken into consideration, then sequence I can
be substantially extended toward the N-terminus, as shown in
Fig. 13. Of course, if the sequence information from the mild acid
cleavage and *o*-iodosobenzoic acid cleavage experiments was not
already available, then these procedures would have been used at
this stage by the second HGSH team.

A tally of the clostripain and *Staphylococcal* protease peptides
shows that only one peptide from each digest is unaccounted for.
These are the two peptides that gave no sequence, R2 and E5, and
can therefore be presumed to be from the N-terminus of HGSH.
Peptide R2 is probably 23 residues long and is likely to end at the
arginine just N-terminal to peptide R1. If the N-terminal portion of
the sequence in Fig. 13 is subtracted from the composition of R2,
then the following residues remain: B, S, A. These are therefore
likely to be the N-terminal three residues of HGSH. This hypothe-
sis has added weight when the composition of E5 is taken into consideration. The "left over" residues from this peptide are also B, S, A.

At this stage there were therefore three regions of HGSH that
required additional sequence evidence:

1. The N-terminal few residues;
2. The overlap between sequence I and sequence II; and
3. The overlap between sequence II and sequence III.

As is often the case, the amount of work required to finish off a
sequence can be disproportionately great.

2.2.3. Cleavage Procedures (Part 2)

2.2.3.1. Cyanogen Bromide

Inspection of the nearly complete sequence of HGSH (Figs. 12 and 13) indicated that cleavage at methionine residues with CNBr might provide the necessary fragments to complete the sequence. One CNBr fragment would start at residue 18 of Sequence I, and a good sequencer run of about 30 residues would clearly establish the overlap. Another CNBr fragment would start at residue 10 of Sequence II, and it should be quite straightforward to obtain sufficient information from a reasonable sequencer run. Two of the remaining CNBr fragments would be small peptides from the N-terminal region, one of which would be useful for mass spectrometric determination of the N-terminal blocking group and the N-terminal residues (see Section 2.2.4.). The fifth CNBr fragment resulting from cleavage at the methionyl–tryptophan sequence shown in Fig. 13 would not be required because there was already ample sequence information from this region.

A sample of 500 pmol of HGSH was dissolved in 70% formic acid and treated with CNBr. The sample was dried, resuspended in dilute trifluoroacetic acid, and the fragments were separated by reverse phase HPLC (Fig. 14). The results were not encouraging, since only three peaks were observed instead of the expected five. Nevertheless, fractions were collected, concentrated, and subjected to amino acid analysis (see Table 5). The results clearly show that fragment CN1 corresponds to the tripeptide near the N-terminus and that CN2 is the N-terminal peptide itself. Fragment CN3 had two of the three phenylalanine residues and must therefore correspond to the C-terminal CNBr fragment (see Fig. 12). Additional evidence in support of CN3 being derived from the C-terminus was the absence of any homoserine in the amino acid composition (Table 5). Fragment CN3 was subjected to automated sequencing and gave a good sequence of 21 residues. These results are included in the final assembled sequence shown in Fig. 15. The other two expected CNBr fragments were not observed at all and were presumed to be insoluble. This is a common problem with CNBr fragments and is the main reason why this procedure is not recommended as an early step in sequencing strategies.

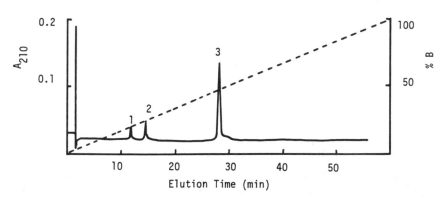

Fig. 14. Separation of CNBr fragments from HGSH by reverse phase HPLC. The peptides were separated with a linear gradient of Solvent A (0.1% trifluoroacetic acid) and Solvent B (80% acetonitrile).

The results from the CNBr cleavage therefore provided good evidence for the overlap between Sequence II and Sequence III (*see* Fig. 12), but, disappointingly, failed to establish the overlap between Sequence I and Sequence II. A number of alternative strategies were considered to try to solve this problem, and basically there were two possibilities: 1. digestion of HGSH with a different proteolytic enzyme such as chymotrypsin or thermolysin, and 2. subdigestion of a peptide already isolated from the clostripain or *Staphylococcal* protease digests. Both of these possible approaches were likely to yield the required information, but it was considered that the work involved to find the desired peptide from a digest of HGSH would be much greater than to use the subdigestion strategy. Accordingly, it was decided to treat clostripain peptide R3 with *Staphylococcal* protease, and to characterize the expected three peptides.

2.2.3.2. Subdigest of Clostripain Peptide R3

A sample of about 200 pmol of peptide R3 was digested with *Staphylococcal* protease as described in Fig. 10. The peptides were separated by reverse phase HPLC (not shown), and three peaks were collected that were characterized by amino acid analysis (Table 6). The compositions clearly showed that peptide R3E2 corresponded to the N-terminal portion of R3 (*see*, for example, the methionine and pyridylethylcysteine content). Peptide R3E1 was

Table 5
Amino Acid Compositions of CNBr Fragments from HGSH[a]

Amino acid	Peptide					
	CN1		CN2		CN3	
Asx	—	—	0.9	(1)	0.9	(1)
Glx	1.1	(1)	—	—	2.2	(2)
Ser	—	—	0.8	(1)	—	—
Gly	—	—	—	—	2.8	(3)
His	1.0	(1)	—	—	0.9	(1)
Arg	—	—	—	—	1.0	(1)
Thr	—	—	—	—	0.8	(1)
Ala	—	—	1.0	(1)	1.0	(1)
Pro	—	—	1.0	(1)	2.1	(2)
Tyr	—	—	—	—	0.9	(1)
Val	—	—	—	—	2.0	(2)
Met	—	—	—	—	—	—
PE-Cys[b]	—	—	—	—	0.9	(1)
Ile	—	—	—	—	1.7	(2)
Leu	—	—	—	—	3.8	(5)
Phe	—	—	—	—	1.9	(2)
Lys	—	—	—	—	2.1	(2)
Trp	ND	—	ND	—	ND	—
Hse	+	(1)	+	(1)	—	—
Total		(3)		(5)		(27)

[a]The results are uncorrected for destruction during hydrolysis, or incomplete hydrolysis of peptide bonds. Tryptophan was not determined. Methionine is converted into homoserine (Hse) and homoserine lactone by CNBr treatment. The recovery of these amino acids is variable, and quantitation was not attempted. The results are expressed as residues/mole of peptide, with the values in brackets corresponding to those established by the sequence determination. The dash indicates less than 0.3 residues/mole in the composition, or not present in the sequence.

[b]Abbreviations: PE-Cys, pyridylethyl-cysteine. Hse, homoserine.

a dipeptide (Thr, Arg) that had to be derived from the C-terminus of R3 as it had no glutamic acid residue. The presence of this dipeptide in the *Staphylococcal* protease digest confirmed the overlap between Sequence I and Sequence II. Peptides R3E1 and R3E3 were sequenced to provide additional evidence for the sequence in this region of HSGH. The results are included in Fig. 15.

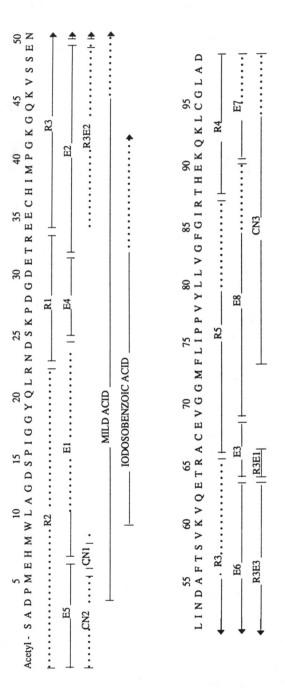

Fig. 15. Assembly of the complete sequence of HGSH. The sequenced portions of the peptides are indicated by the continuous lines, and the unsequenced portions by the dotted lines.

Table 6
Amino Acid Compositions of Peptides
from *Staphylococcal* Protease Subdigestion of Peptide R3[a]

Amino acid	Peptide					
	R3E1		R3E2		R3E3	
Asx	—	—	—	—	2.9	(3)
Glx	—	—	4.2	(4)	2.2	(2)
Ser	—	—	1.8	(2)	0.9	(1)
Gly	—	—	1.9	(2)	—	—
His	—	—	1.0	(1)	—	—
Arg	1.0	(1)	—	—	—	—
Thr	0.9	(1)	—	—	0.9	(1)
Ala	—	—	—	—	1.0	(1)
Pro	—	—	1.1	(1)	—	—
Tyr	—	—	—	—	—	—
Val	—	—	1.0	(1)	1.9	(2)
Met	—	—	0.9	(1)	—	—
PE-Cys[b]	—	—	0.9	(1)	—	—
Ile	—	—	0.9	(1)	0.9	(1)
Leu	—	—	—	—	0.9	(1)
Phe	—	—	—	—	1.0	(1)
Suc-Lys[b]	—	—	2.2	(2)	1.1	(1)
Trp	ND	—	ND	—	ND	—
Total		(2)		(16)		(14)

[a]The results are uncorrected for destruction during hydrolysis, or incomplete hydrolysis of peptide bonds. Tryptophan was not determined. The results are expressed as residues/mole of peptide, with the values in brackets corresponding to those established by the sequence determination. The dash indicates less than 0.3 residues/mole in the composition, or not present in the sequence.

[b]Abbreviations: PE-Cys, pyridylethyl-cysteine. Suc-Lys, succinyl-lysine.

2.2.4. Mass Spectrometry
to Characterize the N-Terminal Residues

At this stage, there was only one remaining portion of HGSH that required additional evidence to establish the sequence. The order of the amino acids near to the N-terminus could not be determined by Edman degradation because of the presence of some as yet unidentified chemical moiety blocking the α-amino group.

Table 7
Mass Spectrometric Analysis of Peptide E5*

Amino acid	Residue M_r
Ser	87
Ala	71
Asp	115
Pro	97
Met	131
Glu	129
OH at C-terminus	17
H at N-terminus	1
Subtotal	648
Total—subtotal	43

*Total M_r of peptide E5 = 691.

The HGSH research team decided that the best approach to elucidate the N-terminal sequence and to determine the nature of the
blocking group would be to use mass spectrometry. They knew
from the amino acid compositions of peptides E5 and CN2, and
from the site of the mild acid cleavage that the N-terminal region
was likely to be:

$$X (S,A) D P M \ldots$$

where X is the blocking group.

A 10 pmol sample of peptide E5 was taken for time-of-flight
laser-desorption mass spectrometric analysis. This method yields
a very accurate measure of the mol mass of a peptide or protein.
The results show that the relative mol mass of peptide E5 is 691. As
shown in Table 7, the relative mol mass of the N-terminal blocking
group must by difference be 43. This is very strong evidence that
the N-terminus is blocked by an acetyl group (CH_3CO), because
this is the only commonly occurring N-terminal blocking group
with a mass of 43.

The technique of fast-atom-bombardment (FAB) mass spectrometry was then used to deduce the order of the amino acid residues in peptide E5. The peptide fragmentation pattern observed
in the FAB mass spectrum showed the presence of an ion corre-

sponding to acetylserine. The other fragment ions were consistent with the following sequence: Acetyl-S A D P M E. These results completed the determination of the sequence of HGSH.

2.2.5. Assembly of Complete Sequence (Part 2)

The entire sequence of HGSH deduced from enzymic cleavage, chemical cleavage, and mass spectrometry is shown in Fig. 15. It can be seen that every residue has been sequenced in at least two different fragments, with the exception of the N-terminal three residues. The HGSH team were therefore very confident of the sequence of residues 4–99, and were prepared to stand by their sequence results even if any discrepancies occurred with the DNA sequencing of the cDNA. The evidence for residues 1–3 is weaker because it relied on a single mass spectrometry experiment and on amino acid compositions. In this case, the protein sequencing team would be prepared to reexamine their data if some disagreement with the DNA sequencing should arise.

2.2.6. Uses and Pitfalls of Homologous Sequences

DNA and protein sequences are being determined at ever increasing rates, and the sequence data base now encompasses many thousands of sequences. There are thus many occasions when the sequence of a "new" molecule will be similar to a previously sequenced molecule. It is obviously sensible to take advantage of the knowledge of the structure of the homologous DNA or protein, but this must be done with caution.

One aspect of protein structure determination that can be helped by homologous sequences is the design of the sequencing strategy. For example, it is likely that active site and functionally important residues will be conserved among homologous proteins, and this information can be exploited for the isolation of particular peptides. Examples could include the choice of peptide separation methods to purify a peptide on the basis of its likely size, its net charge, or its hydrophobicity. Similarly, hydroxylamine or mild acid cleavage sites may be conserved, and these methods would thus be sensible initial steps. In general, a cleavage method that has been useful for one member of the family will be useful for another, and the pattern of peptides obtained will be similar.

			10									20	
			*		*							*	
Yeast	P K L V L V R H G Q S E W N E K N L F T
Human RBC	S K Y K L I M L R H G E G A W N K E N R F C
Human muscle	A T H R L V M V R H G E S T W N Q E N R F C
Human brain	A A Y K L V L I R H G E S A W N L E N R F S

Fig. 16. Comparison of the N-terminal sequences of phosphoglycerate mutases. The residues implicated as being important for activity are indicated by an asterisk. The sequences are taken from (16).

It is very tempting to use homologous sequence information to assign the order of peptides or even to make assumptions about the sequence of a particular peptide. It is probably quite reasonable to use homologous sequences to help to align peptides of say six residues or more in smallish proteins such as HGSH. For example, in the case of HGSH, if several homologous protein sequences had been available, it would have been acceptable to have decided that Sequence I and Sequence II in Fig. 12 overlapped, as did Sequences II and III. But the larger the protein and the smaller and less distinctive the peptide, the more likely it is to make incorrect alignments, and it is obviously necessary to be cautious.

The use of homologous sequence information to deduce sequences of individual peptides should be avoided. The literature is full of examples where this temptation has not been resisted, and incorrect sequences have been proposed, often with consequential problems concerning an understanding of the function and evolution of the protein in question. The sequence of cytochrome c from rattlesnake provides a notorious example of a protein missequenced by homology (for a discussion, see [15]). An examination of the active site sequences of members of the phosphoglycerate mutase family (Fig. 16) serves to illustrate the type of problem that could result from incautious use of sequence homology. The first mutase in this family to be sequenced was the yeast enzyme with the sequence N E K N at residues 14–17. The second enzyme to be sequenced was the human erythrocyte enzyme that has the sequence N K E N at the corresponding residues. Clearly the sequence would have been incorrect if it had been assigned by homology.

2.3. Summary of Strategies
for Sequence Determination

This Section is an attempt to provide a summary of the strategies that may be used to obtain protein sequence information. The summary is presented as a flow chart in Fig. 17.

Note: The hormone HGSH and all the scientists who worked on it are *fictitious* and any resemblance to actual proteins and people is *coincidental*.

3. Chemistry of Edman Degradation

Previous sections and chapters have described the strategies that can be adopted to determine amino acid sequences, and have described how to purify and characterize proteins and peptides in preparation for sequencing and amino acid analysis. It should be apparent that a crucially important aspect of protein sequencing is successful sample preparation. It is now appropriate to consider the actual processes for the determination of amino acid compositions and amino acid sequences. The following sections will describe the chemistry and the equipment, and will indicate some of the limitations and potential problems.

3.1. The Reactions

The procedure for the sequential degradation of a protein from its N-terminus was developed by Per Edman *(17–19)*, and in his honor the reactions involved are called "Edman degradation." Three major steps can be recognized (Fig. 18):

1. The *coupling* of phenylisothiocyanate (PITC) to the free N-terminal amino group (plus all the ε-amino groups of lysine residues during the first coupling). This coupling takes place in alkaline conditions because the free, uncharged amino group is required. The average pKa of the terminal amino group of peptides is 7.8, so the pH of the coupling conditions should be above 8. Very alkaline conditions (pH > 10) should be avoided, because competing side reactions, such as hydrolysis of PITC, are base catalyzed.

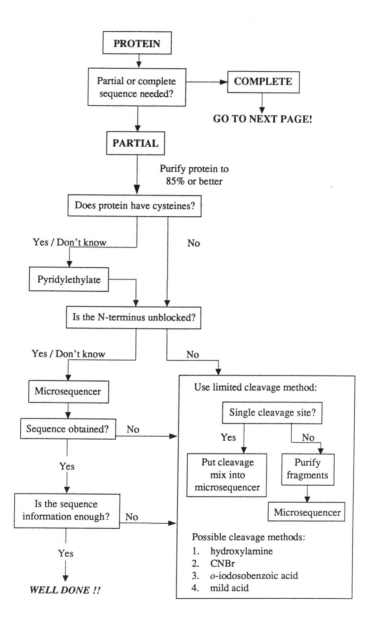

Fig. 17. Strategies for protein sequence determination.

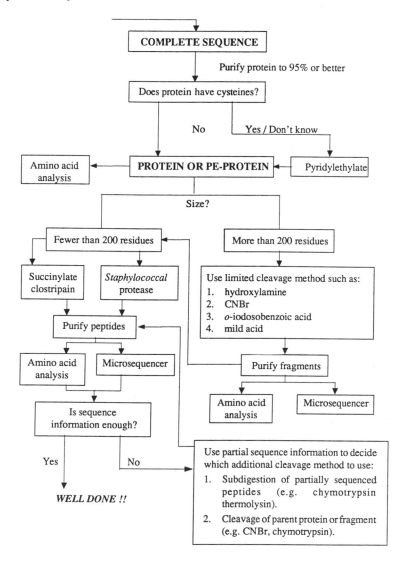

Fig. 17. (continued)

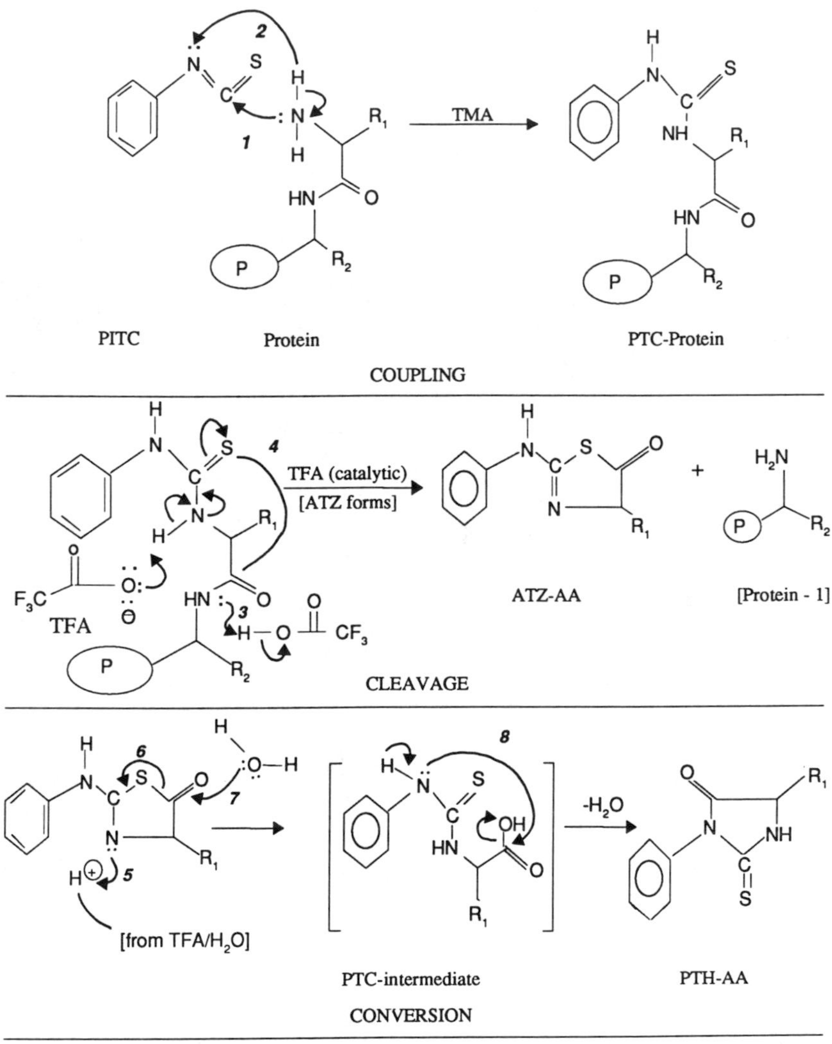

Fig. 18. Mechanistic guide to Edman chemistry. 1. Amino-terminus of protein, which is deprotonated attacks the isothiocyanate carbon of PITC. 2. The nitrogen atom picks up a proton from the amino terminus. 3. TFA supplies its exchangeable proton, causing the release of [protein-1]. 4. The carbonyl carbon atom is attacked by the sulfur atom leading to cyclization and reformation of protonated TFA. 5. A proton from TFA/H_2O is picked up by the ATZ ring nitrogen atom. 6. The ATZ ring is opened with the formation of a carbon–sulfur double bond. 7. The carbonyl carbon atom is hydroxylated by water, which then releases a proton. 8. Recyclization to the PTH derivative occurs when the isothiocyanate nitrogen atom attacks the carbonyl carbon atom, releasing water.

Table 8
Evolution of Edman Degradation

1956	Manual
1967	Spinning cup
1971	Solid phase
1981	Gas phase
1986	Pulsed liquid

2. The *cleavage* takes place under the influence of anhydrous strong acid. Edman used HCl dissolved in acetic acid. These days perfluorinated acids (trifluoroacetic acid [TFA] and heptafluorobutyric acid [HFBA]) are most popular. Cleavage yields an anilinothiazolinone (ATZ) amino acid residue and a new N-terminal amino acid that can be coupled again.

3. The ATZ residue is unfortunately unstable and cannot be identified easily as such. Therefore, *conversion* to the more stable phenylthiohydantoin (PTH) amino acid has to be carried out using aqueous or methanolic acid. Various procedures for the identification of PTH amino acids have been described, including back-hydrolysis, gas liquid chromatography, thin-layer chromatography, mass spectrometry, and HPLC. Currently, the most popular methods are HPLC with reverse phase columns *(20,21)*, and thin-layer chromatography on polyamide plates for manual sequencing *(22,23)*.

3.2. Development of Automated Sequencers

Edman soon realized that the repeating set of reactions simply asked for automation, and together with Begg *(24)* constructed the "spinning cup sequenator" (a summary of the evolution of Edman degradation is given in Table 8). A fast spinning thick walled beaker with a content of about 30 mL, contained in a housing that could be heated and evacuated, was loaded with a protein sample (several hundred nmol). Reagents and solvents were introduced into the cup via several centrally located lines of Teflon® tubing. Excess liquid could be removed by a "scoop" at the top of the cup, and any remaining liquid could be completely dried under vacuum. Conversion was not done automatically, so the ATZ residue was collected into a refrigerated fraction collector for further manual workup and identification. Later on, other workers added automatic conversion (25) and refined the procedures used in controlling the chemical events in the machine. One of the biggest problems observed with this type of equipment was washout of protein from the cup. One of the methods to prevent washout was to bind the protein covalently to a solid support (26). Several reagents and strategies were tried and solid-phase sequencing was born. This method is potentially the best sequencing approach available, mostly because inefficient washing and losses at washing steps are no longer troublesome. However, not all peptides are chemically suitable for covalent coupling to inert supports. Moreover, the expertise of a skilled protein chemist is required. As a result, very few laboratories have the capability for doing high quality solid-phase sequencing (27).

Although the initial enthusiasm for solid-phase sequencing has subsided, the need for highly sensitive sequencing machines has grown continuously. At Caltech, Leroy Hood stimulated some of his researchers to try to develop convenient high sensitivity methods in sequence determination. First, a commercial spinning cup sequencer was modified (28). The hardware was based on developments carried out in the group of Wittmann-Leibold (29) in Berlin. Special miniature valves were used, argon was substituted for nitrogen, an on-line PTH conversion device was included, and last, but not least, all the chemicals were extensively purified. The results were very promising, and the next step in the development was an entirely new machine (30). Here, the sample was no longer loaded into a spinning cup but instead was held on a glass fiber filter. In order to prevent washout, the highly polar coupling base (trimethylamine, TMA) and the cleavage acid (trifluoroacetic acid, TFA) were blown through the filter in the gas phase. All dead volumes were minimized, and the only liquids used were solvents that were good at removing reaction byproducts, but poor at extracting the protein or peptide sample. Since 1982, the resulting machine was commercially available from Applied Biosystems as the "gas-phase sequencer." At the end of 1986, the successor of this highly popular machine was introduced as the "pulsed-liquid-phase sequencer." In this last machine, the cleavage is no longer performed by blowing TFA-saturated argon through the glass fiber filter, but by precisely measuring the exact volume of acid needed to saturate the filter, without washing out the sample. One of the main advantages of this method is a considerable decrease in cleavage time. Also, the same measuring technique can be used to clean out the valve block manifolds, thus giving cleaner results. This machine is also supplied by Applied Biosystems as the Model 477A Microsequencer, and a schematic of the instrument is given in Fig. 19.

3.3. Manual Edman Degradation

The instruments described in the preceding section are undoubtedly costly. A major investment of $250,000–500,000 is required to establish a well equipped protein sequencing facility. This level of investment is clearly beyond the scope of many individual laboratories, and it is important to include a mention of manual sequencing techniques that can be set up at relatively low cost. The capital and operating costs of a manual system are about two orders of magnitude lower than those of the automated systems. Moreover, and probably more important, the manual systems are very suitable for doing batchwise sequencing, which is especially helpful in screening or characterizing HPLC fractions. In this type of application, the need is for just a few cycles of sequence from a very large number of samples. A manual sequencing approach would be most suitable for this purpose. If, on the other hand, the requirement is for an extended sequence, then an automatic sequencer is to be preferred.

The essence of successful and efficient manual sequencing is attention to detail, and the arrangement of a convenient and well-thought-out workstation. George Tarr at the University of Michigan has been particularly instrumental in improving the approaches to manual sequencing techniques, and detailed descriptions of the procedures and workstation are given in recent publications (see, for example, 23,31,32). The Tarr procedures are routinely useful for determining 10–30 residues with 500–2000 pmol of peptide or protein, and can conveniently be done in batches of 10 samples. The time required is about 30–60 min/cycle, depending on the type of detection system used. The repetitive yields (see next section) obtained are of the order of 90% for proteins and long peptides, and about 80% for short peptides.

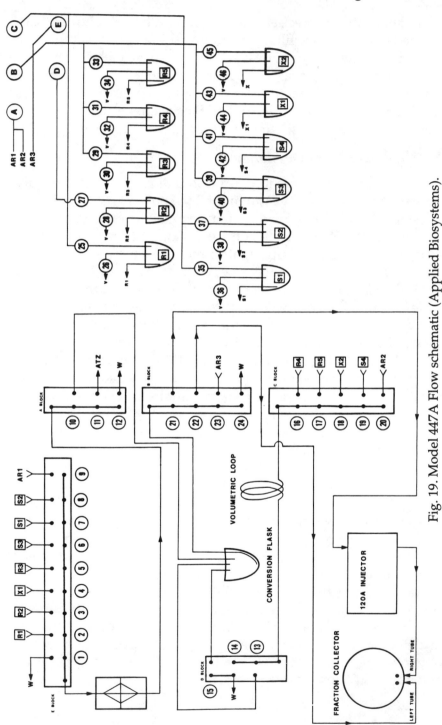

Fig. 19. Model 447A Flow schematic (Applied Biosystems).

3.4. Limitations, Artifacts, and Potential Problems

3.4.1. Repetitive Yield

The steady decrease in signal during a sequencing experiment arises from physical loss of sample from many factors, including loss of sample from the reaction vessel, side reactions that block a portion of the amino-terminal residue at each cycle, acid-catalyzed cleavage of the peptide chain, and incomplete coupling or cleavage that result in carryover (lag) of amino acid into cycles subsequent to that residue's actual position in the sequence. The overall efficiency, limited by these losses, is represented by the repetitive yield (RY) defined as:

$$RY = (Y_n/Y_m)^{1/n-m}$$

where Y_m = yield at cycle m, and Y_n = yield at a later cycle n. The repetitive yield is best calculated between pairs of the same amino acid occurring at different positions, because of the variable recovery of some amino acids during the conversion and PTH analysis stages. For example, a sequence that has a valine at position 2 with a yield of 55 pmol, and a valine at position 10 with a yield of 40 pmol, would have a repetitive yield of $(40/55)^{1/8} = 0.96$ or 96%.

It is obvious that the greater the efficiency of the Edman chemistry, the longer the sequence that can be determined. How many Edman degradation steps are possible for any given repetitive yield? It can reasonably be assumed that a sequence determination ceases to provide useful information when the overall yield has fallen to about 25%, because at this level the PTH-amino acid becomes difficult to identify against the growing background. Table 9 illustrates the number of steps possible for repetitive yields of 99, 95, and 90%.

Thus, it would appear from this table that with a repetitive yield of 99% at least 100 residues could be sequenced, whereas with a yield of 90% this is drastically reduced to little more than 10 steps. The length of sequence that can be determined will also depend on the amount of sample and the sensitivity of detection of the PTH-amino acids. Thus, if the limit of confident detection is about 10

Table 9
Percentage Overall Yield After n Steps
in the Sequential Degradation of a Protein
Assuming Different Repetitive Yields

	Repetitive yield		
n	99%	95%	90%
10	90	60	35
50	60	8	0.5
100	36	0.6	0.003

pmol, then a sample with 50 pmol initial yield could only give useful results for about 30 cycles, assuming a repetitive yield of 95%. Edman originally reported a repetitive yield of slightly greater than 98%, and this has really never been improved. With modern instruments such as the Applied Biosystems pulsed-liquid sequencer, it is possible routinely to attain repetitive yields of 97–98% with sample loadings of about 1000 pmol. In the case of very high sensitivity (100 pmol or less), the repetitive yield is usually 93–95%.

3.4.2. Artifacts of Edman Chemistry

The Edman reagent, phenylisothiocyanate, readily reacts not only with amino groups in proteins, but with a wide variety of amino groups on other molecules. Some of these molecules are routinely present during protein sequencing, and several artifacts are usually observed in the HPLC chromatograms. The presence of the two most common artifacts is not actually a disadvantage because they can be resolved from the PTH-amino acids (see Section 5.1.). Moreover, the quantities of the artifacts can give an indication of the quality of the sequencer performance.

The most abundant artifact is diphenylthiourea (DPTU), which is formed from aniline and PITC (see Fig. 20). The aniline is derived from PITC itself in the presence of water, and abnormally high quantities of DPTU indicate that the PITC reagent has suffered excessive exposure to water. Another artifact, dimethylphenylthiourea (DMPTU), is also readily apparent in most PTH chromatograms. This compound is derived from the reaction of PITC and dimethylamine (see Fig. 20). Dimethylamine gradually forms from the cou-

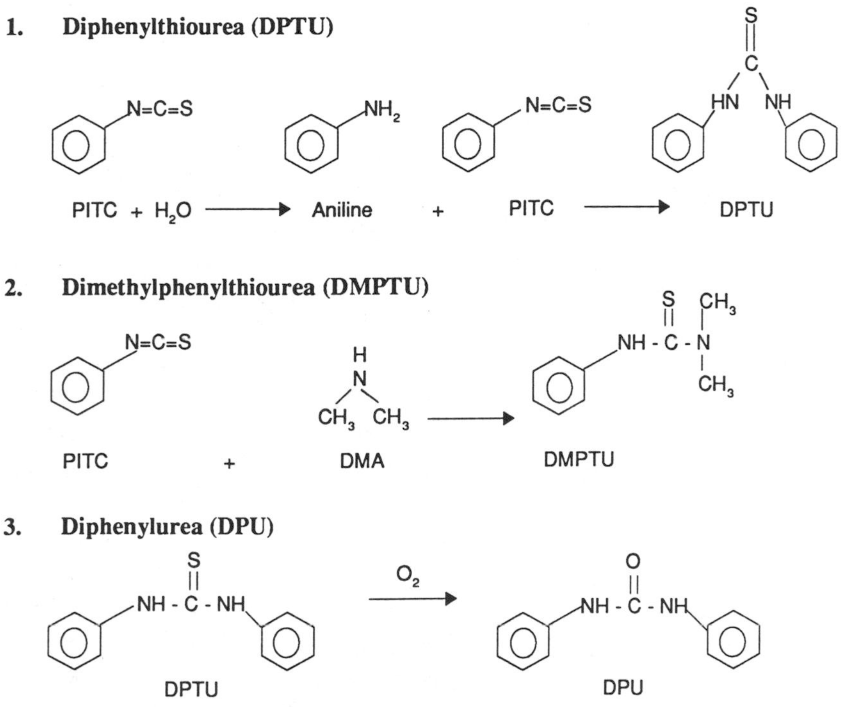

1. **Diphenylthiourea (DPTU)**

PITC + H₂O ⟶ Aniline + PITC ⟶ DPTU

2. **Dimethylphenylthiourea (DMPTU)**

PITC + DMA ⟶ DMPTU

3. **Diphenylurea (DPU)**

DPTU ⟶ DPU

Fig. 20. Edman chemistry byproducts.

pling base, trimethylamine (TMA), and for this reason the TMA must be carefully stored and relatively frequently replaced on an automated sequencer. A third artifact inherent in Edman degradation is more of a problem because it coelutes with PTH-Trp in a commonly employed elution system. This third type of molecule is diphenylurea (DPU), which forms from diphenylthiourea in the presence of oxygen (*see* Fig. 20). Extensive precautions are taken to exclude oxygen from all of the Edman chemistry, and abnormally high levels of DPU are a good indication that a leak has developed somewhere in the system.

Other PITC-related artifacts are derived from contaminants present in samples. Thus, PITC will react with Tris and Coomassie Blue to give uv-absorbing products. Of course, any free amino acids present in the sample (say, from the use of glycine buffers!)

will react with PITC and can give an enormous contaminating peak. It is clearly very important to ensure that all samples are free of these contaminating molecules.

3.4.3. "Difficult" Sequences

Some sequences and some amino acid residues cause problems during Edman degradation, either because they block degradation or because they are unstable or otherwise difficult to recover. Probably the most common problem is with proline residues. The fact that proline differs from all the other amino acids in having a secondary amino group means that it requires longer treatment with TFA at the cleavage step to achieve complete cleavage. Thus, during Edman degradation with standard conditions every time a proline is encountered, the yield drops and a substantial amount of the proline is detected in the next cycle. This out-of-phase sequencing persists, so that in effect two sequences are obtained that are one residue out of step. Of course, if a second proline is encountered, the problem becomes worse. It is possible to solve the problem to a large extent by increasing the cleavage time. Unfortunately, the increased cleavage time cannot be used routinely because it in turn causes unacceptable losses of other residues, especially of the labile amino acids, Ser, Thr, and Trp. Excess treatment with the cleavage acid can also cause N-terminal glutamine residues to cyclize to pyrrolidonecarboxylic acid residues that block Edman degradation. In effect, the only way to have good results with sequences containing several prolines is to sequence the sample twice: once with standard cleavage times, and a second time with increased cleavage cycles at the prolines.

Another commonly encountered difficulty is with recovery of the labile amino acids Ser, Thr, and Trp. The PTH derivatives of these amino acids are notoriously difficult to recover in high yield, and in fact before the use of automatic conversion were frequently missed altogether. The poor recovery of these residues means that if the sequencing is being done toward the limits of detection, then it is all too possible to have a blank cycle where it is not feasible to assign the residue with any confidence. Probably the only way to solve this problem is to obtain a larger amount of sample, or to purify another peptide with the labile residues nearer to the beginning of the sequence.

Unfortunately, there are several other possibilities for having a blank cycle. Thus glycosylated residues will not be identified, and may even block subsequent Edman degradation. Phosphorylated residues are also not recovered because they are so hydrophilic that they are not transferred from the glass fiber disk to the conversion flask with the solvents normally used for this purpose. In addition, samples that have disulfide bridges still intact may give a blank in the cycle that involves the Cys residue. All these problems may be solved by appropriate enzymic or chemical modifications prior to the sequencing. It should be apparent that careful attention to all these potential problems before a sample is actually subjected to Edman degradation will go a long way to ensuring that useful results are obtained.

4. Identification
of N-Terminal Amino Acids

Essentially two different methods exist for the identification of amino acid residues during Edman degradation. One method involves the direct identification of the amino acid residue as it is cleaved from the parent protein or peptide. This procedure is adopted by automated protein sequencers. The other method identifies the N-terminal residue of a protein or peptide before each round of Edman degradation takes place, and is commonly used for manual sequencing.

4.1. PTH-Amino Acids

As mentioned previously (Section 4.), automated protein sequencers usually incorporate online conversion of the thiazolinone amino acid derivatives to the more stable thiohydantoins. These PTH-derivatives are relatively hydrophobic molecules that can conveniently be separated by reverse phase HPLC. The standard procedure employed by the Applied Biosystems Microsequencer involves automatic injection of the PIH-derivative at each cycle onto a C_{18} reverse phase column that is eluted with a gradient of acetate buffer and acetonitrile. A typical separation is shown in Fig. 21. It can be seen that all of the commonly occurring PTH-amino acid residues are resolved, and are separated from the Edman

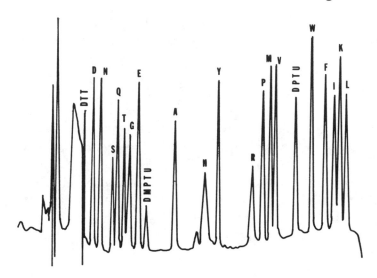

Fig. 21. Elution of a 25 pmol standard mixture of PTH-amino acids according to the procedure employed by the Applied Biosystems Microsequencer.

degradation byproducts DMPTU and DPTU. The pyridylethyl derivative of cysteine elutes between PTH-Val and DPTU, and can thus be readily identified.

A number of different successful reverse phase HPLC procedures have been reported for the identification of PTH-amino acid residues. Both gradient and isocratic elutions can be used, and these can be incorporated with both manual and automated sequencing methods.

4.2. Dansyl and DABITC

A large number of methods have been described for the determination of N-terminal residues in peptides and proteins. Two of these methods are particularly convenient and sensitive, and are used routinely. Dansyl chloride (1-dimethylaminonaphthalene-5-sulfonyl chloride, Dns-Cl) reacts with amino groups in proteins under alkaline conditions to give sulfonamide derivatives that are resistant to acid hydrolysis under conditions appropriate to hydrolyze peptide bonds. The method thus involves initial reaction with dansyl chloride, followed by hydrolysis of the peptide bonds, and identification of the α-N-Dns-amino acid derivative by thin layer chromatography on polyamide plates (33). Dansylation used in conjunction with Edman degradation allows the N-terminal sequence of a peptide or protein to be determined. It is relevant to note that the dansyl method requires relatively large quantities of sample because portions have to be removed at each cycle of Edman degradation for identification.

Another sensitive and convenient method for N-terminal identification involves an alternative Edman reagent: 4-dimethyl-amino-azobenzene-4'-isothiocyanate or DABITC (22,34). The method is essentially that of Edman degradation except that DABITC is substituted for PITC. The main advantage of this method is that the thiohydantoin derivatives are brightly colored, and can thus be readily identified by thin-layer chromatography. A detailed protocol for this method is given in Wittmann-Liebold et al. (35) and in Yarwood (23).

5. Amino Acid Analysis

Probably one of the most difficult analyses routinely required by biological laboratories is the quantitative determination of amino acids in a protein or peptide sample. There are a number of reasons for the inherent difficulties:

1. Proteins have 20 commonly occurring amino acids, some of which are chemically very similar. Any separation technique must therefore be able to resolve complex mixtures of similar compounds.
2. The relative amounts of the different amino acids can vary widely from just one or two residues/mol of a nonabundant amino acid like histidine or tyrosine, to 20 or more residues of abundant amino acids like glutamic acid or leucine. It is inherently difficult to quantify compounds that are present in substantially different amounts in a single sample.
3. The amino acid residues vary in their lability to acid hydrolysis. Tryptophan is completely destroyed, asparagine and glutamine are completely deamidated, and serine and threonine are partially destroyed under the conditions required to achieve complete hydrolysis of the peptide bonds. In addition, some peptide bonds are relatively resistant to acid hydrolysis.

4. Cysteine and methionine residues can be oxidized during the hydrolysis and separation conditions. It is thus essential that cysteine residues be deriviatized prior to analysis, and that reducing agents be present to prevent oxidation of methionine.
5. High sensitivity amino acid analysis demands very careful purification of reagents and the adoption of many precautions to avoid contamination with amino acids, which are present everywhere.

This list of problems gives some impression of the care with which amino acid analyses must be performed and interpreted. For accurate analyses, it is essential that several different times of hydrolysis be used (say 20, 48, and 96 h at 110°C in 6N HCl), and that each sample be done in triplicate.

5.1. Hydrolysis

Until recently, acid hydrolysis was usually done by adding 6N HCl to a sample that had been dried in a glass tube suitable for sealing with a torch under vacuum. The evacuated tube was then usually heated at 110°C for 20–96 h. The advent of very high sensitivity methods has meant that the levels of contamination inherent in the above procedure are unacceptable. The use of vapor phase hydrolysis has been adopted to try to minimize the contamination problem. A convenient method involves placing samples in small glass tubes that have been scrupulously cleaned and carefully stored prior to use. The samples are then dried in a vacuum centrifuge, and then placed in a small desiccator over high purity 6N HCl (such as that supplied by Pierce Chemicals). The desiccator is evacuated and heated at 110°C for an appropriate length of time.

An alternative procedure, which is undoubtedly the most satisfactory for appropriate samples, is to do the hydrolysis automatically within the instrument that will then do the derivatization of the amino acids prior to their chromatographic separation. This approach is available in the Applied Biosystems 420A Derivatizer.

5.2. Derivatization Pre- or Postseparation

Amino acids in general cannot be easily detected directly, and some sort of derivatization method is required. The procedure pioneered by Stein and Moore (36) involves reaction of the amino acids with ninhydrin, which yields a chromophore absorbing at 550 nm (and 440 nm for the proline derivative). The derivatization is done after the amino acids have been separated by ion exchange chromatography. An alternative detection system to ninhydrin is the use of o-phthalaldehyde (OPA), which produces fluorescent derivatives with primary amines. A complication with OPA is that proline does not give a fluorophor unless it is first oxidized in a specific manner. This adds to the complexity of the instrumentation.

The early amino acid analyzers took about 24 h to achieve a separation of all the amino acids, and required 10s of nmols of sample. There have been substantial improvements over the years, especially with regard to ion-exchange resins and plumbing systems, such that separation times of several hours are possible, and detection with ninhydrin has a sensitivity of 100–1000 pmol. The OPA detection system is somewhat more sensitive, although the on-line conversion of proline to a primary amine poses some problems.

Until recently, most commercially available analyzers employed ion-exchange chromatography and detection by ninhydrin or OPA. The advent of reverse phase HPLC instruments, however, has opened the door to methods employing precolumn derivatisation of the amino acids to form chromogenic or fluorogenic derivatives that are relatively hydrophobic and thus suitable for reverse phase separations. These methods show a substantial speeding up of separation times with 20–30 min being typical. The sensitivity has increased to the low pmol range.

A number of precolumn methodologies have been reported and these are compared by Jones (37). The procedure adopted by Applied Biosystems exploits the reaction of amino acids with PITC, the same reagent as used in Edman degradation. The product of the reaction is the phenylthiocarbamyl (PTC) derivative, shown in Fig. 18. All the PTC-amino acid derivatives commonly present in protein hydrolyzates can be readily resolved in about 15 min on a reverse phase column with a gradient of acetate buffer and acetonitrile (Fig. 22). The uv detection system is sufficiently sensitive (a few pmol) that the limitations are in sample preparation, and reagent and water quality (*see* Section 6.3.). One of the main reasons why the PITC methodology has not previously been much used, is that

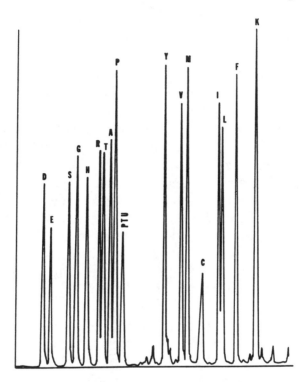

Fig. 22. Elution of a 500 pmol standard mixture of PTC-amino acids according to the procedure employed by the Applied Biosystems Amino Acid Analyzer.

the PTC-amino acid derivatives vary in stability, with PTC-glycine being particularly labile. This problem has been circumvented in the Applied Biosystems instrument by having automated derivatization immediately prior to separation on the HPLC system.

5.3. Limitations and Potential Problems

Undoubtedly the major limitations of amino acid analysis are those imposed by the need to hydrolyze the peptide bonds to release the free amino acids. The customary hydrochloric acid hydrolysis conditions result in complete destruction of tryptophan residues, and complete deamidation of asparagine and glutamine. Moreover, serine and threonine are partially destroyed, with about 10% of serine and 5% of threonine lost in 20 h hydrolysis. Tyrosine is also poorly recovered mostly because of chlorination of the aromatic ring during hydrolysis.

Various procedures can be adopted to circumvent some of the limitations just mentioned. Thus, hydrolysis with methanesulfonic acid (38) can sometimes give good recoveries of tryptophan. However, this acid is not volatile like HCl, and must therefore be neutralized prior to separation of the amino acids. This hydrolysis method is probably not suitable for precolumn derivatization techniques. Alternatively, tryptophan can be determined spectrophotometrically, providing no unusual chromophoric groups are present in the sample (the method is given in Allen [12]). The usual procedure to obtain a reliable estimate of serine and threonine contents is to do several different times of hydrolysis, and to extrapolate back to the amount at zero time. The recovery of tyrosine can be improved by including phenol in the HCl used for hydrolysis.

Unfortunately, there is really nothing that can be done to prevent hydrolysis of asparagine and glutamine. The only sure way to determine the amounts of these amino acids is by sequencing. It is relevant to mention here that amino acid analysis results are often presented in a misleading way by failing to indicate that the amounts of aspartic acid and glutamic acid are derived from both the free acid *and* the amide. The results should be given as Asx and Glx, to make it quite clear what is meant, especially to those who may not be very familiar with the technique of amino acid analysis.

The successful operation of an ultrasensitive amino acid analyzer demands extreme care in sample and reagent preparation. Amino acids and other amines that can react with PITC are everywhere and can completely invalidate the amino acid analysis of a sample unless suitable precautions are taken. For example, it can be shown that the amount of some amino acids transferred when a finger touches a glass surface can be as much as several hundred pmol. Dust, saliva, and glassware washing procedures can all introduce substantial contamination. It is probably necessary to clean all glassware by pyrolysis (heating at 500°C for 6 h). The quality of the water used to prepare the reagents can be a major problem, especially with regard to ammonia content. The ELGA water purification system has been found to be generally satisfactory, in contrast to most other systems, including Waters MilliQ.

Section 6. started with the assertion that amino acid analysis is probably one of the most difficult analyses routinely required by biological laboratories. The justification for the statement should now be apparent!

Bibliography

*References 1–6 and 12 can be recommended as valuable sources of information regarding all aspects of protein sequencing and amino acid analysis.

1. Findlay J. B. C. and Geisow M. J., eds. (1989), *Protein Sequencing: A Practical Approach* (1989), IRL, Oxford.
2. Hugli T. E., ed. *Techniques in Protein Chemistry* (1989), Academic, London.
3. Shively J. E., ed. *Methods of Protein Microcharacterization* (1986), Humana, Clifton, NJ.
4. Wittman-Liebold B., Salnikow J., and Erdmann V. A., eds. *Advanced Methods in Protein Microsequence Analysis* (1986), Springer-Verlag, Berlin.
5. Villafranca J. J., ed. *Current Research in Protein Chemistry: Techniques, Structure and Function* (1990), Academic, London.
6. Hirs C. H. W. and Timasheff S. N., eds. *Methods in Enzymology vol. 91.* (1983), Academic, London.
7. Richardson J. S. (1981), *Adv. Prot. Chem.* **34,** 147–339.
8. Noguchi T., Inoue H., and Tanaka T. (1986), *J. Biol. Chem.* **261,** 13,807–13,812.
9. Takenaka M., Noguchi T., Inoue H., Yamada K., Matsuda T., and Tanaka T. (1989), *J. Biol. Chem.* **264,** 2362–2367.
10. Noguchi T., Yamada K., Inoue H., Matsuda T., and Tanaka T. (1987), *J. Biol. Chem.* **262,** 14,366–14,371.
11. Omenn G. S. and Hermodson M. A. (1975), *Isozymes III. Developmental Biology* (Markert C. L., ed.) Academic, New York, pp. 1005–1018.
12. Allen G. (1989), *Sequencing of Proteins and Peptides* 2nd ed., Elsevier, Amsterdam, pp. 56–59.
13. Mitchell W. M. (1977), *Methods Enzymol.* **47,** 165–170.
14. Drapeau G. R., Boily Y., and Houmard J. (1972), *J. Biol. Chem.* **247,** 6720–6726.
15. Ambler R. P., *Computer-Assisted Bacterial Biosystematics* (Jones D., Goodfellow M., and Priest F. G., eds.) (1985), Academic, San Diego, pp. 307–335.
16. Fothergill-Gilmore L. A. and Watson H. C. (1989), *Adv. Enzymol.* **62,** 227–313.
17. Edman P. (1950), *Acta Chem. Scand.* **4,** 283–293.
18. Edman P. (1953), *Acta Chem. Scand.* **7,** 700–701.
19. Edman P. (1956), *Acta Chem. Scand.* **10,** 761–768.
20. Hunkapiller M. W., Granlund-Moyer K., and Whiteley N. W. *Methods of Protein Microcharacterization* (Shively J. E., ed.) (1986), Humana, Clifton, NJ, pp. 315–327.

21. Geisow M. J. and Aitken A., *Protein Sequencing: A Practical Approach* (Findlay J. B. C. and Geisow M. J., eds.) (1989), IRL, Oxford pp. 94–97.

22. Chang J. Y. and Creaser E. H. (1976), *Biochem. J.* **157,** 77–85.

23. Yarwood A., *Protein Sequencing: A Practical Approach* (Findlay J. B. C. and Geisow M. J., eds.) (1989), IRL, Oxford, pp. 119–145.

24. Edman P. and Begg G. (1967), *Eur. J. Biochem.* **1,** 80–91.

25. Wittmann-Liebold B. (1973), *Hoppe Seyler's Z. Physiol. Chem.* **354,** 1415–1531.

26. Laursen R. A. (1971), *Eur. J. Biochem.* **20,** 89–102.

27. Findlay J. B. C., Pappin D. J. C., and Keen J. N., *Protein Sequencing: A Practical Approach* (Findlay J. B. C. and Geisow M. J., eds.) (1989), IRL, Oxford, pp. 69–84.

28. Hunkapiller M. W. and Hood L. E. (1978), *Biochemistry* **17,** 2124–2133.

29. Wittmann-Liebold B., Grafunder H., and Kohls H. (1976), *Anal. Biochem.* **7,** 621–633.

30. Hewick R. M., Hunkapiller M. W., Hood L. E., and Dreyer W. J. (1981), *J. Biol. Chem.* **256,** 7990–7997.

31. Tarr G. E., *Methods of Protein Microcharacterization* (Shively J. E., ed.) (1986), Humana, Clifton, NJ, pp. 155–194.

32. Kuhn C. C. and Crabb J. W. *Advanced Methods in Protein Microsequence Analysis* (Wittmann-Liebold B., Salnikow J., and Erdmann V. A., eds.) (1986), Springer-Verlag, Berlin, pp. 64–76.

33. Gray W. R. (1972), *Methods Enzymol.* **25,** 326–332.

34. Chang J. Y. (1977), *Biochem. J.* **163,** 517–520.

35. Wittmann-Liebold B., Hirano H., and Kimura M. *Advanced Methods in Protein Microsequence Analysis* (Wittmann-Liebold B., Salnikow J., and Erdmann V. A., eds.) (1986), Springer-Verlag, Berlin, pp. 77–90.

36. Spackman D. H., Stein W. H., and Moore S. (1958), *Anal. Chem.* **42,** 1190–1206.

37. Jones B. N. *Methods of Protein Microcharacterization* (Shively J. E., ed.) (1986), Humana, Clifton, NJ, pp. 121–151.

38. Inglis A. S. (1983), *Methods Enzymol.* **91,** 26–36.

Nine

Electrophoretic Techniques of Analysis and Isolation

James L. Dwyer

1. Introduction

Electrophoresis is a technique whereby components move through an electrically conductive medium in response to an applied voltage gradient. All electrophoretic procedures resolve analytes on the basis of the analyte's effective charge and/or molecular dynamic radius, but there are many modes of operation that provide separations exploiting hydrophobicity, chirality, complexation, specific affinity, and so on. By altering the physical and chemical interactions between analytes and the system, one has access to a surprisingly broad range of separation techniques.

Electrophoresis is a versatile fundamental laboratory tool, universally used by peptide and protein chemists. It is predominantly an analytical tool, although small-scale preparative techniques may be employed. It is beyond the scope of this Chapter to be able to provide the details of the many procedures presently employed. The principal electrophoretic methods in present use are described here, and the reader is referred to comprehensive manuals and literature sources for details of methodology.

The original electrophoretic zone electrophoresis separations developed by Tiselius had very limited resolving power (comparable to ultracentrifuge methods). At present, electrophoresis has

From: *Protein Biotechnology* • F. Franks, ed. © 1993 The Humana Press Inc.

evolved to an array of methods capable of very high resolution, some of which exceed the resolution afforded by HPLC.

Many of the electrophoretic procedures possess the advantage of operating while providing an environment that avoids denaturation of proteins and preserves biological activity. Electrophoretic techniques are equally adapted to the separation of denatured species.

2. Basis of Separations

Consider a conduit (such as a tube) filled with an electrolyte, and connected to reservoirs at either end. These reservoirs contain electrodes. When an ionizable solute is placed in an electrically conductive solution and an electrical field is imposed, the solute will migrate toward the oppositely charged electrode. The rate of this migration is a direct function of the voltage gradient and the effective charge of the solute. Mobility is inversely proportional to the hydraulic drag of the solute. The velocity in free solution as a first approximation is expressed in Eq. (1):

$$V = Eq/x\,6\pi\,r\,\mu \tag{1}$$

Here, E is the voltage difference between electrodes and x is the distance between the electrodes. $\cap$ is the charge on the solute, r is its effective hydraulic radius, and W is the electrolyte viscosity.

Proteins and peptides are ampholytes, possessing both anionic and cationic ionizable groups. Their net charge, hence mobility, are a function of the surrounding pH. A protein can possess either net positive or negative charge, depending on pH. Electrophoretic migration can thus be made to migrate at varying rates, and toward either the anode or cathode as a function of pH.

Ionic strength also has a marked effect in electrophoresis. Increasing ionic strength of the electrolyte reduces the net effective charge of a protein. Generally, the mobility of a charged particle is inversely proportional to the square root of the ionic strength.

An electrophoretic medium thus must possess a controlled ionic strength, conductivity, and pH. A wide variety of buffer systems has been developed to meet the needs of electrophoretic separation. Generally speaking, buffers of low ionic strength yield higher rates of separation, whereas high strengths will produce somewhat sharper zones of separation.

The passage of current during electrophoresis generates heat. This will give rise to convective mixing, destroying the separation. Because of convective mixing, liquid electrolytes are practical only in certain situations, and most electrophoretic procedures are carried out in anticonvective media such as gels. These gels are composed of porous, hydrophilic polymer matrices. These gels are mechanical solids, but electrolytes and solutes can diffuse and move through their molecular porous structure. Conceptually, one can regard the gel as a fixed lattice of pores of macromolecular dimension. Liquids and solutes are free to flow or migrate through this lattice in response to hydraulic or other energy gradients. Diffusive processes can similarly occur. All of these mass transport processes are mediated by the gel lattice.

In many situations the gel is expected to serve solely as an anticonvective agent. Where molecular size separations are desired, the formulation of a gel is selected that will provide retardation of analytes within a desired molecular size range.

If samples are insoluble or tend to aggregation in the electrophoretic buffer system, adding solubilizing agents must be incorporated in the sample solvent and gel buffer. Compounds such as urea (up to $8M$), glycerol, nonionic detergents (Triton X-100, Brij 35), and ionic detergents (SDS) are some of the commonly employed agents. Urea slowly forms cyanates, which can react with proteins to form spurious bands. To avoid this problem,

1. Remove cyanate from urea solutions by ion exchange;
2. Preelectrophorese the gel before adding sample; or
3. Use tris buffers that react with cyanate and protect proteins.

3. Media

A variety of electrophoretic media are employed. Because of the need to prevent convective mixing, most procedures utilize solid media. Mainly cast hydrogels are used. Alternative media include gel beads (such as are used in chromatography), paper, and microporous membranes.

The prime function of the media is anticonvective. In addition, media should be:

- Inert.
- Uniform, possessing no physical or electrical irregularities in use.
- Reproducible from experiment to experiment.
- Shelf stable.
- Stable during electrophoretic, fixation, staining procedures.
- Devoid of undesired interactions with analytes.
- Capable of mechanical manipulation and handling.
- Noninterfering with visualization/quantitation of analytes.

3.1. Electroendosmosis

Electroendosmosis (EEO) is another important characteristic of media. Consider a gel or solid surface in the electrophoretic field that possesses ionizable sites. Such fixed ions give rise to mobile counter-ions in their immediate vicinity. These mobile counter-ions will migrate in the voltage field; and in so doing, they create a hydraulic drag, carrying all liquid components through the gel lattice. Most of the gel and solid media used will exhibit a net negative charge, giving rise to counter-ions that migrate toward the cathode. In most electrophoretic procedures is an unwelcome side-effect, but there are capillary and immunoelectrophoretic methods that rely on controlled EEO to provide the desired migration effects. Just as the mobility of a charged molecule is a direct function of the voltage gradient, so is EEO.

Uncharged molecules are used to measure EEO, materials such as urea, dextran, sucrose, and deoxyribose being employed. Blue dextran is frequently used as it can be directly visualized. To measure the amount of EEO in a system, prepare a sample consisting of the uncharged solute and one of known electrophoretic mobility. Allow electrophoresis to proceed, and measure the position of both species from the origin. The EEO is calculated as follows:

$$EEO = X_{EEO}/(X_{EEO} + X_{MIG}) \qquad (2)$$

Where X_{EEO} and X_{MIG} are the (absolute) migration distances of the two species. Note that the molecular weight of the species used must be low enough that there is no migration hinderance occasioned by the gel matrix. In high concentration gels, use low molecular weight solutes such as urea or sucrose and a dye such as Bromphenol blue.

3.2. Liquid Systems

Liquid systems have been largely superseded by gels. They are employed in capillary electrophoresis (where convection is suppressed by virtue of capillary dimensions), and in some methods that employ vertical electrophoresis, tubes and density gradient liquid compositions.

3.3. Gels—Agarose

A variety of natural and synthetic hydrogel materials have been used in electrophoretic separations. The principal materials presently used are agarose and polyacrylamide.

Agarose, a purified polysaccharide derived from seaweed, has many properties that commend its use. Formulation and gel casting are simple. It has good mechanical properties. A 1% agarose gel is quite strong and able to withstand handling. It is also nontoxic. It is excellent for the migration of very large protein molecules, but it cannot sieve proteins. It also possesses some EEO and generally is not preferred for isoelectric focusing. There are, however, specially purified agaroses that can be used for focusing.

Agarose solubilizes at 90–100°C, and gels when cooled to approx 60°C. To form a gel, heated agarose solution and buffer components are mixed, cast on a plate or in a form, and allowed to cool.

Agarose is a preferred medium for most immunoelectrophoretic methods. The large pore size of agarose permits diffusion of large molecules. Large proteins with hydrodynamic radius greater than 5–10 nm are readily accommodated.

3.4. Gels—Acrylamide

Polyacrylamide gels dominate electrophoretic techniques for peptides and proteins. They are formed by polymerization of acrylamide monomers of structure:

$$CH_2 = CH - CO - NH_2$$

Acrylamide of itself will form a linear water-soluble polymer. To form a gel, it is admixed with the crosslinker N,N'-methylene-bis-acrylamide before polymerization:

$$CH_2 = CH - CO - NH - CO - CH = CH_2$$

The result is a transparent three-dimensional structure. Solution concentrations used for polymerization determine the gel molecular pore size, elasticity, and mechanical strength.

Gels are generated by compounding all components plus buffer. Catalyst is added just prior to casting. Acrylamide gels are formed by free radical reaction. Ammonium persulfate is used to produce free-oxygen radicals via a base-catalyzed mechanism. The catalysts frequently used are N,N,N,N''-tetramethylethylene-diamine(TEMED) or 3-dimethylaminopropionitrile (DMAPN). Alternately, one may generate free radicals photochemically by using small amounts of riboflavin in conjunction with TEMED.

You must deaerate these systems, because O_2 inhibits polymerization; and you want to avoid bubble formation in gel. Acrylamide monomer is neurotoxic and requires careful lab handling procedures, both of unreacted monomer and of finished gels (unreacted monomer will be found in gels). Unpolymerized monomer and catalysts can react with samples, and one may have to preelectrophorese a gel to remove undesired components.

3.4.1. Pore Size Effects

One controls gel pore size and exclusion properties by monomer concentration, and secondarily crosslinker concentration. Concentrations as high as 15–20% can be cast, which will exhibit exclusion properties for peptides of several thousand Daltons size. Below1% gel acrylamides lack sufficient gel strength to be practical in most procedures. Composites of agarose-acrylamide gels can extend capability to higher mw ranges.

3.4.2. Gradient Gels

A gel of given concentration will have some limited range of utility for molecular sieving. Molecules larger than this limit will not exhibit significant retardation differences, whereas small molecules move through the gel without retardation. To extend the dynamic range of sieving, and to manipulate separation of closely migrating species, one can employ gel gradients. These are gels of increasing acrylamide concentration in the direction of electrophoretic migration. They are formed by a flowing a mixture of low-T (acrylamide concentration) and high-T into the bottom of a casting form. Relative proportions are continually varied as the gel solution flows into the casting form, resulting in a continuous gel concentration gradient in the polymerized gel. Gradients can be linear, concave, or convex; but linear gradients are generally sufficient for most work.

4. Apparatus and Equipment

Equipment for performance of the electrophoretic separation is relatively simple. A major piece is the electrophoresis chamber itself, and a power supply. Most other elements used are common laboratory items (syringes, mixers, trays).

Qualitative viewing of finished patterns requires simple transilluminators and/or dark field viewers. In some situations, UV cabinets are used to view fluorescence patterns.

Quantitation of fixed patterns calls for densitometers, which have grown increasingly sophisticated. In its most basic form, a densitometer is simply a spectrophotometer with a narrow sample aperture. The electrophoresis pattern is driven past this aperture, illuminating the separated bands and recording their adsorption at the detector.

The advent of 2-dimensional electrophoresis methods has occasioned the need for densitometers that can record high resolution information from a complex 2-dimensional array. Data acquisition rates, total data storage, and dynamic range of these systems has given rise to a different class of equipment (discussed in the section covering 2-D electrophoresis).

4.1. Electrophoresis Chambers

Most gel electrophoresis takes place in one of three formats: horizontal or vertical rectangular slabs and vertical tubes. Slab dimensions range from a few centimeters on a side to gels that are

25–30 cm/side. Slab thickness ranges from submillimeter to a few millimeters. With tubes diameters, 1 mm to 1 cm are employed, with lengths of 10–30 cm.

The vertical forms remain encased in their casting forms, and samples must be applied at the upper end that contacts the electrode buffer chamber. In some methods, slots or wells are formed into the gel, and samples are injected into these wells. Agarose or acrylamide may be incorporated into the sample to prevent dispersion. If the electrode buffer can alter the sample, a cushion layer of gel buffer is interposed between the sample and electrode buffer.

Horizontal slab gels have the benefit of accessibility of the entire upper gel surface. This provides more flexibility in sample application procedures. Connection to electrode compartments is made with blotter paper wicks, or with gel blocks.

All gel systems will carry a significant amount of current during the separation. This results in a significant amount of electrolysis at anode and cathode, with corresponding oxidation and reduction byproducts. Electrode chambers must be large enough to buffer the running gel region from these effects.

Gel systems should be uniform in their migration characteristics. This requires careful control of gel and buffer formulations, uniformly held casting conditions, and control of heat generation.

The control of ohmic heating in gel electrophoresis is poorly addressed in most apparatus. It is not uncommon for slab gels to carry 50 watts of power. This heat must be dissipated, preferably in as uniform a fashion as possible. With no provision for heat dissipation, a gel can readily heat to the boiling point, destroying the sample, causing bubble formation, and destroying the separation. On a less catastrophic level, one should keep in mind that heat will increase the mobility of species by ≈2.4%/°C (1). With higher temperature the buffer electrical conductance increases. Both of these factors conspire in a gel with nonuniform cooling to produce distortion of the pattern, and destroying the ability to make precise measurement of mobility.

Thin slabs and small diameter tubes reduce heating, but it is highly desirable to maintain a uniform temperature throughout the electrical conductor (the gel) during electrophoresis, and to dissipate the heat generated. Vertical gel tube and slab systems that rely on air cooling do this poorly. By immersing the tube/slab in the lower buffer, one can provide better heat removal, particularly if there is some mechanical circulation in this chamber. This transferred energy will cause a continual heat buildup in the electrode buffer, and such systems also incorporate some form of heat exchanger. Many of these are inadequate plastic tubes connected to a cold water bath circulator.

Heat transfer of glass and plastics is quite poor. Gel slabs on tubes are commonly formed in glass. (One can determine a significantly higher temperature in a gel than in the adjacent cooling media.) Because of electrical conductivity, metal heat transfer surfaces cannot be used. Fortunately, ceramics such as aluminum oxide possess both good thermal conductivity and electrical insulating capability. In some of the better apparatus, ceramic heat transfer surfaces are used. One can also replace one of the glass plates with a ceramic plate to effect uniform heat transfer out of gels. By proper use of heat transfer, one can achieve uniform electrophoretic patterns at high power loads.

One of the benefits of horizontal slab equipment is the ability to provide a good heat transfer surface on which the gel slab rests. This surface in turn is connected to a thermoelectric cooling system. The result is the ability to maintain uniform, selected gel temperature during electrophoresis.

4.2. Power Supplies

Zone electrophoresis power supplies must produce potentials of several hundred volts and wattages of 50–100 watts/slab for large slab systems. Ripple is not critical in electrophoresis of proteins and a modest filtering from line frequencies suffices. The voltages and currents employed in electrophoresis can be lethal, and power and ground fault interrupts must be incorporated into the cell/power supply system. Generally, these power supplies provide both (selectable) constant voltage and constant current regulation. These features address the above issue of ohmic heating. For discontinuous buffer systems, it may be preferable to use a constant power regulated supply.

For isoelectric focusing, several thousand volts may be employed, but at much lower currents. For this reason, focusing power supplies are used separately from those used for zone electrophoresis. Controls on the focusing power supplies used for two-dimensional isoelectric focusing permit one to program a complex control of voltage and power during the course of the focusing run. (There are major shifts in conductivity during establishment of a focus gradient.)

The recent rise of capillary electrophoresis has given rise to another class of power supplies. These units provide only microamp current capacity, but operate at voltages as high as 30,000 V.

5. Visualization and Quantitation

Most research electrophoretic procedures have a qualitative end. Examples:

- Is a particular protein present/absent?
- Is there discernible change in a fraction?
- Comparison of two or more materials.
- Was a particular protein expressed/repressed?
- What is the electrophoretic mobility of a protein?

In these situations, direct visualization of the chromatographic pattern can supply the desired information. An array of techniques are available to provide such information.

5.1. Staining

Many different chromophoric stains are used. Both general protein stains and protein-specific stains are useful. A dye such as amido black or Ponceau S will bind to all proteins (though not necessarily in equal mass equivalents), rendering all proteins of sufficient concentration in a separation visible. A lipid stain such as Sudan Black fixes primarily to the lipoproteins in a separation, whereas periodic acid-Schiff stains are directed to those proteins possessing sites of glycosylation.

Some widely used general protein stains include Amido Black, Xylene Brilliant G, and Coomassie Blue R250. Bands containing a fraction of a microgram can be detected with such stains. Stains can frequently be used in conjunction with other subsequent methods. One can find a protein in a gel by a rapid stain procedure; then proceed to excise and recover that material or take other appropriate steps.

For most situations proteins should be fixed, since staining-destaining can take a number of hours. Diffusion of unfixed proteins will cause both loss of resolution and loss of material as the protein diffuses out of the gel. Proteins are commonly fixed with acids, such as trichloroacetic acid (TCA), sulfosalicyclic acid, or acetic acid-methanol. Alternately, materials such as formaldehyde can be used as fixatives. Many stain formulations are combination fixative-stain.

Fluorescent stains are useful. They are quick, (no fixation) but not as sensitive. o-Phthaldialdehyde (OPA) can be used to visualize proteins in a separation. One can use anilonaphthalene sulfonate (ANS) without denaturing enzymes.

The colloidal metallic stains (silver and gold) are at least two orders of magnitude more sensitive than the dyes (2) (they approach the detection levels of radiolabeled proteins). A dye stained separation, if processed with one of these metallic stains, will frequently demonstrate a number of low concentration bands that escaped detection by the dye stain. Colloidal silver is often used in procedures such as 2-d gel separations. The mechanism of staining remains unclear. Silver ions apparently bind to protein zones, and then serve as nucleation sites for deposition of more silver. This is a surface reaction. (Look through a silver stained gel and tilt the gel obliquely to your line of view. Two stain spots will be observed for each fraction, owing to silver deposition on each surface of the gel.) With these colloidal stains, the degree of staining is time dependent. By increasing staining time, low concentration materials can be brought up; but a background color will develop. One can destain overstained silver gels with photographic reducers. Reliable commercial kits are available for silver and gold stains.

The metal stains are extremely sensitive, and meticulous attention must be paid to equipment preparation and procedures to avoid nonspecific metal staining.

5.2. Destaining

A whole gel takes up stain and imbibes the stain, which must be removed before the proteins can be seen. Generally speaking, there are two methods for stain removal: wash out gels or electrophoretically destain them. Where gels are formed on an adherent

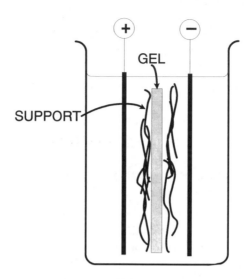

Fig. 1. Electrophoretic destaining of gels.

medium, press drying can considerably accelerate staining and destaining. The gels are first fixed to immobilize protein, and then pressed against an adsorbent medium, either by mechanical pressure or by vacuum. Most of the liquid is removed in this process, and a brief warm air drying process completes the collapse of the gel to a very thin film. Press-dried gels permit very fast stain uptake and destaining.

With unpressed gels, staining and destaining procedures generally take several hours. Rocker table agitation is used, along with multiple changes of the destaining solutions. Dye adsorption with a charcoal canister is sometimes employed. Dyes may be weakly bound, in which case one must watch out for excessive destaining.

Destaining can take several hours, because unbound dye is removed from the gel by diffusion, the only driving force being the concentration gradient.

Electrophoretic destaining greatly improves the situation. In general, a gel is held in a vessel that has electrodes parallel to the gel faces (Fig. 1). When a voltage is applied, the dye will rapidly migrate out of the gel in the electric field. The power supply requirement for destaining is low voltage, but high current (owing to the large cross-sectional area of the gel). The procedure is rapid enough that a cooling source does not need to be part of the system. Take care not to run too long and remove stain fixed to protein bands.

5.3. Radiolabeling

Radiolabeling is the most sensitive detection technique, but it is time-consuming and laborious. It is generally carried out by labeling the sample before electrophoresis with appropriate radiolabels. This labeling can often be effected without changing biological activity or electrophoretic mobility. Radio-tagging is used both for localization and quantitation of small amounts of material. Measurement techniques include scintillation counting, autoradiography, and fluorography.

Scintillation counting is performed by cutting bands of interest from a gel, drying the gel on a piece of filter paper, placing in a counting vial with a scintillant, and counting. For low energy tags, gels can be solubilized or combusted.

Autoradiography can be used both to visualize and quantitate an electrophoretic pattern. It is highly sensitive and possesses a broad dynamic range. Isotopes of C, P, S, Fe, Cu, and I, are commonly used. With the high energy isotopes, one can simply cover the gel with a thin sheet of polyethylene, contact with X-ray film, expose and develop. ^{14}C Gels are commonly dried before film exposure. One can use several film exposures on the same gel to accommodate a wide dynamic range of concentrations. Developed films are scanned with densitometers. For calibration standards, wedges are made by incorporating known quantities of tagged material into adjacent gel strips, and placing one of these wedges next to a sample gel before film exposure.

Fluorography is also used to increase sensitivity of autoradiographs. It is particularly useful for a low energy emitter such as ^3H. The gel is impregnated with a scintillator before exposing to X-ray film. The procedure is laborious and entails the handling of hazardous chemicals such as dimethyl sulfoxide (DMSO) and 2,5-diphenyloxazole. Recently available liquid fluor preparations (EN³HANCE\, New England Nuclear Corp.) simplifies fluorographic procedures and reduces laboratory hazards.

More recently introduced two-dimensional array counting equipment permits direct visualization of a pattern of radioactivity in a gel, without the need for autoradiography, or cutting of gels. Some resolution is lost with such equipment, and autoradiography remains the highest performance method.

5.4. Conjugate Reagents

Enzyme conjugates can provide sensitivity equivalent to radioassay procedures. They are most commonly employed in immunochemical procedures, and are described in that section of this Chapter.

5.5. Densitometry

Densitometry is used for quantitation of an electrophoretic pattern. It is used both on stained and unstained (UV) proteins. One can use spectrophotometers with scanning accessories for densitometry, but the resolution achievable in gels by current methods may require high resolution densitometers. Generally, slabs provide higher scan resolution than do tube gels. The thinner slab permits better optical resolution, without the cylindrical lens effects generated by tube gels.

The issues associated with densitometry of electrophoretograms include:

- Nonlinear takeup of dyes by various proteins.
- Departure from Beer's law of optical adsorption.
- Differential adsorption by different proteins: Measuring the area under one peak and dividing by total area under all peaks can give very misleading results. One must employ standards and standards curves ala HPLC. Alternately, one can include internal standard into an electrophoresis sample. This adjusts for staining variability, gel-to-gel.

As mentioned above, one can UV scan unstained gels. A caution: below 250 nm, gel materials will adsorb UV radiation. Buffers can also exhibit adsorption. One can also scan unstained gels by utilizing fluorescence of tryptophan residues (excitation = 280 nm, emission = 340 nm). Gels to be scanned by UV or fluorescence must be run in quartz tubes.

Other quantitation methods include spectrophotometry of dye eluted from a gel. Cut bands from the gels, chop, elute (Amido Black—use basic methanol solution. Coomassie Blue R250: Use 25% dioxane or pyridine (3) and measure dye adsorption. Be aware that this procedure measures the amount of dye, not the amount of protein. One must develop dye uptake calibration curves for precise determination of specific proteins.

6. Separation Modes

6.1. Homogeneous Gel Systems

In homogeneous systems, there is a uniform composition of buffer in the electrode chambers and across the gel. Protein samples may be native or denatured, and migration velocity of each species is dependent on its charge and effective molecular radius. Resolution of components is highly dependent on achieving a sample application that is not dispersed along the migration axis.

In vertical gel systems, the electrode buffers are loaded such that buffer is in direct contact with the top of the gel (the application site). The sample is applied by inserting a microsyringe through the buffer and to the top of the gel, and injecting sample. To prevent mixing, it is advisable to increase the liquid density of the sample by premixing the sample with sucrose, urea, or glycerol. A tracking dye (such as Bromphenol Blue for alkaline buffers, methylene green for acid buffers) is also added to the sample. The tracking dye migrates ahead of the sample components, and provides a visual indication of when to stop the run. One also uses tracking dye position to calculate relative mobilities of sample components.

When horizontal gels are used, sample application is made either via small troughs formed during gel casting, or by using narrow paper wicks soaked with sample.

Homogeneous gels systems are generally simpler to prepare and run than discontinuous systems. Their use is favored in the following situations:

- Simple sample mixtures not requiring the highest resolution.
- Single component sample comparisons.

- Separations where pH must be maintained rigorously so as to retain bioactivity.
- Experiments requiring prerunning of gels (to remove catalysts, and so on).

    Common applications include:

1. Determining purity of a homogeneous compound (single band);
2. Molecular weight determinations; and
3. Identity-run sample in a lane next to a known and verified mobility.

6.2. Discontinuous and Gradient Systems

Ornstein (4) and Davis (5) defined a methodology that provided a quantum advance in the resolving capability of electrophoresis. Such systems incorporate step changes in composition and pH between the cathode buffer, sample application region, and the separation gel. Homogeneous gel systems are resolution limited by the practical achievable width of the sample application band. Discontinuous systems exploit mobility differences of constituents to generate transient pH and conductivity gradients. The net result is to spatially compress (along the migration axis) the various sample components, and therefore increase resolution. In regularly used form, the system consists of a "stacking" gel and a "running" gel.

The function of the stacking gel is to compress applied sample components to extremely thin bands before it enters the running gel. Proteins then migrate at rates determined by their charge and molecular size. The electrophoretic separation is fast, and the gel media hinders simple diffusion of protein molecules as they migrate. As a result, it is possible to achieve resolution comparable to that obtained in HPLC.

Gels are cast in two stages: the running gel and then the stacking gel. The stacking gel is a relatively short segment interposed between the running gel and the sample application zone. These two zones have different pH values, and the cathode buffer still another composition. The stacking gel is highly porous, but the running gel is cast at a higher solids content, to provide some sieving .

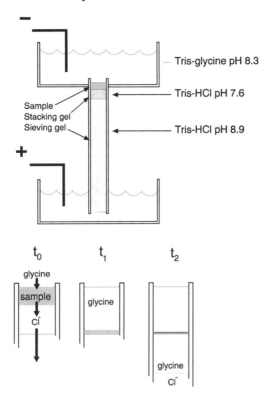

Fig. 2. Zone sharpening with discontinuous buffer systems.

Figure 2 depicts a discontinuous system. The tris-HCl buffer has a pH of 7.6 in the stacking gel, and 8.9 in the running gel. The upper electrode buffer is a trisglycine system. In this example, the electrophoretic mobilities of components are:

Chloride > sample protein > glycine

The chloride and glycinate ions are called the "leading" and "trailing" ions, respectively. Sample is applied on top of the stacking gel and voltage is applied across the system (time t_0). The leading ions, having a high mobility, start to move toward the anode. In so doing, they deplete the ionic strength, hence conductivity of the stacking gel region. This low conductivity generates a sharper voltage gradient in the stacking zone (time t_1). This causes the slower trailing ions to accelerate their migration. Sample components also accelerate, in order of their mobility. The trailing ions migrate immediately behind the leading ions, creating a very thin zone, with sample proteins compressed in this zone. Individual protein components in this boundary region are stacked into layers only a few microns in thickness. This boundary region is called the Kohlrausch boundary.

At the stacking/running gel interface, there is an abrupt change in both gel porosity and pH. When the Kohlrausch boundary reaches this interface, the higher pH in the running gel increases the ionization of the glycine. The proteins, by contrast, are retarded by sieving effects. The net result is that the trailing ions move through the tightly stacked protein layers and migrate immediately behind the leading ions (a Kohlrausch front—time t_2) through the running gel. Sample components move more slowly through the running gel; migration rates are determined by charge, size, and shape. With sodium dodecyl sulfate (SDS), denatured proteins migration rate is effectively just based on size.

This process is, in fact, a form of isotachaphoresis. Although the above description calls out a gel for the stacking region, the stacking zone can be a liquid with the sample admixed. A tracking dye is commonly included, either with the sample or in the trailing ion buffer, and will migrate at the Kohlrausch front. When the tracking dye is seen at the anode end of the gel, the electrophoresis run is terminated. Systems may be designed for selective stacking by selecting a trailing ion of proper mobility.

Many electrophoresis buffer systems using this principle have been described in literature. It is most convenient to use vertical gels, either slabs or tubes. The elimination of gel for the stacking region simplifies preparation, although there is some loss in resolution. Most systems incorporate cooling. Because of varying conductivity during the run, constant power supplies are generally preferred.

The major benefits of such discontinuous systems are very high resolution and the ability to apply large volumes of dilute sample without narrow application zone constraints. Balanced against this is the increased complexity of preparing such systems.

Common applications include:

- Study of microheterogeneity and compositional variations
- Chemical, enzymatic, or genetic modifications of proteins
- Distinguishing of conformational isomeric forms
- Resolution of isoenzymes
- General: closely similar protein species

6.3. Molecular Weight Determinations

Separation is based on both size and charge, but one can eliminate charge differences by the use of SDS. The detergent coats protein molecules, and results in a constant charge/mass ratio for the various sample components. Molecular weight (mw) can thus be determined from the mobility of the various proteins in a sieving gel. When running samples in slab, one or more lanes are reserved for known mw standards prepared in a similar fashion. After electrophoresis, sample and standards migration distances are measured. The migration distances are plotted against log (mw) of standards for a calibration curve. Finally, the molecular weight of sample components are determined, based on their migration distances.

In a typical procedure, treat sample protein with 1.4 g SDS/g protein. This denatures proteins, providing them all with a complete coating of the SDS. Note that you must also reduce disulfide bonds (2-mercaptoethanol is commonly used) to eliminate conformational effects.

For standards, one can either make a mixture of known proteins, or use a standards kit from a supplier. The table below lists various proteins that are used as molecular weight standards.

Molecular Weights of Selected Proteins (6)

Bacitracin	1480
Glucagon	3500
Insulin(reduced)	6600
Trypsin inhibitor (lima bean)	9000
Cytochrome C (muscle)	11,700
α–Lactalbumin	14,176
Lysozyme	14,314
Ribonuclease B	14,700

Hemoglobin	15,500
Avidin	16,000
Myoglobin	17,200
β-Lactoglobin	18,363
Soybean trypsin inhibitor	20,095
Trypsin	23,300
Chymotrypsinogen	25,666
Carbonic anhydrase B	28,739
Carboxypeptidase A	34,409
Pepsin	34,700
Lactate dehydrogenase	36,180
Aldolase (muscle)	38,994
Alcohol dehydrogenase (liver)	39,805
Enolase (muscle)	41,000
Ovalbumin	43,000
Fumarase (muscle)	49,000
IgG heavy chain	50,000
Glutamate dehydrogenase (muscle)	53,000
Pyruvate kinase (muscle)	57,000
Catalase (liver)	57,500
Albumin (bovine serum)	66,290
Transferrin	76,000
Plasminogen	81,000
Lactoperoxidase	93,000
Phosphorylase (muscle)	100,000
Ceruloplasmin	124,000
β-galactosidase (*E. coli*)	130,000
Serum albumin dimer	132,580
IgG (unreduced)	150,000
IgA (unreduced)	160,000
α_2-macroglobulin (reduced)	190,000
Myosin (heavy chain)	220,000
Thyroglobulin	335,000
α_2-macroglobulin (unreduced)	380,000
IgM (unreduced)	950,000

Electrophoretic procedures conform to the methodologies previously described. Multiphasic buffer systems can be employed for higher resolution. It is advisable to pretreat proteins with 2-mercaptoethanol to ensure consistent binding of SDS. Any proteolytic enzymes must be inactivated in the pretreatment process.

Gel concentrations used will depend on the molecular weight range of the sample. Suggested concentrations (7) are shown below.

Gel Concentrations for Molecular Weight Determinations

Protein mw range	Gel concentration
10,000–60,000	15%
10,000–100,000	10%
20,000–350,000	5%

After electrophoresis, proteins are visualized by staining, fluography, or autoradiography. The use of SDS in gels interferes with Amido Black, and Coomassie Blue R250 is the most commonly used stain.

Molecular weight is related to protein mobility as follows:

$$\log mw = a \log T' + b$$

Here mw is the molecular weight and T' is the concentration of the acrylamide. The values of the slope (a) and intercept (b) are determined from a plot of the protein standards. At the lower molecular weight range for the suggested gel concentrations, the mobilities will depart from this log-linear relationship. Below 6000 Daltons, all proteins will tend to migrate at the same velocity, regardless of gel concentration. The addition of 8M urea and the use of more crosslinker for the gel can improve results for small molecules.

Highly basic proteins tend to precipitate in the presence of SDS. Also highly acidic proteins may exhibit anomalous migration characteristics. The use of cationic detergents such as cetyltrimethylammonium bromide (CTAB) or N-Cetylpyridinium chloride (CPC) can be used to address such solubility problems.

An alternate approach to molecular weight determination is the use of Ferguson (8,9) plots, generated by measurements of mobility in gels of varying monomer concentration. A protein will exhibit mobility relative to gel concentration in the following manner:

$$\log R_f = \log Y_0 - K_R T$$

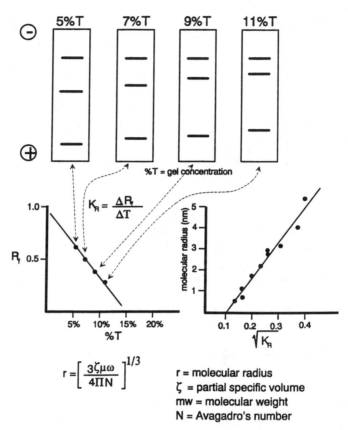

Fig. 3. Determination of molecular weights by electrophoretic mobility.

Here R_f is the sample relative mobility (ratio of mobility of the protein to the mobility of a tracking dye). Y_0 is the relative free mobility (extrapolated mobility in a gel of zero concentration). T is the gel monomer concentration, and K_R is the retardation coefficient of the protein. In practice, one runs a sample in acrylamide gels of several concentrations, and determines relative mobilities for each gel. The mobilities are then plotted against concentration as per Fig. 3. The retardation coefficient, K_R, is the slope of such a plot. This determined retardation coefficient is a linear function of molecular weight of the sample protein, and by plotting sample coefficient against coefficients of known standards, one can obtain molecular weight determinations for samples. Using this approach, one can obtain molecular weight estimations without denaturing a protein.

6.4. Gel Gradients

With simple separations (a few components of similar molecular size) gels of selected constant composition will produce optimum resolution. Complex samples may contain a number of proteins of widely varying molecular weight. A constant composition gel may not adequately resolve all components of such a sample. In this case, a gradient gel is often used. Gradient gels are cast such that there is an increase in gel concentration along the migration path. Electrophoresis proceeds with smaller molecules moving more quickly than the larger molecules. All molecules are increasingly impeded as they migrate through the gel concentration gradient. In a properly selected gradient, the smaller molecules will not migrate off the end of a gradient gel, but rather, will migrate increasingly slowly as time proceeds. The net result is to increase the dynamic range of separation that can be accommodated, as compared to a uniform concentration gel.

As in LC, use of a gradient expands dynamic range of resolution. Gradients can be linear, convex, or concave. Gradients are preferred for complex multicomponent samples, or samples with components having a wide range of mw. In gradient gels, there occurs a transient formation of relative migration rates, after which—as all components will move more slowly into increasing gel concentration—their relative pattern remains unchanged. The use of gradients tends to simplify the experimental method. Time of migration or voltage used is not critical. Migrating components tend to pore-limit positions where they are diffusionally highly constrained, and there is little danger in having components migrate off the end of the gel if time/voltage is not carefully controlled. One also tends to get sharper zones in gradient gels. One can let gels sit unfixed for extended periods, because components migrate into pore limit regions where the diffusional mobility is highly constrained.

A gradient gel is formed by filling a gel mold with acrylamide solution of regularly varying composition.

In a linear gradient gel, the logarithm of molecular weight of a component is proportional to the logarithm of distance traveled. Data for standards and samples can be plotted on log-log coordinates to determine molecular weight. At the extremes, such plots are sigmoid, but they are linear over a broad range of molecular weight. A useful characteristic of gradient gels is their independence from protein charge differences. These charge effects of proteins are diminished as proteins reach their pore limit location in the gel.

SDS complexed samples are often used, but nonionic detergents such as Tween 80 and Triton X-100 have also been successfully employed. These solubilizing agents can avoid the denaturing effects of the ionic detergents. They cannot be used for mw determination in homogeneous gels, because the proteins still exhibit their intrinsic charge characteristics; and mobility is thus a function of both molecular weight and charge.

6.5. Isoelectric Focusing

In isoelectric focusing, fractionation is based solely on differences in net charge of analytes. The mechanism of these separations is essentially electrophoresis within a stabilized pH gradient. This is achieved as follows:

An electrophoresis system is established in which the anode solution is a strong acid and the cathode a strong base. The separation zone is charged with a mixture of special amphoteric molecules (ampholytes—generally synthetic polyaminopolycarboxylic acid oligomers (300–900 mw, containing a number of ionization sites) that will be anionic or cationic, depending on the surrounding pH. The electrode solutions create a pH gradient in the separation zone, and when voltage is applied, a negatively charged ampholyte will migrate toward the anode. As it approaches this more acidic region, carboxylic acid sites ionization will decrease, whereas the amino groups increase ionization. At some location, positive and negative ionization is exactly balanced, i.e., it is isoelectric; and migration of the ampholyte molecule ceases.

A mixture of such ampholytes of varying isoelectric points can form a continuous, stable pH gradient. The ampholytes have two principal functions in isoelectric focusing: they provide buffering of the pH gradient and they provide adequate and uniform electrical conductivity.

Early focusing experiments were done on liquid ampholyte columns. These are complex glassware systems on which first forms a density gradient. Ampholytes and sample are added to the system, voltage is applied, and focus is obtained. When the system is focused and there is no further protein migration, power is shut off, and the resulting focused system is carefully decanted, by means of a valve at the bottom of the column. Fractions are collected.

Liquid column systems are useful for preparative work. There are several problems associated with control of thermal convective mixing, gradient profile uniformity, and sample recovery in liquid columns. Total run times are long. Most separations presently are done in tube or slab gels. Note that preparative work can be done by employing thick gels. There are also gel bead systems wherein one loads gel beads with ampholyte buffers, and casts this slurry on a horizontal plate. After focusing, one can scoop out fraction areas to recover the proteins of interest. Location can be determined by performing a quick membrane blot on the gel, and staining the blot.

When isoelectric focusing is performed in cylindrical gels, the common practice is to prefocus the gradient in the gel (the upper electrode is the cathode), shut off power, and apply the sample on top of the focused gel. A spacer buffer is added to isolate the sample from the cathode electrolyte. At the end of the focusing run, the gel rod can be extruded from the tube, and processed as desired.

Horizontal slabs offer more experimental flexibility than tubes. Multiple samples can be run on a slab. For horizontal slab gels, the position of application is not critical—the sample can be added anywhere along the focused gel gradient. One way to make sample application is to apply the sample to paper wicks that are laid on the gel.

Horizontal slab gels provide an additional advantage, in that they can be run on electrophoresis chambers that provide cooling to the gel. The focusing gel is cast on a glass plate. The plate is placed on a flat platform in the electrophoresis chamber. This platform is made of a thermally conductive cooling material, and is contacted on its lower side by Peltier coolers. A cooled system will typically generate a more uniform gradient, focus up in less time, and allow somewhat higher voltage gradients. Inasmuch as the isoelectric point of a protein is temperature dependent, the use of cooling temperature control aids precision of pI determinations.

Various pH range ampholyte mixtures are commercially available. For precise work, one should not assume that a gradient is linear. One can determine a gradient profile by including known marker proteins as internal standards. The gradient can be determined after focusing by either applying pH indicator papers to the gel, or by contacting the gel with a pH microelectrode, to obtain a plot of pH as a function of location. When analyzing unknowns, use a broad (3–10) pH range gradient, then use a narrower range to resolve similar species.

Samples can be quite dilute, as the focusing process concentrates the sample components. Also, as long as the gradient is maintained by the power supply, there will be no band spreading because of diffusion.

Common applications of focusing include isoelectric point (pI) characterization of proteins. Isoelectric point data is very useful in the development of ion exchange chromatography protocols.

The high resolution achieved (one can resolve 0.001 pH differences) is based on charge. As a consequence, focusing is useful in the determination of microheterogeneity, small genetic differences, and in the resolution of isoenzymes.

One must use a nonsieving medium, commonly acrylamide gels of high porosity. Since such a low gel content acrylamide has very poor mechanical strength, special agarose can be added to provide mechanical integrity. Conventional agarose is unsatisfactory owing to the electroendosmosis possessed by agarose. There are specially prepared low EEO agarose materials available that permit formulation of acrylamide/agarose gels for focusing of large proteins.

Good resolution is favored by a low diffusion coefficient (high molecular weight) and a high mobility slope at the pI of the sample proteins. Low molecular weight peptides are more difficult to focus than proteins because of their higher diffusivity.

Recent developments in isoelectric focusing include the formulation of gels containing immobilized pH gradients (10).

These gels are polyacrylamides with carboxylic acid or tertiary amino groups (–R).

$$CH_2 = CH - \overset{\overset{\displaystyle O}{\displaystyle \|}}{C} - \overset{\overset{\displaystyle H}{\displaystyle |}}{N} - R$$

They are copolymerized into a conventional gel matrix. You form the gradient by casting a continuously varying ratio of acid and base acrylamide monomers. One of the components has its pK close to the desired pH range (buffering component), and the other is used to adjust the pH. Benefits include:

1. No ampholyte in extract.
2. No need to desalt sample.
3. Stability of the gradient permits narrow gradients; thus high resolution (0.001 pH unit).
4. Relatively higher gel loading capacity.

If sample solubility is a problem, one can use nonionic detergents. Salts will interfere with focus and must be removed from the sample by dialysis or ultrafiltration. Should this pose a solubility problem, one can add ampholytes or glycine to aid protein solubility in a low salt medium.

Focused samples can be visualized by staining. Sample proteins in the gel are first fixed, and then the ampholyte is removed. One can directly stain with Amido Black or Coomassie Blue R250.

6.6. Isotachophoresis

Isotachophoresis (ITP) is a solute stacking method utilizing the same principles employed in a multiphasic buffer PAGE separation. Buffer systems are formulated containing leading and trailing ion species, such that sample mobilities are intermediate. When electrophoresis commences, the sample solutes will be sandwiched between the leading and trailing solutes, in a narrow, high concentration zone. Mixed solutes in this zone will migrate in their order of relative mobility, but with each solute contiguous to its neighbors. By employing additional buffer salts of mobility that is intermediate between two solutes, these salts will interpose themselves between the solute pair, with the solutes forming discrete separated zones. ITP has been called displacement electrophoresis, in an analogy to dispacement chromatography.

ITP can be useful as a sample preparation method, in that it allows one to segregate relatively high concentration zones of solutes. One can readily collect such prepared analytes. The technique can be performed in a liquid system or in a gel, provided that the gel offers no sieving of the sample. The technique is useful mainly for samples containing relatively few components of low molecular weight. One must know *a priori* the mobility of the solutes, so as to select the proper spacer salts. There has been very limited adaptation of this technique by the protein chemists, because alternate methods of separation can be used for preparative work, and the analytical resolution of ITP is inferior to other electrophoretic methods.

6.7. Two-Dimensional Electrophoresis

A high resolution electrophoretic separation may separate as many as 50 components. In general, lesser resolution is achieved. Samples may contain a much wider array of proteins that are not detected/visualized because: The detection system cannot cover widely varying solute concentrations. One must overload to detect minor components. Minor components next to major bands are difficult to detect.

By using two different methods of separation, the resolving power is multiplicative rather than additive. Two-dimensional electrophoresis (2-D) uses two high resolution separations (electrofocusing and molecular weight) sequentially to obtain a rectangular 2-D map of proteins. In combination with visualization/detection methods of a broad dynamic range, 2-D can detect literally thousands of proteins present in a sample. O'Farrell (11) demonstrated 1100 resolved proteins from lysed *E. coli* cells. Several thousand proteins in prokaryotic cell lysates can be readily detected.

Two-D is conceptually quite simple. A sample is separated in a small diameter cylindrical gel by isoelectric focusing. Following focus, this gel is laid across the top (cathode end) of a vertical slab gel, and separation of the focused solutes is made based on molecular size.

Fig. 4. Two-dimensional electrophoresis pattern. Reprinted with the permission of the MILLEPORE Corp.

The result (Fig. 4) is a slab gel containing an array of protein spots that are located on a grid of isoelectic charge and molecular size.

The information obtained is:

1. Approximate value of isoelectric point and size of each protein.
2. The expression of proteins not normally expressed.
3. The absence of proteins normally expressed.
4. Changes in relative quantities of proteins.
5. A "fingerprint" of a sample in the form of a protein map.

The technique can provide enormous amounts of information regarding the constellation of proteins of a cellular process, its response to environmental change, and the genetic expression/ repression operative mechanisms of a cellular system. The complex signature pattern of a 2-D experiment can serve to type variants of microbial, plant, and animal tissue samples.

The drawbacks to use of this extremely powerful technique include:

1. Labor intensive, long experiments.
2. Demanding manual skills.
3. Rigorous control of system chemistries.
4. Formidable data processing requirements.

Regarding the last point, the method produces what can be viewed as a glut of information. In most situations, the data field is so rich as to preclude gaining information solely on the basis of visual observations of patterns. To be used effectively in most situations, 2-D patterns must be densitometrically scanned by image processing systems that can interface with resident protein databases. Consider a eukaryotic cell system that may express 2000–3000 detectable proteins, only a small portion of which have been identified. Heat stress or exposure to a toxic material may alter relative concentrations of several hundreds of these proteins, cause genetic expression to normally nonexpressed proteins, and similarly cause repression. Faced with this vast amount of information, drawing accurate and comprehensive conclusions without data processing systems would be difficult at best.

Fortunately, such scanning/data handling systems are available today. Further, one can use 2-D electrophoresis to isolate individual proteins for sequence analysis. This in turn allows one to identify the parent genome from the cellular system. Working from the other end, genetic modifications can be assessed by virtue of alterations in protein expression of a cell system.

Typical sample application is of the order of 1–10 μg. Protein stains are not sufficiently sensitive, and silver stains are preferred. Still more sensitivity is obtained from autoradiography. Silver staining is quicker and simpler than autoradiography, but provides only semiquantitative information.

Autoradiography or fluorography are more demanding in terms of time and equipment. Balanced against this, these techniques have higher sensitivity and are amenable to quantitative densitometry. By making a series of varying film exposures to a tagged gel, autoradiography allows a very broad dynamic range of quantitation of the various proteins present.

Denaturing agents are used in both dimensions of separation. Typically, urea and nonionic surfactants are used in the first dimension separation. SDS is employed for the second dimension. Frequently, one will observe a horizontal row of multiple protein spots, corresponding to subunits or microheterogeneity caused by glycosylation variations.

Applications include:

1. Universal study of cellular changes occurring at the molecular level of expression. These changes may be:
   a. Genetic translation/expression variations
   b. Disease-induced
   c. Results of therapy
   d. Cellular response to chemicals, environment
   e. Cellular differentiation
   f. Monitoring time-course changes
2. Isolation of proteins suspected of causal relation, followed by sequencing, with identification of related genes. This is a powerful tool even though identity of isolated protein is not presently established.
3. Fingerprinting of plant or animal tissues. Taxonomic study. Forensics.

6.8. Capillary Electrophoresis

Most electrophoresis methods are manual and labor intensive, requiring considerable dexterity. Capillary electrophoresis (CE) has recently developed into a mechanized and automated technique. In many respects, the characteristics of CE are comparable in operation and interpretation to HPLC.

Capillary electrophoresis, as its name implies, is carried out in small glass capillary tubes, such as are used for capillary gas chromatography. A buffer system or a gel system is drawn into a capillary 20–100 cm long with an internal diameter of 100μ or less. Sample is next drawn into the prepared capillary. When the ends of the capillary are placed in anode and cathode buffer chambers, electrophoresis will proceed. A detector (typically UV) is placed at a point along the capillary axis, and protein bands are detected as they migrate through the detector zone.

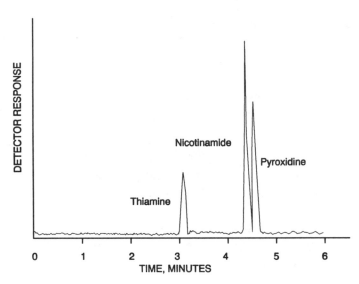

Fig. 5. Rapid separation of a vitamin mixture by capillary electrophoresis.

The small diameter of the capillaries used produces two salutary effects. They have good heat dissipation relative to the ohmic heat generated in their small cross section. As a result, very high voltage gradients can be employed.

A second factor also arises from the small diameter. Radial eddy diffusion is eliminated, even though there are radial temperature gradients.

These two factors have the following consequence: the very high voltage gradients that can be used result in very rapid electrophoretic separations, frequently several minutes (Fig. 5). The short separation times coupled with the lack of eddy diffusion results in excellent resolution. Defined in terms of theoretical plate heights, it is not uncommon to operate capillary electrophoresis systems with plate counts of several hundred thousand (Fig. 6). Separation efficiency exceeds that obtained by HPLC.

The technique lends itself well to mechanization and automation. Procedures and data handling closely mimic chromatography, and in fact, it is chromatographers that in large part have contributed to the development of CE technology and applications.

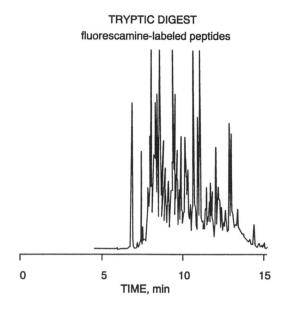

TRYPTIC DIGEST
fluorescamine-labeled peptides

Fig. 6. The very high resolving power of capillary elelctrophoresis enables separations of very complex samples.

Capillary electrophoresis is surprisingly versatile. One thinks of electrophoretic separation based on either molecule charge or size. Methods of capillary electrophoresis have been developed that are counterparts to virtually all the major modes of chromatography. Alteration of the surface characteristics of the capillary wall can modify solute migration. By the use of surfactants that form micelles, separations analogous to reverse phase chromatography can be achieved. There are various schemes of chiral separation that can be employed. The incorporation of gels in capillaries can form the basis of molecular weight separation. Electrofocusing corresponds to the control of separation by ion exchange chromatography.

Electroendosmosis is an important factor in the operation of a capillary system. Any ionization of the interior capillary surface results in free, mobile counter ions in the zone adjacent to the capillary wall. When an electric gradient is applied, these mobile counter ions migrate toward the electrode of opposite charge. In so doing, they generate a hydraulic drag, that results in a nonselective "plug flow" through the capillary. In most systems, conditions are such that electroendosmotic flow is toward the cathode, counter to the electrophoretic migration of polypeptide solutes. Various techniques of surface treatment are used to alter or modify the endosmotic migration.

Sample loading to the capillary is performed either by electroinjection or by hydraulic loading. Electroinjection is achieved by immersing one end of a filled capillary into a sample containing receptacle. An electrode is also dipped into this container, and voltage is applied across the capillary. Endosmotic flow into the capillary draws in sample. By controlling time and voltage, the amount of sample uptake can be regulated.

Hydraulic loading is achieved by immersing one end of the filled capillary into the sample, and then applying a small pressure gradient. Sample is inducted by the resulting viscous flow. The pressure gradient can be applied by pressure or suction at the ends of the capillary, or by siphoning (lowering the distal end of the capillary several centimeters below the sampling end to create a pressure gradient). Typical sample load volumes are a few nanoliters.

Various detectors are employed, the most common being UV absorbance. The capillary is stripped of its protective polymer coating and inserted into the optical path, thus becoming an optical cell. Other detection modes utilize miniaturization of fluorescence, conductivity, electrochemical redox, radioisotope detection, and refractive index. Mass detection limits are extraordinarily low, some detection schemes claiming to have attamole sensitivity. From the standpoint of solutes concentration, however, detection sensitivity is somewhat less than comparable to chromatography detectors. A UV capillary detector, for example, operates in the domain of 10^{-3}–10^{-6} molar.

The speed and mechanization of CE make it an attractive laboratory companion to chromatography. Because of differing separation characteristics, the capillary system can be quite useful in confirming that a chromatographic solute is indeed homogeneous. A capillary separation can confirm the composition of HPLC polypeptide digest fractions. The ability to work with very small fluid volumes of solutes favors CE interface to mass spectroscopy.

7. Protein Blotting

Staining methods utilize insolubilization of proteins to "fix" them in place within an electrophoresis gel. Alternately, there is an array of blotting techniques that involve transfer of proteins from gel to a microporous membrane matrix where they are immobilized, commonly by adsorption. Originally developed for analysis of nucleic acids, blotting techniques have become broadly used by protein chemists. Protein is transferred from internal bulk of gel matrix onto the surface of a membrane. The material is concentrated on the blot membrane, but preserves the geometry of the electrophoretic separation, and is readily accessible to various stains and reagents. Blotting is also a sample cleanup operation, whereby proteins transferred from a gel can be washed free of gel materials such as SDS and glycine.

The membrane materials used are pliable, and readily manipulated. Isolated protein bands or spots are readily cut out of a blot membrane. A blot sample can be stored for long periods of time. Multiple procedures can be carried out on a single membrane. Gas phase sequencing of peptides can be done directly from membrane blots. A membrane piece with isolated polypeptide is excised and placed directly in a sequencer chamber.

7.1. Blot Materials

A variety of membrane materials are used. Originally, the membranes employed were nitrocellulose, but more recently, membranes made of polyamides and polyvinylidine difluoride (PVDF) have displaced nitrocellulose. Highly cationic charged Nylon 66 membranes function well in transfers from SDS-PAGE gels. Nylon and PVDF are mechanically stronger, with better hydrolytic stability than NC. Binding mechanisms remain somewhat obscure. A combination of hydrophobic and dipole protein interactions is generally accepted as the binding mechanism. The binding capacity of membranes varies with membrane type, but is on the order of 100 $\mu g/cm^2$ of membrane. Blot immobilized proteins can be eluted from membranes with detergents, chaotropic reagents such as urea or NasCN, or with acid pH.

Low molecular weight peptides elute well from a gel, but exhibit poor binding efficiency to the blot matrix. To overcome this, blot on matrices that covalently attach to peptide. Alwine et al. *(12)* introduced diazobenzyloxymethyl (DBM) derivatized paper. Negatively charged peptides adsorb to the positively charged diazonium groups, then form covalent bonds via azo derivatives. Other chemistries include diazophenylthioether (DPT). These reactive papers demonstrate efficient capture, but have low binding capacity ($\mu g/cm^2$). Activated membrane materials, with their high specific surface, have proven to be quite useful in this regard. Membranes activated with a variety of chemistries (cyanogen bromide, diisothiocynate, arylamine) are available for covalent transfers.

7.2. Blot Transfer Mechanisms

Blot methods involve transfer of proteins from a gel to the blot matrix by convective flow, diffusion, or electrophoretic transfer. Southern *(13)* introduced the first blot transfer procedure for transferring DNA to nitrocellulose. It was perhaps inevitable that alternate developments were described as Northern (convective transfer to DBM paper), Western (electrophoretic transfer to nitrocellulose), and Eastern (convective transfer from isoelectric focusing gels) blots; thus adding levity but little clarity to the art.

All of these procedures have been adapted from DNA handling techniques to the transfer of polypeptides. Diffusive transfer is the simplest but slowest method. Here a gel is sandwiched between two blot membranes. The membranes are backed by some sheets of blotter paper or foam, and contained between two sheets of plastic or metal mesh. This "club sandwich" is immersed in a buffer for 24–48 h , allowing proteins to diffuse from the gel to the membrane matrix, where they are captured. Because of the long diffusion times, a highly resolved electrophoretic separation will show some resolution loss on the resulting blots.

The long transfer time arises primarily from the diffusion of protein molecules within the gel matrix. The Southern blot procedure considerably accelerates transfer by convective flow.

Here, a stack of materials is used, as shown in Fig. 7. A buffer saturated pad is placed on a glass plate. This pad may consist of a long strip such that the ends immerse in buffer solution, and serve as wicks.

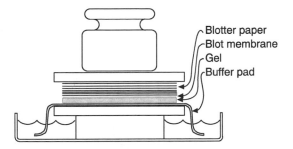

Blotter paper
Blot membrane
Gel
Buffer pad

Fig. 7. Arrangement of materials for Southern blotting.

The gel is placed on top of this pad, and a wet blot membrane is in turn placed on the gel. A stack of dry blotter papers, 2–5 cm thick is next placed on the membrane. Topping this all is another glass plate and a weight.

Capillary wetting of the dry blotter stack creates a convective flow of buffer through the gel and through the blot membrane. The result is a more rapid and efficient transfer.

Much the same result can be achieved in a vacuum drying apparatus. Here, a horizontal vacuum chamber has a porous plastic frit as its upper surface. A transfer membrane can be placed on this surface, with the gel stacked on top. Additional buffer soaked blotter papers are then overlaid. An upper seal plate or flexible silicone rubber membrane tops this stack. When a moderate vacuum is applied, liquid is expressed from the gel, and passes through the blot membrane, effecting transfer.

By virtue of speed and efficiency, electroblotting is the most widely used protein transfer technique. A variety of protocols are described in the literature, and a number of electroblotting instruments are commercially available.

Figure 8 shows the essential elements of an electroblotting apparatus. The gel is contained in a sandwich similar to the diffusion apparatus, with a blot membrane placed in contact with one gel face. Electrode grids in the chamber produce a uniform electric field normal to the gel. At voltage gradients of 5–10 V cm⁻¹, transfers can be effected in 1–3 hours time. A useful equipment is the "semidry" electroblotting unit. In these devices, the gel/blot matrix is contained between flat porous graphite plates, which serve

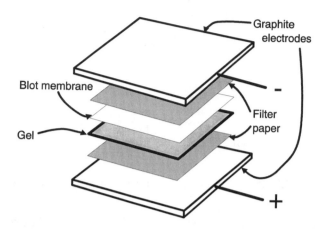

Fig. 8. Principal elements of a "semidry" electroblotting apparatus.

both as electrodes and heat sinks. Because of the short electrical path, an adequate gradient can be obtained at very low voltage, resulting in lower ohmic heating.

Ohmic heating must be controlled. Low ionic strength buffers are used, and low voltage, high current capacity power supplies are utilized. Depending on pH, gel composition, and pI of proteins, migration can be either anodic or cathodic, so care must be taken to place the blot membrane in the electrophoretic migration path of the electroeluted proteins.

With any of the blotting techniques, one should be aware of the transfer efficiency of the procedure. Transfer efficiency from gels to blot membranes is generally inversely related to protein molecular weight. Conditions that are optimized for low molecular weight peptides may result in poor transfer efficiency for high molecular weight polypeptides. Conversely, more aggressive transfer conditions may cause low molecular weight peptides to be only partially captured by the blot membrane, the rest migrating through to the anode.

Buffer composition can influence transfer and binding characteristics. Adding SDS, for example, to a gel will improve the release of protein, but will also reduce the binding efficiency of the membrane. Blot results therefore should be regarded as semiquantitative.

The addition of prestained quantitative protein standards to a blot experiment can improve the quantitative aspects of blot procedures and confirm adequate transfer.

Transfer buffers typically contain methanol (MeOH). This helps in stripping SDS from proteins transferred from SDS gels. Tris-glycine buffers are often used, but if one is to do peptide sequencing on blot transferred materials, this buffer must be completely washed from the blotted material. An alternate is the use of the CAPS (3–[cyclohexyl-amino]-1–propanesulfonic acid) for sequence samples.

Proteins on the blot membranes can be visualized and quantified by a variety of detection schemes. Materials blotted to nitrocellulose or PVDF membranes can be stained with the common protein stains such as Amido Black or Coomassie. These stains will strongly bind to the ionic transfer matrices, giving an unacceptably high background. Detection limits of Coomassie blue stained blots is approx 0.1 µg. Dye stained membrane blots can be quantitated either by elution or by direct densitometry of the membrane.

Silver and colloidal gold stains can be used with great sensitivity (low ng range) on blot membranes. Colloidal carbon, in the form of india ink, is a highly sensitive stain. In general, blot membranes present superior visualization. All of the transferred material is present as a thin monolayer on the membrane surface, maximizing the visual image. Because of the thin layer of material and lack of diffusive resistance typical of a gel, staining procedures with blot membranes take only a few minutes.

Blot membranes also provide a good substrate for autoradiography. The sample is present as a very thin layer, which can be placed in close contact with the photographic film. This not only aids sensitivity, but also increases spatial resolution when compared to the long emission paths of radio decay in a gel slab. Membrane materials rapidly take up fluors for detection of ^{3}H-labeled proteins.

7.3. Recovery of Proteins from Blots

The blot is an effective way to isolate a protein from an electrophoretic gel, to concentrate it, and to free the protein from electrophoresis matrix materials that might interfere with subsequent procedures. Further, blot isolated proteins show excellent stability, frequently retaining activity after months of storage on a blot membrane. Note that many procedures such as sequencing or immunogenic stimulation can be performed directly on a blot isolate, without elution. Cleavage of n-terminal blocked peptides can be performed directly on the blot isolate. If one wishes to elute the protein from the blot matrix, one generally uses either detergent systems or organic/aqueous solvent mixtures. By selection of optimum elution conditions one should expect recoveries on the order of 75–90%.

Selection of an elution system is dependent on the protein itself, the type of membrane used, and the subsequent use of the eluant. There are some generalizations to be observed: High molecular weight species will show lower elution efficiencies. Heating or exposure to pH extremes during blotting can reduce elution efficiency.

8. Immunomethods

Immunoelectrophoretic methods provide a very useful array of techniques for the research laboratory. Techniques are fast and very sensitive, yet have very modest laboratory requirements. One can specifically detect an antigen present in very complex samples, without any need for prepurification or isolation techniques. Both qualitative and quantitative immunoelectrophoretic procedures are available.

Quantitative immunoassays are regularly employed that are not based on electrophoretic principles. These employ various spectrophotometric and radiometric instruments. Corresponding reagent systems are commercially available. Standard immunoreagents may not be available, and the immunoelectrophoretic methods provide a versatile set of techniques requiring simple equipment.

Agarose is frequently used in immunomethods. Detection by immunochemical methods involves mass contact of antigen and antibody species. These materials must be free to migrate in the gel media. The large pore size of agarose gels permit unhindered diffusion of large molecules. Agarose gels of 0.5–1.5% gel content provide good mechanical strength with little or no size sieving.

8.1. Detection and Visualization

Immunoelectrophoretic methods rely on direct visualization of species, and quantitation by densitometric methods. Visualization of immunoreactants can be direct. Antigen–antibody complexes form a precipitate within the transparent gel that can be directly observed. Visualization is aided by dark field illumination. A dark field illuminator is a horizontal viewer box with a glass surface onto which the gel is placed. The observed field below the glass is a nonreflective black surface. Fluorescent tubes are located under the perimeter of the glass view plates, such that they cannot be seen when looking down at the view field.

When a gel is placed on the glass surface, light from the tubes enters the gel at an oblique angle. The precipitins in the gel scatter this light, and so become quite visible to the observer, whose view position is directly above the gel. The effect is entirely analogous to microscope dark field illumination. Detection limits with direct dark field illumination are on the order of 5–10 µg/mL. The use of polyethylene glycol (Carbowax 4000®) can intensify precipitins and aid visualization.

The use of protein stains will increase sensitivity by more than an order of magnitude, and a dried stained gel provides a permanent record. Staining requires that unreacted protein first be washed out of the gel. Fluorescent tags/stains can also be employed. The use of a second fluorescent antibody (reactive to the complex) avoids the need for washing of the gel, and provides good sensitivity when used with a UV transilluminator.

The washing of gels to remove unfixed protein can be greatly accelerated by the use of press-drying. Here the gel must be cast on an adherent surface. (Mylar® films, precoated with an extremely thin layer of agarose are commercially available for such adherent casting surfaces. They take advantage of the temperature hysteresis of agarose gelation/solubilization. A hot agarose solution poured onto such a surface will cause the agarose chains of the two systems to intertwine without releasing the dried gel film.) Pressing the gel as in the blotting techniques will quickly remove most of the water, along with unreacted protein. The resulting preparation is hot air dried. When soaked in aqueous systems, this collapsed film will not reswell: Hydrogen bonding of the gel chains prevents this. As a result, such press-dried preparations are rinsed, stained, and destained in a matter of minutes.

Radioimmune assays (RIA) and enzyme linked immunoassays (ELISA) can further increase detection sensitivity to the region of 10–20 ng/mL. In radioimmune methods, one can use a labeled antibody, either as a direct reactant, or as a secondary reactant to a preformed immunoconjugate. Resultant gels are press-dried, rinsed, and autoradiographed. The ELISA methods employ antibody-enzyme conjugates, and a chromophoric substrate for detection. Sensitivity is generally comparable to what can be achieved by RIA.

SDS is used in many electrophoretic procedures, primarily as a solubilization reagent for proteins. Unfortunately, SDS interferes with immunoprecipitin formation. SDS will also perturb the electrophoretic mobilities of species, and can frustrate techniques that rely on electrophoretic contact of analytes with reactant antibodies. For immunoelectrophoretic methods then, SDS should not be used. Non-ionic surfactants can be used as solubilizers, often in conjunction with moderate concentrations of urea. Where separations have been performed with SDS, one must fix the proteins in the gel and then wash out the surfactant.

8.2. Immunoblots

Immunological reagents can be employed to visualize extremely small amounts of specific proteins. A sample is electrophoretically separated and transferred to a membrane by one of the blot techniques described in a preceding section. After blot transfer, the membrane is contacted with a blocking protein to cover all non-specific protein adsorption sites. Following blocking, the membrane is contacted with an antibody specific for the protein(s) of interest. The membrane is rinsed and contacted with an antibody conjugate directed against the first reagent antibody. Using enzyme conjugates or avidin–biotin systems, as little as 100 pg of material is readily detected. Nespolo et al. (14) achieved detection sensitivity for transferred bands of >1 pg by employing streptavidin colloidal gold conjugate and silver enhancement.

The use of radioimmune assay reagents extends sensitivity still further. The blot transfer methods are fast and sensitive. Additionally, they exhibit one other benefit relative to direct immunofixation of protein in the electrophoresis gel. Reagents such as SDS present in the electrophoresis buffer can inhibit some of the enzyme visualization systems. A protein transferred to a blot membrane can readily be washed free of such interfering substances.

An alternate to dealing with radiolabels is to use the various enzyme labeled conjugates, peroxidase and alkaline phosphatase systems being the most commonly employed. Sensitivities of such systems can rival the radiolabel assays.

In addition to immune reagents, other systems can be used effectively on blot proteins. Glycoproteins, for example, can be visualized by contact of the blot with the appropriate lectins, followed by conjugate visualization reagents. With the use of specific lectins and glycosidases, detailed information about glycoprotein structure can be obtained (15).

8.3. Immunodiffusion

A useful starting point for understanding immunoelectrophoretic methods is a description of the predecessor, immunodiffusion. The immunodiffusion system as developed by Ouchterlony (16) is illustrative. Here a gel is cast into a small Petri dish and circular wells are punched into the gel, a few millimeters apart. If an antigen is placed in one well and its antibody in the other, these species will each start to diffuse from its well of origin. In so doing, they each generate a concentration gradient, centered on their well of origin.

As the diffusion boundaries overlap and the molecular species come into contact, an antigen–antibody complex will form. Such complexes are soluble when either antigen or antibody is in marked excess. At the proper combining ratios, however, the complex is a molecular aggregate of very high molecular weight. It is insoluble, and immobile in the gel. At some region of the gel between the two wells, therefore, a precipitin gel forms as a visible line. Its position relative to the two wells is a function of the relative amounts (hence concentration profiles) of antigen and antibody. At a high antigen/antibody ratio, the precipitin line will form closer to the antibody well, and vice versa.

By forming patterns of multiple wells in a gel, one can determine identity or nonidentity of the immunodeterminant sites of samples. A common pattern is a hexagonal array of wells, with a sample well in the middle of the hexagon. An antigen can be placed in the center well with various sample antibodies at the hexagonal vertices. Each sample well containing an antibody that reacts with epitope sites of the antigen will form a line somewhere between that sample well and the antigen well. If two adjacent antibodies react with identical sites, their two precipitin lines will form a continuous fused juncture (identity reactions). By contrast, antibodies each reacting with the antigen, but at different epitopes, will form two independent precipitin lines, crossed through each other (nonidentity).

One can thus readily screen materials and by employing strategies of well patterns and sample locations, deduce a great deal of the immunochemical characteristics of the materials.

An analogous quantitative technique is radial immunodiffusion, as developed by Mancini (17). Here, a gel containing a determined antibody concentration is cast. Single wells are cut into the gel. Antigenic samples are placed into these wells. As antigen diffuses outward from a well, it forms a radial concentration gradient. A precipitin will form as a circle at the equivalence concentration of antigen. Measuring the diameter of such precipitin circles provides a ready quantitation of the antigen sample.

8.4. Immunoelectrophoresis Methods

The immunodiffusion methods are simple and flexible, but they are slow, and do not have the ability to separate species. The various immunoelectrophoretic methods make use of electric fields to position antigens and antibodies for reaction.

This section covers several of the most frequently employed immunoelectrophoretic methods. There are a variety of less frequently used additional methods (fused rockets, tandem crossed iep, line iep, rocket-line iep) omitted here for brevity.

8.4.1. Immunoelectrophoresis

Grabar and Williams (18,19) developed the basic method of immunoelectrophoresis (IEP) whereby a mixture of proteins were first separated by zone electrophoresis, and then allowed to form immunoprecipitins by diffusion against nearby antibodies. The technique permitted diagnosis of gammopathies and a number of other plasma protein abnormalities.

IEP is commonly performed in an agarose horizontal slab gel. A series of cylindrical sample wells are punched in the gel, about 1/3 of the way from the cathodic end of the gel. A set of troughs are cut into the gel, the troughs running parallel to the direction of electrophoretic migration. The troughs are spaced midway between the sample wells.

As a first step, samples are placed in the wells and migrated by electrophoresis. The gel is then removed from the electrophoresis chamber. Next, specific and/or polyvalent antisera are dispensed into the troughs and the gel is placed in an incubation chamber for a 12–24-h period.

During this diffusion process, the individual proteins radially diffuse from their positions in the gel, while the antibodies diffuse as linear fronts from the troughs. The result is a set of complex overlapping concentration gradients of antigens and antibodies. Precipitin arcs form for each antigen/antibody combination (Fig. 9). These arcs can be directly visualized or stained. The intensity and position of each arc reflects the relative quantities of species.

By virtue of the electrophoretic separation of sample proteins, one can obtain immunoidentity information from much more complex samples than are accommodated by immunodiffusion. Multiple species can be analyzed simultaneously. It is a useful control procedure for determining the homogeneity of an antibody.

IEP is flexible and requires simple equipment. It is probably universally used in immunochemistry laboratories. IEP is not without shortcomings. These include:

1. Long diffusion times are required to form precipitins.
2. IEP is at best semiquantitative.
3. The user must develop some interpretive skills.

8.4.2. Rockets

The technique of rocket electrophoresis was developed as a noninstrumental method for rapid quantitation of antigens (20–23). Rockets use a slab gel with a series of sample wells placed across the cathode end of the gel. The gel is formed with a

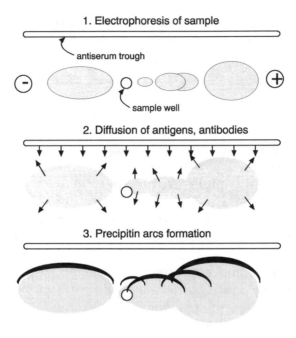

1. Electrophoresis of sample

antiserum trough

sample well

2. Diffusion of antigens, antibodies

3. Precipitin arcs formation

Fig. 9. Steps in the technique of immunoelectrophoresis.

usual buffer to which has been added a specific antibody. Buffer pH is selected to minimize electrophoretic/endosmotic migration of the antibody.

When samples are run in the gel, they migrate toward the anode; and in so doing, sweep through the field of antibody. The result is a triangular precipitin pattern (Fig. 10) originating at the sample well and tapering to a point (hence the name "rocket"). The higher the sample antigen content, the longer the rocket. This is a fast and sensitive method for quantitative assay of an antigen. Stained rockets can be used for quantities as little as 0.1 mg/L. By using ELISA reagents of autoradiography, one can decrease the working range 10–100X.

8.4.3. Crossed Immunoelectrophoresis

This is a variant of the rocket technique for the simultaneous semiquantitative assay of mixed antigens in a sample. In crossed IEP, a gel is formulated with a polyvalent antiserum. The sample is first run in a conventional gel. A strip of this gel containing the

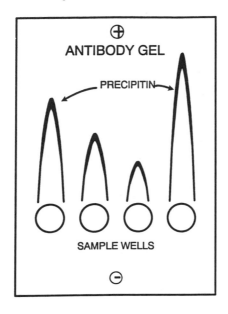

Fig. 10. "Laurell rockets" for immunoquantitation of antigens.

separated sample is cut, and butted at 90° to the antiserum gel (such that the separation of the sample is normal to the electric field in this second gel). When voltage is applied across this composite, the separated antigens migrate through the antibody gel.

The result is a set of rockets in the gel (Fig. 11), positioned one to another in relation to their separation positions in the initial gel. One quantitates by measuring areas under the precipitin lines for each antigen. These areas are compared to those of known standards. By carbamylation of standards (this increases their electrophoretic mobility), they can be employed as internal standards.

8.4.4. Counterimmunoelectrophoresis (CIEP)

This is a rapid and sensitive technique for screening of samples for the presence/absence of an antigen. Pairs of wells are punched in an agarose gel. Gel buffer typically is at a pH that is close to the pI of the antibody. The well pairs are aligned along the axis of electrophoretic migration and a few millimeters apart. Sample (antigen) is placed in the well nearest the cathode, with antibody in the

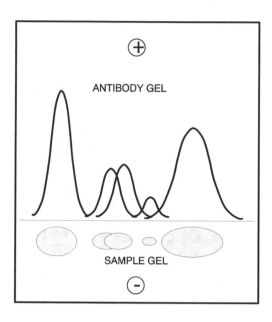

Fig. 11. Crossed or two-dimensional immunoelectrophoresis.

anodic well. When voltage is applied, the antigen starts to migrate toward the anode. The antibody shows little or no electrophoretic mobility, but migrates toward the cathode by virtue of endosmosis. The two species thus migrate through each other, and if the antigen is recognized by the antibody, a precipitin line forms in the zone between the well pair (Fig. 12).

The procedure yields a result in 15–30 minutes. Although nonquantitative, it can provide sensitivity approaching that of RIA, and requires very small amounts of sample/antibody. Basically, the method achieves the same result as obtained in immunodiffusion, but does so with more sensitivity, and much less time.

9. Preparative Electrophoresis and Sample Recovery

Electrophoretic procedures are seldom used for preparative isolation of gram quantities of proteins. The problems of dealing with large current fluxes make scaleup a complex and difficult problem, whereas chromatography, extraction, and crystallization are much more amenable to scaleup.

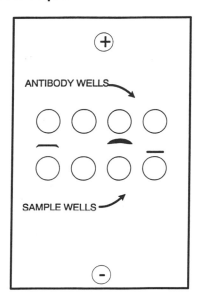

Fig. 12. Detection of antigen–antibody reactions by counterimmuno-electrophoresis.

Electrophoresis is, however, commonly used as a method of preparing or isolating milligram quantities of proteins. In the elementary prep methods, a separation is obtained by one of the preceding techniques. Following separation, one simply cuts out a gel region containing the protein of interest, and extracts the protein from this piece of gel. With horizontal or vertical slab gel equipment, samples can be applied across the whole width of the gel slab. Such gels can often handle up to 50–70 mg of applied protein.

Frequently in preparative electrophoresis, protein bands can be directly visualized. You can see high concentration bands (>100 μg/band) directly by back illumination. SDS-PAGE gel bands will fluoresce by transillumination at 302 nm. Intensity depends on tryptophan content, but is in the range of >10 μg.

Alternately, protein location is determined with guide strips: Apply the sample as a wide band. After electrophoresis, cut off edge strips (and a midstrip); stain to localize protein bands of interest. If gel shrinkage is a problem, notch or punch the gel and guide strips to index.

A protein isolate from a gel fragment can be obtained either by simple liquid extraction or by solubilization of the gel. Gels are usually cut into small pieces to facilitate extraction. One must also maintain good protein solubility, such that it can freely migrate out of the gel. The addition of urea or SDS can be used to this end. These solubilizers are subsequently separated from the protein by trichloroacetic acid (TCA) precipitation, dialysis, chromatography, and so on. Alternately, one can pack a cut up gel into a tube, and electroelute through a packing or a blot membrane, adsorbing the protein of interest.

The protein blot method described earlier is yet another micropreparative technique. A protein pattern from a gel can be electroblotted to a membrane, and the band of interest cut from the membrane. A more elaborate version of electroblotting consists of a vertical slab gel contacted with a moving scroll of membrane at the anodic terminus of the gel. As proteins move out of the gel, they pass through the moving membrane, and are adsorbed as a series of bands corresponding to the mobility order.

Larger preparative quantities can be obtained by applying a sample to a vertical tube gel a few millimeters in diameter. The anodic end of the gel emerges into a chamber, through which buffer can flow transversely. The lower portion of the chamber is covered with a dialysis or ultrafiltration membrane. This provides an electrical path, but prevents protein migration toward the anode. Protein fractions are instead swept out of the chamber by buffer flow, and can be isolated in a fraction collector. Such laboratory devices can accommodate $\approx$100 mg loads.

Isoelectric focusing is a high resolution separation analytical method for proteins. Recently, several commercial systems have emerged that permit isolelectric fractionation of multimilligram and even gram quantities of proteins. They are essentially multicompartmentalized systems that employ liquid rather than gel electrophoresis zones. Material can be focused, and drained off as specific focus zones.

Bibliography

1. Andrews A. T. (1986) *Electrophoresis: Theory, Techniques, and Biochemical and Clinical Applications*, 2nd ed. Clarendon Press, Oxford, UK, p. 3.
2. Switzer R. C., Merril C. R., and Shifrin S. (1979) *Anal. Biochem.* **98,** 231.

3. Fenner C., Traut R. R., Mason D. T., Coffelt-Wikman J. (1975) *Anal. Biochem.* **63**, 595.

4. Ornstein L. (1964) *Ann. NY Acad. Sci.* **121**, 321.

5. Davis B. J. (1964) *Ann. NY Acad. Sci.* **121**, 404.

6. Andrews A. T. (1986) *Electrophoresis*, Oxford University Press.

7. Dunker A. K., Ruekert R. R. (1969) *J. Biol. Chem.* **244**, 5074.

8. Ferguson K. A. (1964) *Metabolism* **13**, 985.

9. Hedrick J. L. and Smith A. J. (1968) *Arch. Biochem. Biophys.* **126**, 155.

10. Bjellquist B., Ek K., Righetti P. G., Gianazza E., Görg A., Postel W., and Westermeier R. (1982) *J. Biochem. Biophys. Methods* **6**, 317.

11. O'Farrell P. H. (1975) *J. Biol. Chem.* **250**, 4007.

12. Alwine J. C., Kemp D. J., and Stark G. R. (1977) *Proc. Natl. Acad. Sci. USA* **74**, 5350.

13. Southern E. M. (1975) *J. Mol. Biol.* **98**, 503.

14. Nespolo A., Bianchi G., Salmaggi A., Lazzaroni M., Cerrato D., and Tajoli L. M. (1989) Immunoblotting techniques with picogram sensitivity in cerebrospinal fluid detection. *Electrophoresis* **10**, 34–40.

15. Gershoni J. M. and Palade G. (1982) *Anal. Biochem.* **124**, 396.

16. Ouchterlony O. (1958) *In Progress In Allergy* (P. Kallós, ed.), vol. 5, Karger, Basel, New York, p. 1.

17. Mancini G., Vaerman J. P., Carbonara A. O., and Heremans J. F. (1963) A single radial diffusion method for the immunologic quantitiation of proteins. *Protiedes Biol. Fluids Proc. Colloq.* **11**, 370–373.

18. Grabar P. and Williams C. A. (1953) *Biochim. Biophys. Acta* **10**, 193.

19. Scheidegger J. J. (1955) Une-Methode de l'Immunoelectrophorese. *Int. Arch. Allergy Appl. Immunol.* **7**, 103.

20. Laurell C. B. (1965) Antigen antibody crossed electrophoresis. *Anal. Biochem.* **10**, 58–361.

21. Laurell C. B. (1966) Quantitative estimation of proteins by electrophoresis in agarose gel containing antibodies. *Anal. Biochem.* **15**, 45–52.

22. Laurell C. B. (1972) Electroimmunoassay. *Scand. J. Clin. Lab. Invest.* **29**, suppl. **124**, 21–37.

23. Ganrot P. O. (1972) Crossed immunoelectrophoresis. *Scand. J. Clin. Lab. Invest.* **29**, **Suppl. 124**, 39–41.

Ten

Production and Application of Polyclonal and Monoclonal Antibodies

Gert Van Duijn and André W. Schram

1. Introduction

During the last 20 years, the use of antibodies has become a self-evident method for biochemists, as well as cell and molecular biologists. The reason for this is that antibodies have proven to be excellent probes for the recognition of proteins, peptides, lipids, and hormones. Particularly, the high specificity of an antibody simplifies studies concerning, for instance:

1. The synthesis, processing, and transport of proteins;
2. The discrimination of different isoforms of an enzyme;
3. The localization of proteins.

Alternatively, antibodies can be used in a wide variety of quantitative studies dealing with concentrations of different types of molecules in body fluids, tissue samples, or food.

Antibodies, also called immunoglobulins, play an essential role in the immune system. They are produced by B-lymphocytes, which circulate in blood and lymph, and which are found in lymphoid organs. One B-lymphocyte is capable of making only one type of antibody in response to the introduction of a foreign molecule into an individual. Antibodies can bind to complementary structures on antigens. This binding is based on multiple noncovalent interactions, such as hydrogen bonds, electrostatic forces, Van der Waals bonds, and hydrophobic interactions.

From: *Protein Biotechnology* • F. Franks, ed.   ©1993 The Humana Press Inc.

An immunoglobulin G molecule consists of two identical heavy chains and two identical light chains that are linked by disulfide bridges. Both heavy and light chains are composed of a constant region and a variable part. The constant domains on the heavy chain determine the isotype of the antibody and thereby its biological activity. Examples of different isotypes are the IgM, IgD, IgG, IgA, and IgE antibodies. The variable regions of a light chain and a heavy chain together form the antigenic binding site of the antibody molecule. The enormous diversity of antibodies is based on the large number of variable regions that can bind to specific antigens. Furthermore, one variable region can associate with various constant regions, resulting in antibodies with the same antigen specificity, but with different biological activities.

In this chapter, the production and use of antibodies will form the central subject. Information will be provided about how to obtain and use a polyclonal antiserum or, alternatively, to make and use monoclonal antibodies. A second important item will be the detection and purification of antibodies. Furthermore, more recent developments in hybridoma technology will be described, i.e., unusual immortalization methods for lymphocytes, the introduction of in vitro immunization procedures, and the production of antibodies directed toward small molecules, the so-called haptens, and peptides. Also the powerful implications of the generation of hybrid hybridomas and antiidiotype antibodies will be discussed. Finally, an overview of the possible applications of antibodies will convince, if necessary, the reader of the enormous impact of immunochemistry.

2. Polyclonal Antibodies

2.1. Immunization Procedure

For raising an antiserum, a wide variety of animals can be used, e.g., goats, sheep, donkeys, rabbits, rats, and mice. A prerequisite for the choice of an animal is that the antigen is recognized by the animal as a foreign molecule.

The immunization procedure itself includes a first injection of the animal with the antigen dissolved in equivalent volumes of complete Freund's adjuvant and an aqueous buffer. The dose of antigen needed for this primary immunization strongly depends on the type of animal. As an example, rabbits require 200–500 µg of antigen, whereas for mice 50 µg are generally sufficient. Two and four weeks after the initial immunization, booster injections with comparable amounts of antigen need to be given. For these booster injections, the antigen is dissolved in a mixture of aqueous buffer and incomplete Freund's adjuvant (1:1 by vol).

Ten days after the last booster injection, blood samples are taken from the animal, and the antibody production is tested (*see* Section 3.) It should be noted that, even after a long period of rest, a simple booster injection could be sufficient to trigger the immunosystem of the animal to obtain another batch of antiserum. After coagulation of the blood sample at room temperature, the blood cells and the fibrin clot are removed by centrifugation. In order to inactivate the complement system, the antiserum is heated for 30 min at 56°C and subsequently dialyzed against an aqueous buffer (Phosphate Buffered Saline, PBS). It is recommended to aliquot and store the antiserum at –20°C, and to prevent repeated freezing and thawing.

2.2. Advantages and Disadvantages

A big advantage of the production of a polyclonal antiserum in comparison with monoclonal antibodies is the cost of production, which is undoubtedly much lower. Furthermore, the production of a polyclonal antiserum is possible within a time frame of 6 wk, whereas the generation of a panel of monoclonal antibodies requires at least 3 mo. Another advantage is that a polyclonal antiserum contains different IgG molecules recognizing several antigenic determinants. An example of such an advantage is in viral or bacterial diagnosis, in which polyclonal antibodies react with several determinants, some of which might not be present on all strains of the pathogen. In contrast, the main disadvantage of the production of a polyclonal antiserum could be the presence of undesirable impurities of the antigen preparation, thereby leading to contaminating antibodies or, even worse, to antibodies that are exclusively directed toward uninteresting immunodominant epitopes. Furthermore, the quality of a polyclonal antiserum obtained from an animal is not constant after various booster injections, whereas during the generation of monoclonal antibodies, stable cell lines can be obtained (*see Section 4.*).

3. Screening Methods

3.1. Enzyme-Linked Immunosorbent Assay (ELISA)

The best known test for the detection of antibodies in a serum or a hybridoma supernatant is the ELISA. In this assay, the antigen of interest is coated onto the surface of the wells of a microtiter plate. These 96-well containing plates are prepared from polystyrene and show a high binding capacity for proteins. In general, the coating of the ELISA plates with antigen is completed within 2 h at 37°C. The unbound antigen is removed by repeated washing of the plates with PBS-Tween (0.05% Tween 20). Next, the wells are incubated with antisera or hybridoma cell supernatant (see Section 4.9.). After an incubation period of 1 h at 37°C, the plates are washed again intensively with the detergent-containing buffer (PBS-Tween). In case antibodies are present that recognize the coated antigen, these antibodies will be bound to the antigen on the assay plate. To visualize bound immunoglobulin molecules, a second antibody, dissolved in PBS-Tween containing 1% of BSA (Bovine Serum Albumine) as blocking agent, is added to the microtiter plates. It should be mentioned that this commercially available second antibody should have been raised against the constant region of immunoglobulin molecules derived from the type of animal used for immunization. Furthermore, the second antibody has to be conjugated to a radioactive isotope or to an enzyme (peroxidase, phosphatase, β-galactosidase) giving rise to a colored or fluorescent reaction product after addition of the substrate. This enzymatic reaction or the immobilization of the radiolabel directly reflects the presence of antigen-bound antibodies in the well.

Alternatively, ELISA can also be used to identify and/or quantify an antigen in solution in the case that antigen-specific polyclonal or monoclonal antibodies are available. In these experiments (so-called sandwich-ELISA) antibodies are first coated to the polystyrene plates, followed by addition of the solution in which the antigen needs to be detected. After incubation and washing, another antibody, raised against the same antigen, is incubated on the microtiter plates. Finally, a suitable enzyme-linked second antibody is added to the ELISA plate. In this way, antibodies can be used to detect and quantify trace amounts of antigen in a sample.

3.2. Enzyme-Binding Assay (E.B.A.)

Although the ELISA is a rapid and sensitive way to screen for the presence of antibodies, it has two major disadvantages. First, the antigen to be used has to be absolutely pure in order to avoid the detection of antibodies directed against contaminants present in the antigen preparation used for immunization; these problems are commonly encountered. Second, binding of a protein to the polystyrene microtiter plates may lead to denaturation of the protein, whereas the generated antibodies may only react with the native protein.

Detection of antibodies reacting with a protein in its native conformation can be achieved by labeling the protein with a radioactive isotope. A drawback, however, is that the labeling procedure may also lead to denaturation of the proteins. In case the antigen is an enzyme, these problems might be overcome by making use of an enzyme binding assay. In this test, a second antibody is first adsorbed to the surface of the well of a microtiter plate. Next, the antiserum of interest or the hybridoma supernatant is added. If antibodies are present in these samples, recognition and binding will take place by the coated second antibodies. The final incubation is done with the native antigenic enzyme in solution. It should be realized that, for this incubation, it is not essential to use the purest enzyme preparation available. Subsequently, a specific substrate for the enzyme is added to the immuno-immobilized enzyme, and the reaction is allowed to occur. If the enzyme has been bound, conversion of the substrate will be detected. Of course, with using this assay, antibodies binding to or blocking the substrate binding site or active center of the enzyme will not be detected. In order to detect such antibodies, an enzyme-inhibition assay can be used.

3.3. Enzyme-Inhibition Assay (E.I.A.)

In this assay, the first step is to coat the wells of a microtiter plate with a specific polyclonal (antienzyme) antiserum. After extensive washing of the plates with PBS-Tween, the purified enzyme preparation or a preparation containing the antigen is added. Next, the antiserum or hybridoma supernatant under investigation is incubated with the immobilized enzyme. In case of inhibition of enzymic activity after this last incubation, one deals with antibodies reacting with the binding site or active center of the enzyme.

3.4. Immunoblotting (Western Blotting)

The immunoblotting technique is an elegant method to determine whether the antibodies obtained are monospecific for the antigen of interest or are crossreacting with other components present in the antigen preparation. As a first step in this procedure, the proteins in the antigen-containing sample have to be separated using, for example, polyacrylamide gel electrophoresis. This separation can be performed under native or protein-denaturing conditions by dissolving the proteins in the absence or presence of a detergent, such as sodium dodecylsulfate. As soon as the separation of the proteins on the gel is complete, proteins are transferred to an adjacently placed nitrocellulose matrix by applying an electric field perpendicular to the gel-nitrocellulose sheet bilayer (1). After this so-called electroblotting procedure, the nitrocellulose sheet, being a replica of the protein gel, is incubated with the antibodies of interest. This incubation has to be performed in the presence of an excess of blocking agent in order to prevent a nonspecific binding of the immunoglobulin molecules to the matrix and the immobilized proteins. After incubation for 5–15 h and after repeated washing of the nitrocellulose matrix, an enzyme-conjugated second antibody is added. Finally, the one or more protein bands, recognized by the first antibody, can be visualized by incubating the nitrocellulose sheet with a soluble substrate, to produce an insoluble colored reaction product by the conjugated enzyme. Additional information, if necessary, can be obtained about the molecular weight of the recognized antigens by including prestained molecular-weight markers in these Western blotting experiments.

3.5. Precipitation of Antigen

Another tool to detect specific antibodies is immunoprecipitation. Since immunoglobulines have two (and in the case of IgM antibodies even ten) identical binding sites, these molecules are able to form a network with the antigen. If so, after incubation of antigen with antiserum, the removal of this network from the supernatant upon centrifugation is an indication of the presence of antigen-specific antibodies. In some cases, despite binding, the obtained antibody is not able to form a precipitable network with the antigen. The first possibility to solve this problem is the use of a second antibody that is still able to form an extended network with the antigen-bound first antibodies. Furthermore, protein A from *Staphylococcus* aureus shows a high affinity for several classes of immunoglobulines, and is thus able to induce comparable complexes. The best result is obtained when the protein A is immobilized to a solid matrix, such as Sepharose or Agarose beads.

An optimal characterization of an antibody can be achieved by applying a combination of screening tests. The choice of antibody detection methods should be based on the final use of antibodies, such as antigen specificity, recognition of native or denatured proteins, and so forth.

4. Monoclonal Antibodies

4.1. Introduction

It has been mentioned before that the use of a polyclonal antiserum may exhibit several disadvantages. These problems generally involve a lack of specificity of the antiserum or disturbing contaminants in the antigen preparation. In order to overcome these problems, one may decide to choose the production of monoclonal antibodies. For instance, in some cases it is desirable to obtain an antibody directed specifically toward one of the more antigenic determinants on a large molecule. This is especially relevant when the antibodies are to be used in studies involving the discrimination between different isoforms of a protein because of posttranslational modification, aggregation of the protein, partial degradation, and so on. Mutant forms of a protein can also be distinguished from a wild-type protein by monoclonal antibodies. The principles of the monoclonal antibody technique are based on the ability of each B lymphocyte to produce and secrete only one type of antibody. However, the main problem is that the antibody producing B lymphocytes do not survive in culture. Kohler and Milstein in 1975 showed that it is possible to produce cloned, stable hybrid cell lines secreting specific monoclonal antibodies by fusing murine B lymphocytes with immortal myelona cells (2). In the next sections, a number of principles of this technology, including immunization procedures, the choice of myeloma cells, fusion methods, and selection and cloning procedures, will be reviewed.

4.2. Immunization In Vivo

Concerning the production of a polyclonal antiserum for the generation of monoclonal antibodies, immunization of an animal is the first step. In general, the choice of the species is restricted to a mouse or a rat, since the availability of suitable myeloma cell lines is limited to these species. For primary immunizations, soluble antigens (20–50 µg) are usually suspended in complete Freund's adjuvant, which is a water in oil emulsion containing heat-killed *Mycobacterium tuberculosis*. The antigen is then introduced into the animal by subcutaneous, intradermal, intrasplenic, or intraperitoneal routes. Booster injections with the antigen dissolved in incomplete Freund's adjuvant (without the bacteria) are given ip 1–3 times within a period of 2–6 wk. Next, a blood sample of the animal is checked for antibody production. If the animal shows a positive titer, then it should be boostered ip with the antigen dissolved in saline or PBS, 3 d before fusion.

4.3. Immunization In Vitro

Although the conventional procedure for the production of monoclonal antibodies includes immunization of the animal in vivo, an alternative and, in some situations, even more attractive approach is the immunization of splenocytes in vitro. The advantages of in vitro immunization procedures are:

1. Only a very small amount of the antigen of interest is required;
2. The immunization itself needs only a very short period;
3. This method can be particularly useful for the generation of antibodies against evolutionary highly conserved antigenic determinants or even self-antigens; and
4. This strategy can be used to produce human monoclonal antibodies.

The in vitro immunization is performed by incubating spleen cells from a nonimmunized animal with antigen for 3–5 d prior to fusion. In addition, in order to guarantee an immunoresponse, growth and differentiation factors need to be added to the splenocyte culture in the form of thymocyte-conditioned medium, supernatant of phorbolmyristate-acetate-activated EL-4 cells, or one or more (recombinant) lymphokines. In some protocols for in vitro immunization, serum is also included in the incubation medium. However, it should be emphasized that some serum components themselves are highly immunogenic.

It has been shown for a considerable number of antigens that it is possible to obtain an antigen-dependent primary immune response in vitro. As an example, antihuman thyroglobulin monoclonal antibodies have been generated using this procedure (3). An optimal immune response was obtained after a cell culture period of 3 d, with only 1 μg of antigen present/mL of culture medium. Increasing the amount of antigen during the in vitro culture resulted in a substantially lower number of antigen-specific hybridomas. Increasing the duration of the in vitro immunization experiment did not lead to a higher yield of antibody-producing hybrid cells. Apparently, besides the presence of antigen, the addition of thymocyte-conditioned medium to the cell culture is also essential in order to obtain (antihuman thyroglobulin) antibody-producing hybridomas. Even in the presence of antigen, the number of specific hybridomas hardly extended the background response when this thymocyte-conditioned medium was omitted from the in vitro cell culture. Although most of the resulting monoclonal antibodies showed a low antigenic specificity, reflected in the pronounced crossreactivity with a large number of nonrelated antigens in various types of screening tests, the observation that one of the murine monoclonal antibodies crossreacted exclusively with human thyroglobulin and not with mouse thyroglobulin provided evidence for an immune response of splenocytes in culture. The observation that most monoclonal antibodies, obtained by in vitro immunization procedures, show a weak antigen specificity can be explained by the fact that this system represents a primary immune response, thereby resulting in the generation of the IgM type of antibodies instead of immunoglobulins of the IgG isotype, which are predominantly obtained after repeated in vivo immunizations. As an alternative to a complete in vitro immunization procedure, it is also possible to immunize an animal in vivo followed by in vitro stimulation of the spleen cells with antigen 3–5 d before fusion occurs. This method has the advantage of producing a higher percentage of antibodies of the IgG isotype. This combination of immunization procedures also allows for a successful primary immunization in vivo with a crude antigen preparation, followed by a secondary immunization in vitro with a very small amount of the pure antigen.

4.4. Immunization with Peptides and Haptens

During the last decade, the production of antibodies directed specifically toward (synthetic) peptides has become a widely used application (4). If a pure protein preparation is not available for immunization, antibodies raised against only a small part of this protein have proven able to recognize the entire protein molecule. Furthermore, this method is even more relevant in those cases where the antibodies are to be used for the discrimination of isoforms of an enzyme. A prerequisite in these situations is that, by using synthetic peptides, the amino acid sequence of the discriminating part of the molecule should be (partly) known or that mild digestion of the protein gives rise to a peptide of interest. The problem in obtaining an immunoresponse against these small peptides or other molecules, such as steroid hormones, is that these so-called haptens are not immunogenic by themselves. The reason for this is that such small molecules lack a T-cell epitope. It is known that these T-cells are indispensable for delivering help to the B lymphocytes, which recognize the hapten. In order to overcome this problem, the small peptide or hapten should be conjugated to a large carrier molecule that does contain a T-cell epitope. This implies that the antibody response is directed toward at least two parts of the antigen. If necessary, the requirement for priming lymphocytes to the carrier can be circumvented in the response to a hapten-carrier conjugate if the animal has previously been primed to the carrier molecule alone or receives spleen cells from a donor that have been primed to that carrier.

Several methods for the conjugation of haptens to carriers are known. Examples of these are conjugations using glutaraldehyde, succinimide, or carbodiimide. Finally, the immunoresponse against a hapten can be further increased by including a spacer molecule between the carrier and the hapten, thereby facilitating the presentation of the hapten part of the complex to the B lymphocytes. In addition, for in vivo immunization experiments, the hapten-carrier conjugates are also usable in in vitro immunization methods.

4.5. Choice of Animal and Cell Lines

Presently, three different systems are used for the production of monoclonal antibodies, namely mouse, rat, and human. Human hybridomas are still at an early stage of development. The main reason is that no non-(immunoglobulin)-secreting human myeloma cell line is available for fusion. The other two systems are widely used for the generation of monoclonal antibody-producing hybridomas, and the choice between the two is fairly open. In principle, this choice is predominantly determined by the nature of the antigen used for immunization. As is true for conventional polyclonal antibody production, the more foreign this antigen, the better the response is likely to be. The rat is generally not employed for the production of antibodies to a rat protein, and vice versa, a mouse is not used for murine antigens. However, in spite of a lack of prediction, with an antigen of neither rat nor mouse origin, the immunoresponse of the two species can still vary greatly in vigor. The murine system was first developed by Kohler and Milstein for the production of monoclonal antibodies (2), and to date, the majority of hybridomas used are still of mouse origin. Nevertheless, the rat system shows a few advantages. One of these is that by using rat splenocytes, about 90% of the obtained hybridomas express spleen immunoglobulins, whereas in the mouse system, this percentage is only 60%. Furthermore, since rats are larger than mice, they can also yield approx 10 times more ascites fluid for the large-scale production of antibodies (see Section 4.11.). One disadvantage of the rat system might be the comparatively low affinity of most rat immunoglobulins for protein A. Thus, purification of rat-derived antibodies using protein A generally results in a lower yield.

To carry out a fusion experiment, it is preferable to use a myeloma cell partner that does not produce antibodies itself. If this is not the case, the random association of heavy and light chains during fusion will result in mixed immunoglobulin molecules containing heavy and light chains from both parental cell lines. If the myeloma cells make exclusively a light chain, then the amount of irrelevant antibodies is limited, and approx 33% of the immunoglobulin molecules produced will be of the desired specificity, including two functional binding sites. However, if both the myeloma cells and the lymphocytes synthesize light chains and heavy chains of the same class, contamination with nonspecific and nonfunctional antibody-producing hybridomas will be considerable. This is particularly relevant where hybridomas secrete low amounts of antibody, since much of the capacity of the cell is directed into the production of useless antibodies. At present, several mouse and rat myeloma cell lines, lacking antibody production or secretion, are available.

4.6. Fusion Procedures

In order to obtain hybridomas, following either in vivo or in vitro immunization, the activated lymphocytes have to be fused with myeloma cells. Using the in vivo immunization procedure, the spleen is removed from the animal, and the splenocytes are teased from the capsule. Before a fusion experiment is started, the splenocytes are washed carefully, and in some cases, contaminating erythrocytes in the cell suspension are lysed by washing with an ammonium chloride solution. Subsequently, the washed lymphocytes are mixed with a myeloma cell population in a ratio that is dependent of the type of fusion protocol used. In this section, two entirely different fusion procedures will be described and discussed. To date, nearly all fusions designed to generate hybridomas are performed with chemical fusogens, such as polyethyleneglycol. However, it has been shown that the most promising alternative fusion procedure appears to be fusion induced by an electric field.

4.6.1. Polyethyleneglycol-Induced Fusion

The mechanism of cell fusion is rather complex, and involves cell aggregation, fusion of the lipid membranes, and swelling of the cells. The optimal conditions for these three processes are frequently at variance. When reviewing published data, it is obvious that wide variations in concentration and molecular weight of polyethyleneglycol (PEG) used have all resulted in successful hybridoma production. Most fusions are performed with PEG in the molecular-weight range between 600–6000. The disadvantage of a low-molecular-weight PEG is that it is more toxic to the lymphocytes. On the contrary, high-molecular-weight PEG is very viscous and therefore difficult to handle. Although the exact mechanism of fusion is not known, it is generally accepted that the hydrophilic PEG binds the free water molecules present between the cells, thereby leading to cell aggregation. The concentration of PEG that appears to result in optimal cell fusion varies between 35–50%. The somatic cell fusion is initiated by the addition of a small vol (1 mL) of the PEG solution in serum-free medium to a loose cell pellet comprised of lymphocytes and myeloma cells in a ratio of 5:1.

After 1 min, the PEG is slowly diluted during 5 min by dropwise addition of 25 mL of serum-free medium, resulting in a final PEG concentration of maximally 2%. Subsequently, the cells are washed and centrifuged twice, and the resulting cell pellet is resuspended in a selective growth medium (*see* Section 4.7.) and divided between the 96 wells of a microtiter plate.

It should be emphasized that the fusion frequency, which is defined as the number of hybridomas divided by the total number of lymphocytes, is low. In the mouse system, this fusion frequency is 1 in 10^5 under so-called optimal conditions, whereas for the human system, this value is even an order of magnitude lower.

4.6.2. Electrofusion

A fundamentally different method for the fusion of lymphocytes with myeloma cells makes use of an electric field (5). Prior to fusion, as a pretreatment, the mixed lymphocyte-myeloma cell population (1:1) is incubated with a proteolytic enzyme, such as pronase, for 3 min. The result of this mild digestion is (without considerable loss of viable cells) the removal of extended parts of proteins at the cell surface, thereby facilitating membrane contact between adjacent cells. Next, the cells are thoroughly washed and suspended in a low-conductance fusion medium containing inositol and histidine.

The first step in the electrofusion procedure itself is the polarization of the mixed-cell population in an alternating electric field (100 V/cm; 2 mHz; 30 s), which is performed in a fusion chamber containing two parallel electrodes. This applied ac field results in dielectrophoresis and pearl-chain formation (perpendicular at the surface of the electrodes) of the cells. One or more pulses of high-field strength (2–4 kV/cm) and short duration (5–25 µs) are then given in order to porate the cell membranes. After dielectrophoresis and electroporation, the alternating electric field is applied again for another 30 s, thereby maintaining the cells in contact and allowing cell fusion. Finally, the cell suspension is removed from the fusion chamber, diluted with a 50-fold excess of a selective medium, and distributed over 96-well tissue culture plates.

It has been shown that both the duration of the electroporation (5–25 µs) and the number of pulses only marginally influence the efficiency of this type of fusion, although a higher number of pulses of high-strength should be avoided. However, an important factor influencing the fusion frequency is the cell density. Optimal results can be obtained when the cell density is between 5×10^6

and 10×10^6 cells/mL. Under optimal conditions, a fusion frequency of 1 in 2000 can be reached in the mouse system. This value is 50 times higher than that obtained by using the more classical PEG fusion method. Interestingly, it has been found that, in comparison to PEG-induced fusion, the percentage of viable antibody-producing hybridomas is the same. The electrofusion technique should be especially suitable for the production of human hybridomas in which fusion frequencies in general are an order of magnitude lower.

4.7. Selection of Hybridomas

Where cell fusion is carried out between lymphocytes (referred to as A cells) and myeloma cells (type B cells), the fused cells will not only represent hybridomas (type AB cells), but also AA and BB cells. Therefore, a positive selection procedure for hybrid cells is necessary. The most commonly used method for carrying out this procedure is the use of a parent myeloma cell line that has been selected for its sensitivity to Hypoxanthine, Aminopterin, and Thymidine (HAT) medium. Aminopterin blocks the main biosynthetic pathway for purine and pyrimidine synthesis by inhibition of the enzyme dihydrofolate reductase. However, salvage pathways do exist that make use of exogenous nucleotides. The pyrimidine pathway, involving the enzyme thymidine kinase, utilizes exogenous thymidine. Alternatively, the purine pathway needs the enzyme hypoxanthine (guanine) phosphoribosyl transferase (HGPRT) with exogenous hypoxanthine. It is possible now to select cell lines that are deficient in either one of these enzymes. To date, nearly all the rodent myeloma cell lines are HGPRT negative. Since splenocytes do not survive in culture media and myeloma cells are mostly HGPRT negative, in HAT medium only the hybrid cells will survive because of complementation.

In general, approx 10 d after fusion, the aminopterin is no longer supplemented to the fused cell population, since at that moment all the lymphocytes and myeloma cells have died. However, since residual trace amounts of the aminopterin might still be present in the cell culture, hypoxanthine and thymidine are still needed for the salvage pathways of the hybridoma cells. Another 10 d later the cell culture will be completely free of aminopterin; therefore, these two nucleotides are no longer added to the hybridoma culture.

4.8. Preselection of Lymphocytes

Preselection of a population of cells secreting antibodies of desired specificity and/or affinity may facilitate the establishment of a continuously growing hybridoma cell line after fusion. This is especially relevant, since even after immunization, the fraction of B lymphocytes producing the antibodies of interest is relatively small. Furthermore, the number of proliferating hybridomas obtained after fusion is low because of the low efficiency of cell fusion, so that the generation of a larger panel of antigen-specific monoclonal antibodies is a time-consuming operation. During the last decade, a variety of methods has been developed in order to increase the number of specific antibody-producing cell lines obtained after fusion of lymphocytes with myeloma cells. It should be mentioned that these selection procedures are all based on the presence of (B-lymphocyte) cell surface immunoglobulins that act as a receptor for the antigen. As an example, the rosetting of lymphocytes with antigen-coated erythrocytes has been shown to be a useful preselection method. These rosetting techniques using erythrocytes have also been used in negative selection procedures. Another method is the use of antigen labeled with a fluorescent compound in conjunction with the fluorescence-activated cell sorter. Furthermore, direct fusion methods have also been described that make use of the high binding affinity between biotin and avidin *(16)*. For this purpose, biotin-coated antigen is added to an activated lymphocyte population and mixed with avidin-coupled myeloma cells. The conjugated lymphocytes and myeloma cells, as a result of biotin-avidin binding, are further fused by means of electrofusion without preceding dielectrophoresis and pearl-chain formation.

Selection of specific antibody-producing hybridomas can also be performed after fusion by the use of antigen-coated magnetic beads *(7)*. After binding of the relevant hybridomas to the beads, the cell rosettes can be isolated via the induction of a magnetic field in the cell suspension.

4.9. Cloning Procedure

Cloning of hybridomas is a very essential step in the procedure for the production of monoclonal antibodies. After fusion, the cells are distributed over one or more 96-well tissue culture plates. The number of plates used is related to the number of lymphocytes participating in the fusion and to the mean fusion frequency. The ideal number of hybridomas per well is one. The main reason is that, in the case that two or more hybridoma clones are present in one well, one of the possibly uninteresting or nonantibody-producing clones may outgrow its neighbors. This is especially relevant, since it is believed that fused cells producing antibodies divert so much energy into this process that they inevitably have less energy left for division and multiplication. Therefore, in the mouse or rat system, the number of lymphocytes seeded in one well of a microtiter plate after fusion with PEG should be limited to 100 000, based on a mean fusion frequency of 10^5. In addition, by using electrofusion, this number of cells per well will even be restricted to 1000 or 10 000, because of the fact that electrofusion is approx 50 times more efficient than PEG fusion.

During the 10 d after fusion of the lymphocytes with the myeloma cells, the wells are examined regularly for growth of hybrid cell clones, using a light microscope. After this period, it is essential to assay the cell culture supernatant for the presence of antibodies. The choice of the screening procedure used is predominantly determined by the desired properties of the antibodies (*see* Section 3.). Those clones producing interesting antibodies are harvested, diluted in cell culture medium, and divided between the wells of a new microtiter plate. The recommended cloning procedure is to aliquot hybridoma cells at approx 1 cell/well and to grow them in the presence of feeder cells (for instance, splenocytes from a nonimmunized animal), or in the presence of purified growth factors, such as lymphokines. Before an antibody can be considered as being monoclonal, this cloning procedure must be repeated several times.

4.10. Purification and Characterization of Antibodies

Purification of antibodies can be performed by different methods. A crude antibody preparation is obtained by precipitation of the antibodies, but also of other proteins, in ammonium sulfate. A further purification procedure involves ion-exchange chromatography using, for instance, DEAE cellulose, followed by a gel permeation step separating proteins in the range of 50–300 kDa. In a last step, high-performance liquid chromatography using hydroxyapatite columns results in a highly purified antibody preparation. To date, a more sophisticated and faster purification can be reached by means of affinity chromatography using columns with immobilized protein A or protein G. With this technique, the antibodies are first specifically bound to the column material. After lowering the pH to 2–3, the antibodies can be recovered from the matrix. Differences between protein G and protein A deal with the ability of protein G to bind only immunoglobulins of the IgG class. Protein A also crossreacts with other isotypes, such as IgM and IgA. It should be stressed that particular classes of Ig are not recognized by protein A, thus limiting the use of protein A to some extent. Further characterization of the purified immunoglobulin molecules can be achieved by polyacrylamide gel electrophoresis under nonreducing or reducing conditions. Using these methods, one gains an insight into the molecular weights of the entire antibody molecule, and of the relative proportions of the heavy and light chains. Mild digestion of the immunoglobulins with pepsin or papain, resulting in the formation of $F(ab^1)2$ fragments, followed by gel electrophoresis, provides additional information about the molecular weight of the antigen-binding part of the molecule. Moreover, these separation techniques can provide proof that the antibodies are indeed monoclonal. More detailed information about the class and, more specifically, the isotype of the antibodies can be obtained by the use of specific antiisotype antibodies in an ELISA system.

4.11. Epitope Analysis

When a panel of monoclonal antibodies has been generated, one will be interested in the question of whether these different antibodies can be distinguished based on the epitope recognized. In order to answer this question, a number of tests are available. The first and most simple assay is to determine whether two monoclonal antibodies compete for the same antigenic determinant. One of the two antibodies to be used should be labeled (for instance with radioactive iodine) and added to an antigen-coated microtiter plate. Subsequently, the second antibody is added to this plate. In the case that the two types of antibodies do not compete for the same or overlapping epitopes, the amount of bound radioactive antibody remains unchanged. Alternatively, if the two antibodies do compete for a common epitope, the second (unlabeled) antibody will displace the first radiolabeled antibody and, consequently, the level of bound radioactivity will be decreased. It should be noted that, in general, the number of epitopes present on an antigen is unknown, but still relevant in such a competition assay. In the presence of an excess of antigen, two principally competing antibodies can bind to the antigen without measurable competition. For this reason, the experiment should be set up with a wide range of dilutions of the monoclonal antibodies and a limited amount of antigen. Another important factor that can disturb a competition assay is the affinity of an antibody for its antigenic epitope. To obtain information about the affinity, different amounts of labeled antigen and a constant amount of antibody are allowed to react. After precipitation of the antigen–antibody complex, the fractions of bound (b) and free (c) antigen can be calculated at all the antigen concentrations employed. The results are plotted according to the Sips equation:

$$\log [b/(Ab\ tot - b)] = a \log k + a \log c$$

where Ab is the antibody concentration and K is the equilibrium constant. If $\log [b/(Ab\ tot - b)]$ is plotted against $\log c$, $K = i/c$ when the ordinate is zero.

A second manner to investigate more precisely the epitope specificity of an antibody is the modification of any of the amino acid residues on the antigen molecule. This can provide further information about the amino acids involved in the structure of the antigenic determinant. A sophisticated modification of this method is the so-called pep-scan technique, in which the peptides are synthesized stepwise and tested for crossreactivity with the antibody.

4.12. Large-Scale Production

Antibody-producing hybridomas can be expanded in vitro and in vivo. For large-scale production of antibodies in vitro, the hybridoma cells are cultured in large flasks with a cell density between 10^5–10^6 cells/mL or, even more conveniently, in roller bottles. This method can yield up to 100 µg/mL of antibody. It is essential to reclone the cells regularly, since otherwise nonproducing hybridomas can start to overgrow producing cells. Cell culture in vitro provides a rather pure preparation of antibodies, since only proteins from the culture medium are present as contaminants. This contamination can be prevented by culturing the hybridoma cells in serum-free medium.

In vivo expansion of hybridomas is possible by ip inoculation of a mouse with 10^6–10^7 hybridoma cells. In order to stimulate the cell growth, it is advisable to prime the animal with mineral oil, 5–10 d previously. In this way, it is possible to obtain 10 mL or more of ascites fluid from a mouse by regular tapping. A disadvantage of the in vivo expansion is that the collected antibody-containing ascites fluid will be contaminated with immunoglobulins and other proteins from the mouse.

5. Antiidiotype Antibodies

A relatively new development in antibody technology is the generation of antiidiotype antibodies *(8)*. The basis for the possible usefulness of these types of antibodies is principally related to prior studies concerning the immune network theory. It is widely accepted that the immune system is composed of a network that is maintained by the interaction of idiotypes and complementary anti-idiotypes. It has been shown that even in a normal immune response antiidiotype antibodies are generated, and that these antibodies play an essential role both in the regulation of antibody synthesis and in the formation of the idiotypic repertoire.

To date, a powerful application of these antiidiotype antibodies is their use as a probe in studies concerning biological receptors. The principle of this strategy is strongly related to the possible mimicry of biological receptors by immunoglobulin idiotypes. In short, an antibody that is raised against a ligand may occasionally resemble parts of the three-dimensional structure of the receptor for this ligand. Consequently, antibodies that are directed against this type of receptor-mimicking antibodies may in turn resemble the ligand itself and thereby recognize the receptor.

The most important advantage of this approach is that receptor-binding antiidiotype antibodies can be generated in the absence of the receptor, using a nonpurified ligand preparation. Furthermore, these so-called antiidiotypes often mimic the biological activity of the ligand upon receptor binding. It should be clear that hybridoma technology facilitates the generation of large amounts of purified and specific antiidiotype antibodies. However, a disadvantage of the antiidiotype approach can be that artifacts arise as a result of a nonspecific crossreactivity of a polyclonal antiidiotype antiserum. This problem is especially relevant, since most of the antiidiotype antibodies are directed against an antigenic determinant on the variable region of the antibody, which does not closely resemble the so-called internal image of the antigen. Therefore, most antiidiotype antibodies are structural antiidiotopes that are not able to recognize a receptor. In order to select the useful internal image antiidiotype antibodies, these antibodies should meet a number of requirements. First, the internal image antiidiotypes should recognize to a substantial degree the (antiligand) antibodies. Second, they should compete with the ligand for binding to the (antiligand) antibody binding site. Furthermore, inoculation of the antiidiotope should elicit an anti-antiidiotope response that resembles the (antiligand) antibody. Finally, the antiidiotype antibody mimics or competes with the ligand function.

In conclusion, in order to obtain a useful preparation of antiidiotype antibodies, the first antibodies (mouse or rat) are preferably monoclonal and directed specifically toward the functional epitope of the ligand. In addition, an optimal monoclonal (mouse or rat) or polyclonal (rabbit) antiidiotype immune response will be achieved using purified (antiligand) monoclonal antibodies, or even better, using F(ab)2 fragments derived from these monoclonals, following pepsin treatment.

6. Hybrid Monoclonal Antibodies

Another new development in hybridoma technology is the production of hybrid bispecific monoclonal antibodies. These hybrid monoclonal antibodies differ from normal monoclonal antibodies by their ability to recognize two different epitopes instead of only one. As has been mentioned previously, a monoclonal antibody is comprised of two identical heavy (H) chains and two identical light (L) chains. The polypeptides in the immunoglobulin molecule are linked via disulfide bridges and form a structure with two identical pairs of H and L chains. Since bispecific monoclonal antibodies consist of two different pairs of H and L chains, they are able to recognize two epitopes simultaneously. As a result, the bispecific hybridomas have applications in, for instance, immunocytochemical research by crosslinking a tissue antigen with a marker molecule. The possible therapeutic use of these antibodies is even more interesting. As an example, the enhanced lysis of tumor cells, crosslinked with a cytotoxic agent via a bispecific antibody, should be mentioned.

For the generation of a bispecific monoclonal antibody, two different strategies are available. The first method includes the fusion between hybridoma cells, directed specifically to a first antigen, with spleen cells activated with a second antigen molecule. At present, most biologically produced hybrid antibodies originate from a fusion between two different monospecific hybridomas. However, it should be realized that a large number of possible recombinants will be obtained after fusion of two different antibody-producing cell types. Whichever cells are used as parents to produce these hybrid antibody-producing cells, the yield of bifunctional hybrid molecules will depend on the degree of heterologous H-H chain and homologous H-L chain pairing. Theoretically, three different types of associations can occur for both the H-H and the H-L recombination, namely:

1. At random;
2. Preferentially homologous; and
3. Preferentially heterologous.

These complex possible recombinations make it obligatory to select and screen the hybrid molecules very carefully. The selection procedure can involve the combination of sensitivity for HAT medium and the resistance for ouabaine in one of the two fusion partners (9). The respective sensitivity and resistance should be the opposite for the other fusion partner. As a result of fusion, some of the hybrid hybridomas will combine the two resistances. Another selection procedure includes the staining of the two parent cell types with two different fluorescent dyes and subsequent sorting of double fluorescent cells after fusion.

After selection, cloning, and growth of hybrid hybridomas, the antibodies produced have to be screened for double reactivity toward the two antigens. In most cases, the detection of bispecific monoclonal antibodies can be performed using an ELISA system with the two relevant antigens.

7. Catalytic Antibodies

Probably the most futuristic development in antibody technology during the last few years is the development of catalytic antibodies. This approach makes it possible to produce antibodies that do not exclusively bind an antigen molecule, but that are also able to act as an enzyme by transforming the antigen (10).

The value of these types of antibodies will unquestionably be their use as therapeutic agents for various diseases, including viral infections and cancer. Furthermore, these catalytic antibodies will provide an insight into a large variety of biological and chemical processes, such as peptide bond cleavage and stereospecific ester hydrolysis. The generation of catalytic antibodies is based on the complementarity of an antibody to its antigenic epitope. Many active sites of enzymes contain nucleophilic, electrophilic, basic, or acidic amino acid side chains that are essentially positioned to react with the substrate. The introduction of these types of side chains into the binding site of an antibody molecule should enable these antibodies to catalyze a chemical reaction.

The question now arises of how to insert such a complementary binding site in the immunoglobulin molecule. As an example, it has been shown that a negatively charged carboxylate can be introduced in the antibody combining site when the antibody is raised against an antigen in which a positively charged ammonium ion replaces the abstractable proton of the substrate. As a result, some antibodies were obtained that could deprotonate the substrate because of the presence of a carboxylate in the antibody within the bonding distance of this abstractable proton. A second important approach for the production of catalytic antibodies makes use of the principal difference between an enzyme and an antibody; an enzyme specifically binds the transition state of a chemical reaction, whereas antibodies prefer binding of molecules in the ground state. Consequently, antibodies raised against a stable analog of a transition state could possibly show enzyme-like activity. Using this idea, antibodies have been generated that can indeed catalyze, for instance, hydrolysis reactions with rate accelerations of 100,000-fold. Furthermore, these catalytic antibodies can even show stereospecificity for a substrate, with selectivities above 98%. Interestingly, antibodies to date are even available that catalyze a Diels-Alder reaction for which no enzyme has yet been isolated. Based on these examples, it can be suggested that antibodies could be generated that catalyze a wide variety of reactions, such as condensations, peptide-cleavage and ligation, and glycosylation.

8. Human Monoclonal Antibodies

Human monoclonal antibodies will unquestionably be very valuable as therapeutic tools. In comparison with mouse monoclonal antibodies, the human monoclonal antibodies will only give a very small antiimmunoglobulin response when used in patients in vivo, and they will be more compatible with human effector cells because of their carbohydrate sequences. However, several studies concerning human hybridoma technology (as an example, see [11]) deal with a number of problems. First, humans are in general not specifically immunized for the production of monoclonal antibodies. Therefore, the application of in vitro immunization should especially be considered for this purpose. The next problem arises from the limited availability of spleens and lymphnodes needed for in vitro immunization procedures. Human peripheral blood is in general the alternative source for lymphocytes in studies concerning the generation of human monoclonal antibodies. Furthermore, the low frequency of antigen-specific B-cells (10^{-5}), in combination with an observed extremely low fusion frequency for these human lymphocytes, will lead to a minimal number of desired antigen-specific hybridomas. Low fusion frequency, which is a common problem in hybridoma technology, is especially applicable to human lymphocytes as long as stable and useful human myeloma cells are not available. With regard to this, the limited stability of human hybridomas, obtained after fusion between human lymphocytes and mouse myeloma cells, is an extra problem. Epstein-Barr virus transformation or oncogenic DNA transfection of human lymphocytes, followed by fusion with murine myeloma cells, seems a promising alternative with respect to fusion frequency and hybridoma stability. As discussed previously, in vitro immunization procedures represent a primary immunoresponse, with the result that the antibodies obtained are predominantly of the IgM isotype, and consequently, show low specificity and affinity for the antigen molecule. This problem could be solved by genetic engineering of an already produced hybridoma. An isotype switch from IgM to IgG can be achieved by changing the gene segments coding for the constant region of the heavy chain. In addition, the affinity of the antibody for its antigen can be increased by site-directed mutagenesis. Despite the various problems reviewed here, the production of human immunoglobulins of defined specificity deserves attention because of its many possible applications.

9. Recombinant Monoclonal Antibodies

Although hybridoma technology has proven to be an extremely successful method for the generation of panels of monoclonal antibodies directed specifically against a wide variety of antigens, this technique is still rather time-consuming. Furthermore, a more pronounced disadvantage can be that the desired monoclonal antibodies are not isolated because of the large diversity of B-lymphocytes in combination with the relatively low efficiency of the immortalization procedures known; as an example, following a fusion experiment with 10^7 mouse lymphocytes, only 10 stable and antibody-producing hybridomas might be obtained. This problem is even more relevant for the production of human monoclonal antibodies, since the clonal frequency for blood lymphocytes is low, the fusion of human lymphocytes is an order of magnitude less efficient in comparison with the mouse system, and many human hybridomas are not stable.

The antibody expression library technology has provided a new pioneering method for the cloning of the immunological repertoire in bacteria for the generation of monoclonal antibodies with desired specificities *(12)*. This technique includes the creation of immunoglobulin cDNA libraries capable of expressing heavy and light chains. This cDNA is first synthesized from mRNA obtained from human or mouse B-lymphocytes derived from peripheral blood, spleen, or lymph nodes. The population of cDNA coding for immunoglobulin sequences can be enriched by PCR (Polymerase Chain Reaction) using valuable primers. These PCR products are first digested and next ligated into heavy and light chain vectors. The antibody molecules are further expressed in *E. coli* and blotted onto nitrocellulose filters. Following incubation of these filters with labeled antigen, the position of phage plaques expressing immunoglobulins recognizing the antigen can be localized. This method enables the screening of millions of clones within a very limited period of time.

Heavy- and light-chain cDNA libraries can, on the one hand, be screened separately and otherwise be coexpressed for the screening of assembled antigen binding sites. Furthermore, this coexpression can be applied for the generation of a diversity of antibody molecules that might be even larger than the one observed in nature. Consequently, new combinations of heavy and light chains may exhibit the desired characteristics that could theoretically not be obtained using the hybridoma technology.

10. Applications of Monoclonal Antibodies

Monoclonal antibodies are exquisitely sensitive reagents that can be used for a wide variety of purposes in experimental practice. It has been mentioned previously in this chapter that monoclonal antibodies can be used for the detection and quantification of proteins using immunoblotting techniques and ELISA systems, respectively. Furthermore, site-specific monoclonal antibodies can be applied for distinguishing between different molecular forms of a protein. In addition, the ultrastructural localization of proteins in cells and tissues can be determined using immunofluorescence cytochemistry and histochemistry or, alternatively, with immunogold labeling electron microscopy. Another common application of monoclonal antibodies is their use for the purification of proteins, based on immunoaffinity chromatography. Finally, monoclonal antibodies can be used for diagnostic and therapeutic purposes in medical practice.

10.1. Immunoaffinity Chromatography

Once an antigen-specific monoclonal antibody is available, it can be applied to the purification of large amounts of this antigen using immunoaffinity chromatography. The principle of this technique is that, first, the monoclonal antibodies are attached to the column material. A well-known procedure is the coupling of monoclonal antibodies to cyanogen-bromide-activated Sepharose 4 B beads (13). Next, the crude protein preparation is layered on top of the column. After rinsing the column with an appropriate buffer, only the antigenic protein is bound to the antibody-coated sepharose, while the contaminants are eluted. Finally, the purified and antibody-bound protein is recovered from the column using an elution buffer facilitating the elution of the antigen.

10.2. Immunocytochemistry

Immunocytochemical and immunohistochemical techniques have the advantage that, in addition to the presence of an antigen, its cellular distribution can also be investigated. In a typical assay, the monoclonal antibodies are first incubated with tissue sections or otherwise with isolated cells. These target cells may also be sliced in order to study the intracellular localization of the antigen. Following this incubation, the bound antibodies are detected by second antibodies which in turn are labeled with a fluorescent dye or with an enzyme, for instance, peroxidase. Finally, the fluorescent dye or diaminobenzidine which is an insoluble substrate for peroxidase, is visualized by microscopy. However, a disadvantage of these methods is that fixatives used for sectioning the tissue might disturb the availability of the antigenic determinants. Therefore, it is advisable to assay tissue or cell sections prepared in different ways.

Where localization of antigens needs to be studied in more detail, the use of electron microscopy can be of great importance. In applying this technique, extremely thin sections of the tissue or cells are required. The antigen-bound antibodies can be visualized under an electron microscope, using a second antibody or protein A, coated with colloidal gold particles.

10.3. Discriminating Between Isoforms Using Monoclonal Antibodies

Monoclonal antibodies can be selected for their capacity to distinguish between different molecular forms of a protein. This is, as an example, confirmed by the properties of a monoclonal antibody directed toward α-glucosidase (14). This enzyme is responsible for the degradation of glycogen in the lysosomes. α-Glucosidase is synthesized as a precursor with a molecular weight of 110 kDa, which is proteolytically processed to the mature protein with a molecular weight of 76 kDa. One of the monoclonal antibodies raised against this enzyme reacts exclusively with the mature form of the protein. Therefore, this antibody could be used to identify the processed form of α-glucosidase specifically.

Human urine contains approximately equal amounts of the precursor and mature forms of α-glucosidase. Whereas a polyclonal antiserum was able to immunoprecipitate almost all of the enzyme activity from urine, the monoclonal antibody precipitated only half of the activity. Therefore, this antibody was especially valuable for a variety of studies concerning the biosynthesis, the transport, and the maturation of α-glucosidase and, furthermore, for immunocytochemical studies on the intracellular localization of the enzyme.

10.4. Diagnostic and Therapeutic Use of Monoclonal Antibodies

In the clinic, monoclonal antibodies can be used both in and outside the patient. This use comprises various diagnostic and preliminary therapeutic applications. An example of an immunotest outside the patient is a serum assay, using monoclonal antibodies, that can detect antigens that are aberrantly present in the serum. Such an assay is in general a sandwich-type ELISA in which a first monoclonal antibody is used to concentrate the antigen from a serum sample onto a solid phase. Next, the antigen can be visualized and quantified using a radiolabeled or enzyme-labeled second monoclonal antibody. A prerequisite for this method is that both monoclonal antibodies recognize different and not overlapping epitopes.

A second application of monoclonal antibodies is their use in immunohistopathology. Although a pathological diagnosis is based on morphological criteria, monoclonal antibodies have proven to be useful additional staining tools for assessing a diagnosis. Staining with monoclonal antibodies in immunohistopathology can be performed using an enzyme-labeled second antibody. In this way, only cells reactive with the monoclonal antibody are stained. Monoclonal antibodies can recognize and identify tumor cells based on the specific expression of all kinds of antigens. In the case that tumor biopsy shows poor morphology, immunohistopathology is the method of choice to confirm a tumor diagnosis. It can be anticipated that monoclonal antibodies will become more important for these purposes, since an increasing number of specific monoclonal antibodies reactive with tumor-associated antigen will be available.

For diagnostic purposes, monoclonal antibodies labeled with a radioactive isotope can also be injected in tumor patients. If the antibody is raised against a tumor-associated antigen, the antibody will bind to the tumor and its metastases. Detection of the label with the aid of a camera will provide an image of the tumor.

The observation that monoclonal antibodies can bind to antigens in a tumor makes it even possible to use them for therapeutical use by coupling a tumor-cell killing agent to the antibodies. Such agents may be radiolabels, cytostatics, toxins, or substances that can trigger the immune system of the patient. An example of this last strategy is the use of hybrid (bispecific) monoclonal antibodies, which can combine, for instance, the reactivity with a tumor-associated antigen with the recognition of cytotoxic T-cells of the immune system. Preliminary data indicate that, in this way, tumor-targeted immune cells are able to destroy tumor cells both in vitro and in vivo. This result indicates that monoclonal antibodies may be very useful tools for a variety of therapies in the future.

Bibliography

1. Towbin H., Staehlin T., and Gordon J. (1979) Electrophoretic transfer of proteins from polyacrylamide gels to nitrocellulose sheets: Procedures and some applications. *Proc. Natl. Acad. Sci. USA* **76,** 4350–4354.
2. Kohler G. and Milstein C. (1975) Continuous cultures of fused cells secreting antibody of predefined specificity. *Nature* **256,** 495–497.
3. de Boer M., ten Voorde G. H. J., Ossendorp F. A., van Duijn G., and Tager J. M. (1989) Optimal conditions for the generation of monoclonal antibodies using primary immunization of mouse splenocytes in vitro under serum-free conditions. *J. Immunol. Methods* **121,** 253–260.
4. Boersma W. J. A., Claassen E., Deen C., Gerritse K., Haagman J. J., and Zegers N. D. (1988) Antibodies to short synthetic peptides for specific recognition of partly denatured protein. *Anal. Chem. Acta* **213,** 187–197.
5. van Duijn G., Langedijk P. M., de Boer M., and Tager J. M. (1989) Height yield of specific hybridomas obtained by electrofusion of murine lymphocytes immunized in vivo or in vitro. *Exp. Cell. Res.* **183,** 463–472.
6. Lo M. M. S., Tsong T. Y., Conrad M. K., Strittmatter S. M., Hester L. D., and Snijder S. H. (1984) Monoclonal antibody production by receptor-mediated electrically induced cell fusion. *Nature* **310,** 792–794.
7. Ossendorp F. A., Bruning P. F., van den Brink J. A. M., and de Boer M. (1989) Efficient selection of high affinity B cell Hybridomas using antigen-coated magnetic beads. *J. Immunol. Methods* **120(2),** 191–200.
8. Gaulton G. N. and Greene M. I. (1986) Idiotypic mimicry of biological receptors. *Ann. Rev. Immunol.* **4,** 253–280.
9. de Lau W. B. M., van Loon A. E., Heije K., Valerio D. G., and Bast E. J. E. G. (1989) Production of hybrid hybridomas based on HAT-neomycin double mutants. *J. Immunol. Methods* **117,** 1–8.
10. Schultz P. G., Lerner R. A., and Benkovic S. J. (1990) Catalytic antibodies. *Chem. and Engineering News* **May 28,** 26–40.
11. Borrebaeck C. A. K. (1987) Human monoclonal antibodies produced from primary in vitro immunized leucine methyl ester-treated peripheral blood lymphocytes, in *In Vitro Immunization in Hybridoma Technology: Progress in Biotechnology* (Borrebaech C. A. K., ed.) Elsevier, Amsterdam, pp. 209–230.
12. Sastry E., Alting-Mees M., Huse W. D., Short J. M., Sorge J. A., Hay B. N. Janda K. D., Benkovic S. J., and Lerner R. A. (1989) Cloning of the immunological repertoire in Escherichia coli for generation of monoclonal catalytic antibodies: Construction of a heavy chain variable region-specific cDNA library. *Proc. Natl. Acad. Sci. USA* **85,** 5728–5732.

13. Aerts J. M. F. G., Donker-Koopman W. E., Murray G. J., Barranger J. A., Tager J. M., and Schram A. W. (1986) A procedure for rapid purification in high yield of human glucocerebrosidase using immunoaffinity chromatography with monoclonal antibodies. *Anal. Biochem.* **154,** 655–663.

14. Oude Elferink R. P. J., Strijland A., Surga I., Brouwer-Kelder E. M., Kroos M., Hilkens J., Hilgers J., Reuser A. J. J., and Tager J. M. (1984) Use of a monoclonal antibody to distinguish between precursor and mature forms of human lysosomal α-glucosidase. *Eur. J. Biochem.* **139,** 497–502.

Eleven

Conformational Stability of Proteins

Felix Franks

1. Definition of Stability

The stability of a protein, i.e., its usefulness as a biologically active molecule, depends on the particular environment and the exposure to conditions that can promote chemical deterioration or conformational changes.

1.1. In Vivo Stability

Under in vivo conditions, protein stability (turnover) is governed mainly by the action of proteolytic enzymes. Protein lifetimes range from milliseconds (hormones, protein kinases) to years (collagen). Of all known proteins, crystallin (eye lens) has the longest half-life: It exhibits zero turnover and remains chemically unchanged throughout the life of the organism. It may, however, become subject to aggregation or undergo conformational changes that manifest themselves as cataracts. Apart from physiological considerations, in vivo stability is important in drug delivery and targeting. Data base searches reveal that certain sequences (PEST: pro-glu-ser-thr) are associated with short half-lives in the living organism. It is also speculated that certain terminal residues are associated with short and long half-lives, respectively (1).

From: *Protein Biotechnology* • F. Franks., ed. © 1993 The Humana Press Inc.

1.2. In Vitro Stability

The stability of an isolated protein in solution, i.e., removed from its native environment, is limited by deleterious chemical reactions and/or conformational changes (denaturation). Destabilization can be caused by chemical reactions, in which case it is permanent (irreversible) or it can result from structural changes induced by changes in the solvent environment, in which case it may be permanent or temporary (reversible). Temporary destabilization is often referred to as denaturation, and the process is utilized in practice during the isolation and purification of proteins. Permanent destabilization is invariably deleterious, because it is accompanied by inactivation. It must, however, be emphasized that few chemical changes can be reversed enzymatically; e.g., thiol disulfide isomerase (TDI) can "repair" incorrectly formed —S—S— bonds.

1.2.1. Chemical (Covalent) Stability

The peptide group is by nature reactive, as are several of the amino acid side chains. They are thus subject to attack by many reagents and can undergo the following reactions:

Hydrolysis (enzymic or chemical);
Oxidation, particularly serine;
Deamidation, particularly asparagine;
Phosphorylation and glycation;
β-elimination;
Isopeptide formation;
Racemization;
S—S interchange and/or thiol/S—S exchange (disulfide scrambling);
Maillard reaction (NH_2 + reducing sugar); and
Chemical modification (immobilization, crosslinking).

Examples of the chemical inactivation pathways are shown in more detail in Scheme 1, and the conditions that promote the various reactions are summarized in Table 1 (1). It is seen that heat and/or high pH are the most damaging conditions. Apart from the above destabilizing reactions, the stability of a protein can also be modified permanently by "engineering," i.e., changes in the primary sequence.

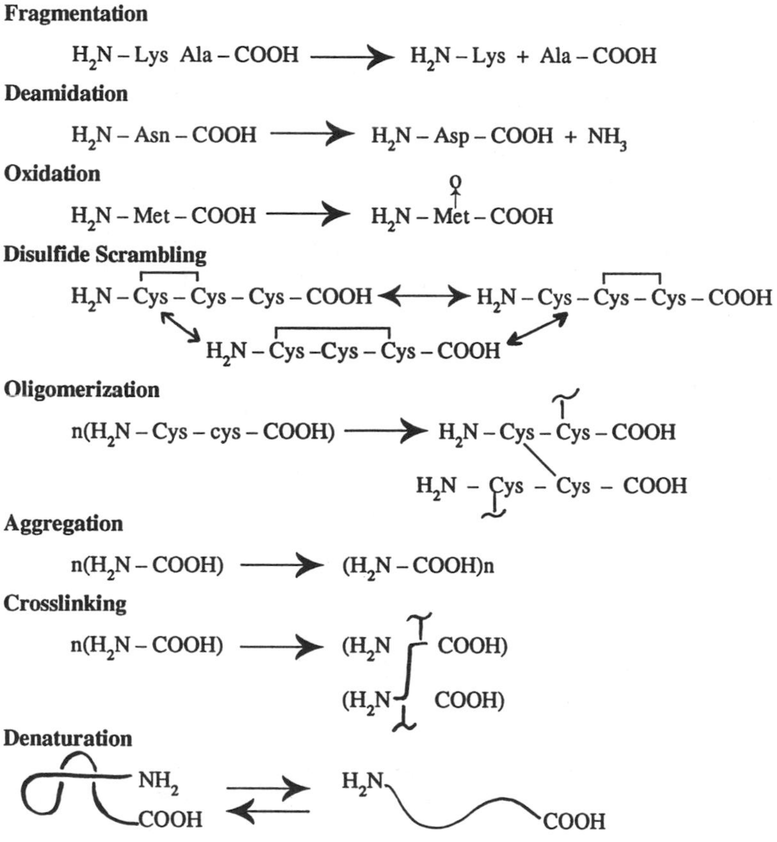

Fragmentation

$$H_2N - Lys \; Ala - COOH \longrightarrow H_2N - Lys + Ala - COOH$$

Deamidation

$$H_2N - Asn - COOH \longrightarrow H_2N - Asp - COOH + NH_3$$

Oxidation

$$H_2N - Met - COOH \longrightarrow H_2N - \overset{\overset{O}{\uparrow}}{Met} - COOH$$

Disulfide Scrambling

$$H_2N - \overline{Cys - Cys} - Cys - COOH \longleftrightarrow H_2N - Cys - \overline{Cys - Cys} - COOH$$

$$H_2N - Cys - \overline{Cys - Cys} - COOH$$

Oligomerization

$$n(H_2N - Cys - cys - COOH) \longrightarrow H_2N - Cys - \overset{\sim}{\underset{|}{Cys}} - COOH$$

$$H_2N - \underset{\sim}{\overset{|}{Cys}} - Cys - COOH$$

Aggregation

$$n(H_2N - COOH) \longrightarrow (H_2N - COOH)n$$

Crosslinking

$$n(H_2N - COOH) \longrightarrow (H_2N \overset{\Upsilon}{\underset{\mathcal{L}}{\int}} COOH)$$

$$(H_2N \int COOH)$$

Denaturation

Scheme 1. Irreversible protein degradative pathways.

1.2.2. Conformational (Noncovalent) Stability in Solution

In addition to the above chemical reactions, which reduce the biological activity, isolated proteins in solution can also be inactivated by changes in their tertiary and higher order structures. Such changes can, under certain circumstances, be reversed and the full activity restored to the protein. Conformational destabilization is produced by the following environmental changes:

Extremes of pH;
Hydrophobic aggregation (e.g., by detergent);
Pressure;
Shear;

Table 1
Mechanisms of Irreversible Inactivation of Enzymes

Inactivation mechanism	Denaturing condition
Aggregation (sometimes followed by the formation of intermolecular S—S bonds)	Heating, guanidine chloride, urea, sodium dodecyl sulfate, shaking
Alteration of primary structure H^+ - and OH^- -catalyzed hydrolysis of peptide bonds, proteolysis, autolysis	Extreme pH values, heating, proteinases
Oxidation of amino acid residues (Cys, Trp, His)	Oxygen (especially during heating), free radicals (during radiation)
Reduction of S—S bonds, intra-molecular thiol-disulfide exchange	Heating, high pH values, thiols, disulfides
Reduction of S—S bonds with subsequent formation of new amino acids (lysinoalanine, lanthionine, ornithinoalanine) or intramolecular rearrangement of S—S bonds	Heating, high pH values
Modification of essential SH-groups	Metal ions, disulfides
Protein phosphorylation and/or glycation	Protein kinases
"Suicide" inactivation in the course of catalysis owing to the action of reactive intermediates	Substrates
Racemization of amino acid residues	Heating, extreme pH values
Deamidation of Asn	Heating, high pH values
Dissociation of coenzyme molecule from active site	Chelating agents, thiols, dialysis, metal ions, heating
Dissociation of oligomeric protein into monomers	Chemical modification, extreme pH values, urea, sur factants, high or low temperatures
Adsorption to the surface of the vessel	Small concentration of protein, heating
"Irreversible" conformational changes or "incorrect refolding"	Heating, extreme pH values, organic solvents, guanidine chloride

Temperature;
Sorption at interfaces (e.g., foaming);
Metal binding; and
Solvent effects, lyotropism (salting in).

Since protein conformation and activity are coupled and are very sensitive to the nature of the solvent environment, then some of the above effects, used under the right conditions, can be employed so as to stabilize or reactivate proteins. Such stabilizing treatments include:

Immobilization, e.g., in gels, on columns, or within cells;
Binding of cofactors/substrates/metals;
Low temperatures (unfrozen);
Crosslinking reactive side chains;
Hydrophilization, i.e., derivatization with hydrophilic groups; and
Lyotropism (salting out).

1.2.3. Operational Stability

Proteins are isolated, fractionated, and purified for the exploitation of their specific activities as therapeutic agents, biocatalysts, food processing aids, clinical diagnostic materials, and so on. Proteins that find their end use as processing aids, e.g., biocatalysts, need to be able to function over considerable periods of time (the longer the better). They must therefore be rendered stable against deteriorating influences, such as those mentioned above and others that might arise from inhibitor binding during the course of a biotransformation. For obvious reasons, the methods that may be used to stabilize an industrial enzyme against chemical attack or high temperatures differ dramatically from those that can be applied to stabilize therapeutic products destined for injection or ingestion. In the latter case, operational stability has no meaning, because each preparation is used just once, and its efficacy is then determined by its in vivo performance.

1.2.4. Storage Stability

As discussed above, their reactive nature makes proteins more or less sensitive to chemical attack and/or conformational inactivation. It has also been speculated that no protein can be purified to the extent that all traces of proteolytic enzymes are removed. This implies that the gradual inactivation in solution can hardly be avoided. In order to obtain a commercially viable product, it is necessary to subject the protein to some form of treatment that renders it stable during processing, distribution, and storage, with an acceptable shelf life, preferably at ambient temperatures. The particular stabilizing treatment to be employed depends on the end use of the product; the choice may be severely limited by regulatory demands, particularly in the case of parenteral products.

Stabilizing treatments of liquid products that find commercial application include salt suspensions, typically in $3M$ ammonium sulfate, and concentrated (50%) glycerol solutions, coupled with shipping and storage at $-20°C$. Solid-state stabilization methods include immobilization, deep freezing, freeze-drying, and vacuum/spray drying. For most applications, stable solutions would be preferred, and some proteins are indeed marketed in this form (e.g., insulin). However, such products are often subject to unexplained losses of activity and changes in physical appearance, e.g., gelling; they have in any case to be stored under refrigerated conditions and have a limited shelf life. Over the past two decades, freeze-drying has established itself as the favored stabilization method, especially in the manufacture of therapeutic products and biochemical reagents. The solid-state stabilization of proteins is discussed more fully in Chapter 14.

2. Conformational Stability in Solution

During their extraction and purification, all proteins are handled in solution; frequently, the purification process includes several precipitation/solubilization stages. An extensive literature deals with folding or "naturation" of nascent proteins on the ribosome or during posttranslational processing. Such topics are beyond the scope of this bcok. This chapter therefore concentrates on the principles that govern so-called denaturation/renaturation processes, with the accent on reversible changes. The simplest model treats protein stability in terms of two states in dynamic equilibrium:

$$\text{N} \quad \text{Native state} \rightleftharpoons \text{Denatured state} \quad \text{D} \quad\quad\quad (1)$$

The native (N) state is uniquely defined in terms of specificity and biological activity. If the detailed tertiary structure is known, this provides yet a further description of the N state. Frequently, the denatured (D) state is equated to the "random coil." This state, commonly used in polymer science, refers to a macromolecule with perfectly flexible links between residues. For proteins, this is an oversimplification because of the rigidity of the peptide C—N bond, which does not allow for complete flexibility. Denaturation can be associated with changes in one or more levels of structure, and can be probed by various spectroscopic and other physical techniques, discussed in an earlier chapter. The most sensitive and definitive monitor of denaturation is change in biological (or immunological) activity. In practice, denaturation is sometimes promoted in order to achieve some desired technological attribute, e.g., solubility.

According to the two-state N/D model, denaturation and renaturation are cooperative, all-or-none processes; no intermediate species exist with appreciable lifetimes or concentrations. Denaturation can therefore often be treated by simple equilibrium thermodynamics, and the process in dilute solution can then be quantitatively decribed by an equilibrium constant, K, of the form

$$K = [D]/[N] \tag{2}$$

where the square brackets denote concentrations.

2.1. Thermodynamic Characterization of Protein Stability

According to the laws of equilibrium thermodynamics, the Gibbs free energy associated with the process in Eq. (2), which is also the measure of protein stability, is given by

$$\Delta G = (G_D - G_N) = -RT \ln K \tag{3}$$

and the temperature coefficient of the equilibrium constant is given by the van't Hoff equation

$$d \ln K/dT = \Delta H/RT^2 \tag{4}$$

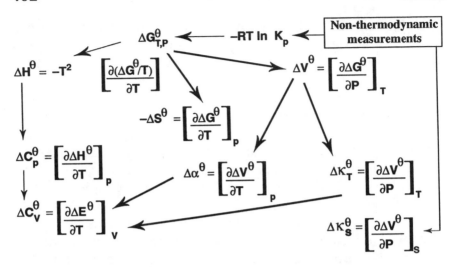

Scheme 2. Interrelationships among thermodynamic properties that are directly accessible to experimental determination. The symbol θ refers to standard conditions.

where ΔH is the enthalpy of denaturation; its temperature dependence is expressed in terms of the isobaric heat capacity change ΔC, which accompanies the denaturation:

$$\Delta H = \int \Delta C dT \qquad (5)$$

ΔC may itself be a function of T, e.g., $\Delta C = A + BT + DT^2$. The relationships among the basic thermodynamic quantities are summarized in Scheme 2.

2.2. Experimental Methods for the Determination of ΔG

The free energy difference is the fundamental measure of protein stability in solution, and it can be measured by so-called indirect methods, according to the above equations or by direct, thermodynamic methods.

2.2.1. Indirect Methods

The measurement of K, as defined in Eq. (2) and, hence, of ΔG is based on the assumption that a two-state N/D equilibrium does in fact exist (2), and this has to be established experimentally. Figure 1

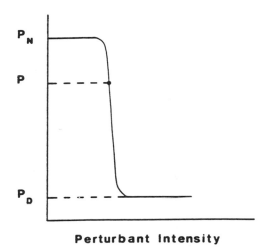

Perturbant Intensity

Fig. 1. The effect of perturbant intensity on the N ↔ D equilibrium; *P* is any physical or biological property that can be measured independently for the N and D states.

is a diagrammatic representation of a cooperative, reversible conformational transition produced by a perturbant, the effect of which is measured by some physical property, P, which is sensitive to the difference between the N and D states. If P_N and P_D can be determined, then for any point on the curve

$$K = f_D/f_N = (P - P_D)/(P_N - P) \qquad (6)$$

where f_D and f_N are the fractions of molecules in state D and N, respectively. In practice, experimental problems limit the measurements to the range $0.1 < K < 0.9$, corresponding to $+5 > \Delta G$ kJ/mol > -5.

The curve in Fig. 1 is highly idealized, because there is no reason why P_D or P_N should be independent of the perturbant intensity. If such a dependence is known and measurable, then Fig. 1 provides a good test for a two-state process.

The experimental techniques that are available for the measurement of ΔG include spectroscopy of various types, light scattering, optical activity, bioassay, immunoassay, enzyme assay, chromatography, sedimentation, electrophoresis, pH titration, and viscosity.

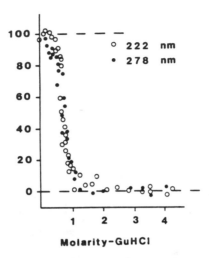

Fig. 2. Reversible denaturation of yeast 3-phosphoglycerate kinase by GuHCl, monitored by CD.

Figure 2 provides an example of such measurements: The protein is phosphoglycerate kinase (PGK), the perturbant is guanidinium hydrochloride (GuHCl), and the measuring technique is CD at two wavelengths (3). The 222-nm band monitors the peptide backbone conformation, whereas the 278-nm band is diagnostic of the aromatic amino acid residues. A comparison with the idealized curve in Fig. 1 shows that the N–>D transition can be adequately represented by the two-state model; i.e., such transition states as exist (and they must of course exist) are not significantly populated and must have very short lifetimes.

When the experimental data are converted to free energies, as shown in Fig. 3, a problem becomes apparent: The quantity of most interest is the PGK stability in the *absence* of GuHCl, $\Delta G°$. Its evaluation requires a long extrapolation. Various techniques have been proposed for such extrapolations, all of them of a semi-empirical nature (3).

One popular approach views denaturation as the equivalent of transferring the amino acid residues from the "native," aqueous solvent to the perturbing solvent. ΔG is then expressed in terms of

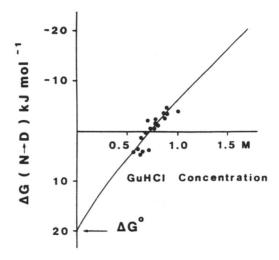

Fig. 3. ΔG data, corresponding to the experimental results shown in Fig. 2 and illustrating the limited range over which the N/D equilibrium can be studied in practice (2).

the sum of the individual amino acid contributions, $\delta g_{tr,i}$ and a mean change in solvent exposure Δa_i to which residue i is subjected during denaturation:

$$\Delta G = \Delta G_0 + \Sigma_i \, \Delta\alpha_i \, \delta g_{tr,i} \qquad (7)$$

Here $\delta g_{tr,i}$ is the free energy required for the transfer of residue i from the aqueous medium to the denaturing medium (aq. GuHCl in Fig. 2). Since $\delta g_{tr,i}$ is not a linear function of the GuHCl concentration, a plot of ΔG vs GuHCl concentration will not be linear (see Fig. 3). The best fit is obtained by adjusting $\Delta\alpha_i$; the curve must also pass through the $\Delta G = 0$ point, as obtained by experiment (see Fig. 2).

The extrapolated ΔG^0 represents the stability of the protein in the absence of the perturbant, i.e., the *intrinsic* stability. However, the higher the concentration of perturbant required for denaturation and the steeper the slope of the ΔG vs concentration curve, the more uncertain the extrapolation becomes. The denaturation of lysozyme by urea provides a good illustration: Experimental

Table 2
Propagation of Errors (SD)
in the van't Hoff Treatment of Spectroscopic Data (4)[a]

Parameter	Experimental design	
	20–30°C in 2-deg steps	5–50°C in 5-deg steps
log K	0.02	0.02
ΔH (J/mol)	4055	794
ΔS (J/mol/K)	14	1.4
ΔC_p (J/mol/K)	2784	117

[a]The accuracy that can be attained by the use of temperature-difference spectroscopy is usually lower, and hence, the errors will be greater than those shown in the table.

measurements are limited to the urea concentration range 4–7M, thus requiring a very long extrapolation to $c = 0$ for the estimation of ΔG^0 (50 kJ/mol).

The advantages of indirect measurements can be summarized as follows:

1. Several of the available techniques are relatively simple to perform and can be automated;
2. Results can be checked by the use of two or more complementary techniques, e.g., spectrophotometry at several wavelengths; and
3. K can be measured over a limited range of perturbant concentrations and/or temperatures for $0.1 < K < 0.9$.

The disadvantages are that the method relies on the validity of the two-state N/D assumption and requires accurate knowledge of P_N or P_D, i.e., an absolute measure of either the N or D state concentration, but preferably both.

ΔH and other derived thermodynamic functions are obtained by differentiation(s) of the van't Hoff equation (Eq. [4]). Each successive differentiation will introduce an increased level of uncertainty into the results, as shown in Table 2 (4). The errors are particularly marked when the experimental temperature range is narrow, a frequent prerequisite for biochemical measurements.

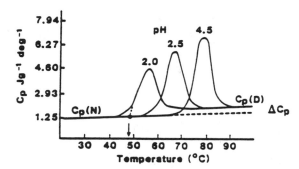

Fig. 4. Specific heat/temperature profiles of lysozyme as a function of pH. The area under C (T) curve is a measure of ΔH (N ↔ D) and the change in the base line is equal to ΔC (N ↔ D) (5).

2.2.2. Direct Methods

All such methods are based on calorimetry; they are termed direct, because they yield thermodynamic results directly, without the need to assume a reaction equilibrium, such as the two-state model. ΔH and ΔC can be measured; ΔG and other thermodynamic functions are obtained by integration:

$$\Delta G_{N \to D} = \Delta H^* (1 - T/T^*) - \int \Delta C dT + T\int (\Delta C/T)dT \qquad (8)$$

where ΔH^* is the heat of denaturation at the transition temperature T^*.

The advantages of direct methods are several: Calorimetry does not rely on the two-state assumption. A further advantage of direct measurements is that they do not require the long extrapolation to zero perturbant concentration discussed above, because experimental determinations are possible on the N and D states independently, and not just in the narrow transition range. Temperature-scanning calorimetry has been developed into a very sensitive experimental tool for the study of protein stability. The instrumental output measures specific heat as a function of temperature, as shown in Fig. 4 for lysozyme at different pH values (5). The following information can be extracted from the data:

1. $C(T)$ of the native protein below the transition temperature T^*;
2. $C(T)$ of the denatured protein above the transition temperature T^*;
3. The heat capacity change ΔC associated with the N $\to$ D transition:
4. The enthalpy change ΔH^* associated with the transition;
5. The temperature T^* of half-conversion, where $K = 1$ ($\Delta G^0 = 0$); and
6. The van't Hoff enthalpy change $\Delta H_{vH} = 2(RT\Delta C)^{1/2}$.

Calorimetry is, however, also associated with certain disadvantages: ΔH can only be measured *at* the transition temperature. In order to overcome this limitation, the assumption is usually made that, at any given temperature, ΔH is independent of pH *(6,7)*, and pH is then used as a means for probing a temperature range. The validity of this assumption remains to be verified; it seems to have no physical basis.

In order to integrate Eq. (8), it is necessary to know the temperature dependence of ΔC. It is usually assumed that ΔC is constant. If not, then it is claimed that ΔC $(= d\Delta C/dT)$ is very small and lies within the experimental error *(7,8)*. Reference to Fig. 4 suggests that such an assumption is reasonable, provided that the temperature range covered is limited. When temperature ranges of the order of 100 are considered, the above approximation leads to severe inaccuracies (*see* Section 3.1.2. *below*). In any case, there is no theoretical reason for putting $C_N(T) = C_D(T)$. Nevertheless, the literature contains numerous statements that ΔC is constant, independent of temperature.

Finally, calorimetry, as an experimental technique, requires a high degree of skill, experience, and knowledge, both in the actual design and performance of the experiment, but also in the interpretation and processing of results. The current trends of interfacing instruments with computers may give the impression that such knowledge and experience are no longer required. Whatever may be the merits of computer control of instruments, in the case of calorimetry, such instrument modifications have not been entirely successful. They can be disastrous, when the programmer is not intimately familiar with the basis of the technique and the significance of the instrumental output *(9)*.

Table 3
Relative Stabilities of Native Globular Proteins
at Physiological Temperatures Relative to the Denatured States[a]

Protein	ΔG kJ mol^{-1}	Method
Eggwhite lysozyme	53	GuHCl equil.
Human lysozyme	117	Do.
Ribonuclease pH 7	62 (39)	Do.
pH6	(54)	
α-Chymotrypsin pH 4	43 33	Heat
Penicillinase pH 7	25	GuHCl equil.
PG kinase	21 (19)	Do.
Peptide 99 → 149	−21	Immunological
Staph. nuclease		
Ferricyt. c pH 7	(64)	GuHCl
pH 4.8	(38)	Heat

[a]Methods of destabilization are given (2).

2.2.3. Experimental Results

Table 3 provides a brief summary of experimentally determined free energy margins of globular proteins. It should be noted that the stability margin is extremely small, seldom exceeding 50 kJ mol^{-1}, equivalent to the energies of two or three hydrogen bonds.

3. Denaturation Mechanisms

We now turn to a discussion of the various types of destabilization processes that are common in protein chemistry or processing. Where they are reversible, they can be treated by thermodynamic equilibrium methods. In practice, most such processes are only partially reversible, unless extremely dilute solutions are used and the environmental changes are imposed slowly.

Thermal denaturation is sensitive to the amino acid composition. This is well illustrated by collagen types derived from different species, where T^* increases with increasing proline (and hydroxyproline) content. In protein engineering, the aim is frequently to increase the thermal stability by single point mutations. It should, however, be remembered that an increase in stability is not always accompanied by a corresponding enhancement of the activity. T^*

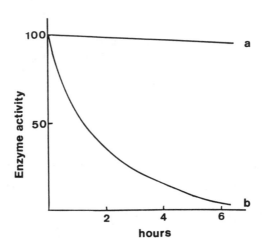

Fig. 5. The thermoinactivation rate of RNase in anhydrous nonane at (a) 110°C and (b) 145°C. The lyophilized enzyme (residual moisture 9.5%) was dispersed in nonane and heated for various periods. The solvent was removed under vacuum, and the dried enzyme redissolved in aqueous buffer and assayed. Redrawn from Volkin D. B., Staubli, A., Langer, R., and Klibanov, A. M. (1991) *Biotechnol. Bioeng.* **37,** 843–853.

is also affected by other stabilization/destabilization treatments, such as pH, salt concentration, and so forth, and is therefore frequently used as a monitor for the effectiveness of such treatments.

The chemical effects of heat on proteins are included in Table 1. From an industrial point of view, the kinetics of inactivation in a reactor are important. Since inactivation is generally slow, it cannot be solely owing to an equilibrium N/D transition. Many studies have been performed with model enzymes, in particular ribonuclease (RNase) and lysozyme (Lz). Figure 5 shows the rate of inactivation, measured at 90–100°C. The reaction responsible is not an aggregation process, nor is it a conformational change, because it is unaffected by denaturants (urea). Gel electrophoresis shows up low-molecular weight fragments, indicating a hydrolytic process. Cleavages take place preferentially at asp residues, but the rate of aspartate hydrolysis can only account for 25% of the measured inactivation rate. The detection of ammonia provides evidence of deamidation of gln and/or asn.

Studies on several enzymes have led to the general result that four reactions contribute to thermal inactivation:

Hydrolysis;
Deamidation (of greatest importance);
—SH oxidation; and
—S—S— rearrangement.

The stability of disulfide bonds is quite constant at 100°C (but not at room temperatures), with $t_{1/2}$ = 10 h for all proteins. Deamidation half-times are also universally of the order of several hours at 100°C.

3.1. Denaturation by Changes in Temperature

3.1.1. Heat Denaturation

The effect of temperature as a means of destabilizing proteins has been studied in detail. High temperatures are of technological importance, but irrelevant to the in vivo functioning of a protein. To study thermally induced denaturation, the treatment outlined in Section 2 can be used. For enzymes, the activity profile is probably the most sensitive monitor. Otherwise any of the spectroscopic, optical rotation, or macroscopic (sedimentation, viscosity, calorimetry) methods find application.

One aim of protein engineering is the production of mutants with superior thermal stability, e.g., enzymes destined for the catalysis of biotransformations on an industrial scale, or for use in products, such as detergents. It is not immediately apparent how amino acid substitutions can enhance resistance against the chemical reactions shown above, unless the sensitive residues, e.g., asn, gln, can be considerably reduced in number or eliminated from the peptide chain altogether.

3.1.2. Cold Denaturation

It was established fairly early on that ΔC accompanying heat denaturation is always positive and large, which then makes ΔG very sensitive to temperature. In practice, ΔG always adopts the shape of a skewed parabola; see Fig. 6 for chymotrypsinogen (6) and Fig. 7 for several lysozyme mutants (10). Franks et al. have demonstrated that ΔG for *any* process that can be described by the

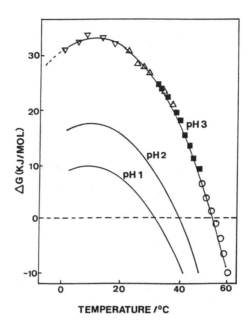

Fig. 6. Stability profiles of chymotrypsinogen, according to Brandts *(6)*: △ pH 1, ■ pH 2, ○ pH 3, ▽ 2.3*M* urea at pH 1.55. The master curve (pH 3) was calculated from data at different pH values and the assumption that, at any given temperature, Δ*H* is independent of pH. *See also* Fig. 9.

two-state equilibrium model will take the shape of a skewed parabola, and the magnitude of Δ*C* only determines the degree of skew *(11)*. This type of Δ*G(T)* relationship implies the existence of a cold denaturation (inactivation?) at some subzero temperature T_L. Since the treatment is based on the existence of only two states (N and D), and since Δ*G* is a continuous function of temperature, the cold- and heat-induced D states must be identical. Brandts speculated on the possibility of cold denaturation as long ago as 1964 *(see* Fig. 6), but could not establish the phenomenon experimentally, because he was unable to prevent the solutions from freezing *(6)*.

More recently, cold denaturation has been established as a real and probably universal phenomenon *(12)*, of great ecological significance. Cold-induced protein transitions play an important role in the natural cold resistance and cold acclimation of many organisms. The biochemistry of acclimation, as applied to insects,

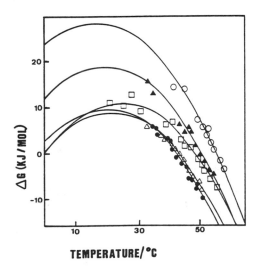

Fig. 7. Stability profiles for lysozyme mutants *(10)*. Circles denote the wild type; other symbols refer to various single-point mutations. Note that all mutations reduce the stability. Redrawn from data in ref. *10*.

has been reviewed in detail by Storey *(13)*. The phenomenon of cold destabilization can be studied directly, provided that freezing is prevented, by one or several of the following means:

1. Partial protein destabilization by pH, salts, or chaotropes, to bring T_L into the experimentally accessible temperature range *(14)*;
2. High concentrations of cosolvent (e.g., 50% methanol) to depress the freezing point of the solution to below T_L *(15)*; *see* Figs. 8A and 8B. Such solvents are referred to as cryosolvents and have been used extensively in studies of enzyme kinetics at subzero temperatures *(16,17)*;
3. Partial destabilization of the protein by point mutations, also coupled with the use of a chaotrope *(18)*;
4. Undercooling, i.e., the inhibition of ice crystal nucleation and growth at subzero temperatures *(19,20)*. This method is the most unambiguous one, because it allows temperature-induced protein transitions to be studied without the necessity for simultaneous destabilization by another perturbant. For further details and applications of this technique, consult Chapter 14.

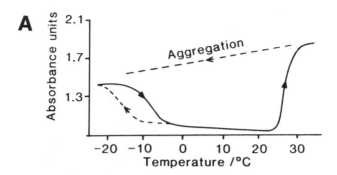

Fig. 8A. The effects of cooling (- - -) and heating (---) on the absorbance of lactate dehydrogenase in 40% aqueous methanol solution, measured at 240 nm. The high temperature denaturation was irreversible; cooling from 30°C resulted in aggregation. Reproduced, with permission, from Hatley, R. H. M. and Franks, F. (1986) *Cryo-Letters* **7**, 226–233.

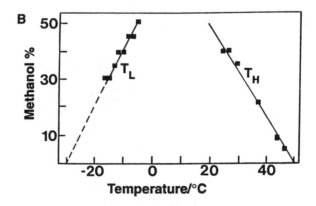

Fig. 8B. The effect of methanol cryosolvent composition on the cold and heat denaturation temperatures of lactate dehydrogenase. Reproduced, with permission, from Hatley R. H. and Franks F. (1989) *Eur. J. Biochem.* **184**, 237–240.

Most recent reports *(14,21,22)*, including several reviews *(23,24)*, claiming to deal with cold denaturation should be treated with caution, because the results described do not refer to actual experimental measurements at subzero temperatures, but are extrapolations of heat denaturation data, arrived at with the aid of unduly simplified versions of Eq. *(8)*. The T_L data and associated

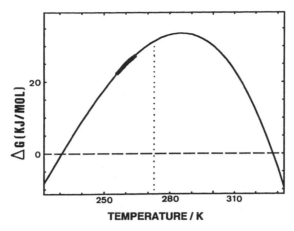

Fig. 9. Stability profile of chymotrypsinogen; *see also* Fig. 6. The dotted line is the limit of previous experimental measurements (freezing point), and the curve corresponds to the best fit *(6)*. The bold portion of the curve denotes experimental data on undercooled solutions *(19)*.

thermodynamic functions do not agree with those determined directly by experiment *(25)*. For recent reviews of cold denaturation and reexaminations of previous work, *see* refs. *26* and *27*.

The basic cause of the discrepancies between direct and extrapolated results lies in the often repeated claim that ΔC is constant, independent of temperature (e.g., refs. *10,14,17,21*). There appears to be no good physical reason why this should be the case, and direct measurements at subzero temperatures have demonstrated that ΔC is not constant. As long ago as 1964, Brandts found that his chymotrypsinogen denaturation data could only be adequately fitted by a function that allowed for a temperature-dependent ΔC *(6)*, and recent direct measurements in the region of T_L have been in excellent agreement with the polynomial function first proposed by Brandts *(19,25)*. The extrapolated and directly measured thermodynamic functions are shown in Fig. 9.

Cold denaturation has been dealt with in some detail, because from the point of view of protein characterization, such studies are more penetrating and informative than the more common heat denaturation investigations. Thus, it is hardly surprising that an ordered structure can be destroyed by increasing its kinetic energy, but the converse, i.e., an order/disorder transition resulting from a decrease in kinetic energy, merits closer study. Although such an analysis is beyond the scope of this discussion, we summarize here some of the important, if not unique, features associated with cold denaturation:

1. The process is completely reversible, even at very high protein concentrations;
2. The denaturation/renaturation cycle appears to exhibit hysteresis, the extent of which is unaffected by the number of cycling processes. It is possible, however, that the hysteresis is an experimental artifact because of a high scanning rate relative to the renaturation rate;
3. The destabilizing effects of cryosolvents (e.g., aqueous methanol) on T_H and T_L are not identical; compare the slopes of the two lines in Fig. 8B;
4. Glycerol and other polyhydroxy compounds (PHCs) are well known to stabilize proteins against heat inactivation, i.e., they raise T_H (28,29). As shown in Fig. 10, glycerol also lowers T_L, and the effect is much more dramatic than at the high temperatures (25); and
5. The thermodynamics of cold inactivation are *not* the mirror image of those associated with heat denaturation. Thus, cold inactivation is exothermic. The exact nature of $C_N(T)$ and $C_D(T)$ remains to be established, although, in principle, there is no reason why ΔC could not change sign at some low temperature (11).

3.1.3. Freeze Denaturation

The main injurious effect of freezing is not the low temperature, but the concomitant concentration of all soluble species while ice separates from the mixture as a pure water phase. Complex relationships exist between the initial protein concentration and the degree of freeze denaturation observed at different subzero temperatures; they are described in detail in the following chapter.

Denaturation is especially marked at high subzero temperatures, damage being reduced as freezing reaches completion (not necessarily the eutectic temperature). Freezing also increases the

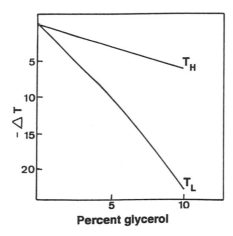

Fig. 10. Relative depression of the denaturation temperatures of lactate dehydrogenase caused by glycerol *(25)*. The reference solvent was a 40% (v/v) methanol cryosolvent in buffer. The aqueous buffer was then replaced by the indicated amount of glycerol. Note that the apparent destabilization at high temperatures is owing to the increasing molar ratio MeOH/(water + glycerol) as glycerol replaces water in the solvent.

concentrations of buffers and other additives present in the solution by orders of magnitude. This can lead to the precipitation of acids and/or salts and large changes in the perceived pH, which in themselves can cause a protein to become inactivated.

3.2. Pressure Denaturation

In vitro protein denaturation by pressure is not of great practical importance, because very high pressures (>2 kbar) are required. The fundamental relationship between stability and pressure is given by

$$d(\Delta G)/dP = \Delta V \tag{9}$$

However, ΔV is itself a function of pressure (compressibility) and temperature (expansibility), as shown in Scheme 2. It has been found that ΔV can change qualitatively with increase in pressure, and that it is also sensitive to pH and concentration *(30,31)*. The relationships between stability and pressure are therefore extremely complex.

3.3. Shear Denaturation

Closely related to pressure denaturation are the effects of shear on protein stability. Since protein purification procedures involve mixing, flow in tubes, ultrafiltration, passage through pumps, and so on, the effects of shear are of utmost importance. Similarly, immobilized proteins with fluid flowing past are also subject to shear degradation, e.g., a coating of anticoagulant or immobilized enzyme on a tube surface. The kinetics of enzyme reactions are also altered by shear. If a macromolecule is so oriented during flow that a critical bond becomes subject to the full shear effect, then under extreme conditions, even covalent bonds can be broken.

The best correlation of the available data over a wide range of shear rates is

$$\gamma t = a(x/x_0)^b \tag{10}$$

where γ is the shear rate, t the exposure time, and x/x_0 the degree of inactivation (32). The coefficients a and b depend on pH and temperature. Some representative data are shown in Fig. 11.

Although the average shear under conditions of turbulent flow cannot be calculated, inactivation increases with increasing Reynolds number. Loss by pumping can be considerable: Fibrinogen in solution, pumped through a finger pump in tygon tubing loses up to 20% of its original activity after 3000 passes at a moderate shear rate ($=145 \text{ s}^{-1}$).

Equation (10) in its integrated form becomes

$$(x/x_0) = [(1/a) \cdot (16/3) \cdot (L/D) \, N]^{1/b} \tag{11}$$

D and L are the diameter and length of the tube and N is the number of passes (33,34).

Conformational changes induced by pressure and/or shear during reversed-phase HPLC have been reported by Katzenstein et al. (35). Intermediate conformations of a range of proteins could be identified, corresponding to different degrees of unfolding, accompanied by partial loss of activity.

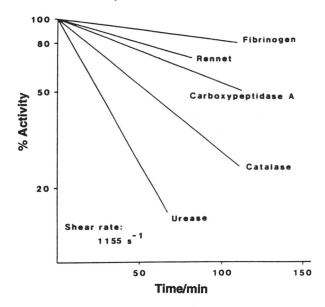

Fig. 11. The residual activities of representative proteins when exposed to constant shear conditions for varying periods.

3.4. Chemical Denaturation

Under this heading are included effects produced on the N–D equilibrium by additives (36). Such effects may rely on electro-static interactions, or they may be caused by salting in/out phenomena. They may also be caused by specific binding phenomena or may be classified as general solvent effects, in that the additive so modifies the structure of water as to render it unable to maintain the properly folded N state. Conversely, some additives are able to enhance the stability of the N state, as evidenced by an increase in the thermal denaturation temperature T^*. It has been found that the stabilization for a given additive concentration observed at low temperatures far exceeds that at high temperatures (see Fig. 10) (25).

3.4.1. pH Denaturation

As expected, the stability of the N state is sensitive to the degree of protonation of the protein. The N –> D transition can therefore be followed by means of a pH titration, as illustrated for lysozyme in Fig. 12 (37). The transition can be considerably sharpened by a measurement of ΔH, instead of ΔG. This is shown in Fig. 13 for carp muscle parvalbumin (38). By combining thermal and pH denaturation data, it is possible to construct the ΔG-pH-T surface for any given protein, and thereby to identify the pH and temperature regions where the native protein is stable. Figure 14 provides an example of such a surface for lysozyme (37). The $\Delta G = 0$ contour marks the limit of stability of the N state for this particular protein. Similar surfaces for other small globular proteins differ in detail, but in no case does the maximum stability of the N state exceed 70 kJ/mol. The stabilities of several proteins at moderate pH values are collected in Table 3, indicating also the method by which denaturation was achieved (2,7).

3.4.2. Salting In/Out and Protein Stability

A clear distinction must be made between salt effects at low concentration (<0.15M) and those produced by higher concentrations. In the former concentration range, the effects on the protein are of a nonspecific, electrostatic nature and can be expressed in terms of the ionic strength. (Note: It may be no coincidence that $I = 0.15$ corresponds to the isotonic salt concentration of many living organisms). At higher concentrations, all the symptoms of the ion-specific effects associated with the lyotropic series become apparent. This is graphically illustrated in Fig. 15 (39). Reference to the earlier discussion on salting in/out shows that those salts that reduce the solubility of proteins also enhance the stability of the N state, whereas salting-in electrolytes weaken the integrity of the N state and favor denaturation. The results for ribonuclease are typical of proteins in general, although the actual degree of stabilization or destabilization varies from one protein to another. The specific effects of ions are made much clearer when we compare the different salts of a common cation. This is done in Fig. 16 for GuH$^+$. Although GuH$^+$ salts are generally taken to be denaturing solvents and GuHCl is the most commonly used denaturing agent, such a generalization is seen to be mistaken, because $(GuH)_2SO_4$, like other sulfates, stabilizes proteins against thermal denaturation (39). This is also true for the phosphate (not shown in Fig. 16).

As indicated earlier, nothing is known about the origin of the lyotropic series. In the protein literature, one finds numerous interpretations relying on "binding" phenomena. However, such

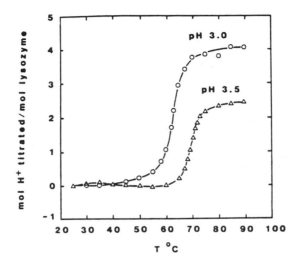

Fig. 12. The proton uptake of lysozyme as a function of temperature *(4)*.

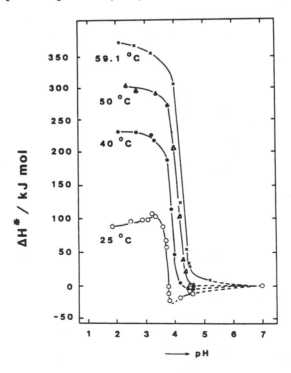

Fig. 13. Isothermal calorimetric pH titration of the N → D transition of carp muscle parvalbumin *(38)*.

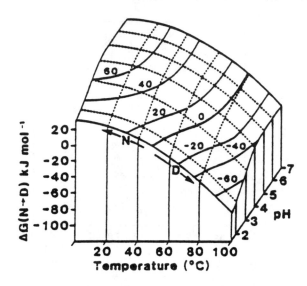

Fig. 14. Temperature-pH-stability profile for lysozyme *(37)*. The $\Delta G = 0$ contour corresponds to the conditions for half-denaturation.

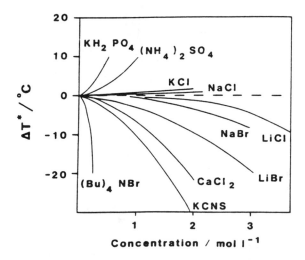

Fig. 15. The effects of salts on the denaturation temperature (T^*) of ribonuclease at pH 7 *(39)*.

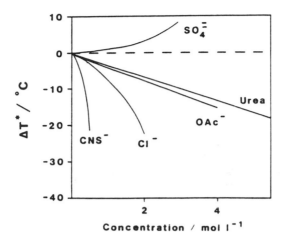

Fig. 16. The effects of urea and guanidinium salts on the denaturation temperature (T^*) of ribonuclease at pH 7 (39).

hypotheses are unrealistic, although they may be made to fit the data for individual experimental studies. Just as ion-specific effects determine the solubility and stability of proteins, so they also influence the solubility of argon gas (and other simple gases) in water (36). The origins of the lyotropic series must be closely related to the details of ion hydration and how such hydrated ions can interact with hydrated argon atoms or protein molecules. There are also indications that the lyotropic series is associated with the phenomenon of hydrophobic hydration , but the details remain obscure.

　　Although nomenclature in terms of *salt* effects may be appropriate to discussions of the effects of ions on protein behavior, such nomenclature appears to be inapplicable to nonelectrolyte effects on protein stability and solubility. Nevertheless, the effects of some organic substances fit in well with those of the ions in the lyotropic series. Thus carbohydrates (sugars and sugar alcohols) tend to stabilize proteins against denaturing treatments, the magnitude of the protection being, to a first approximation, a function of the number of —OH groups (28). This is shown in Fig. 17, where the effects of various organic compounds are compared with those of ions.

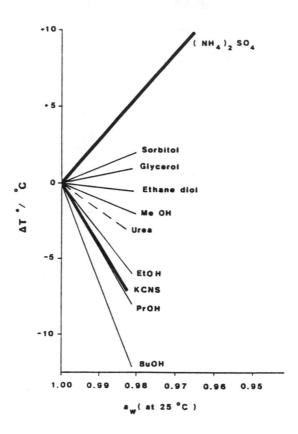

Fig. 17. The effect of organic reagents on the denaturation temperature (T^*) of ribonuclease at (nominal) pH 7. Additive concentrations are expressed in terms of water activity (a_w), mainly to demonstrate that a_w is not a reliable measure of enzyme stability.

The universal destabilizing action (chaotropism) of urea is well known (36). Urea enhances the solubility of nonpolar compounds in water (salting-in?), reduces the tendency of surfactants to form micelles, and destabilizes phospholipid membrane structures. All these effects can be accounted for on a semiquantitative basis by the influence of urea on the intermolecular, hydrogen-bonded structure of liquid water.

The behavior of proteins in multicomponent mixtures should be of technological importance but has not been widely studied. It is believed that in mixtures of salts the observed behavior is the

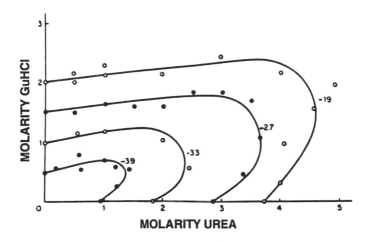

Fig. 18. Concentrations of urea and GuHCl plotted at a constant k_{app} of lysozyme unfolding. Lines represent the average contours of constant $\log_{10}k$ (40).

additive sum of the various ionic contributions to the N–D equilibrium. However, this simple additivity is unlikely to apply in mixtures of urea and GuHCl, which are *worse* protein denaturants than would be expected from their individual effects, as seen in (Fig. 18) (40).

The effects of alcohols and their derivatives (alkoxyalcohols and haloalcohols) on protein stability are particularly complex. This is because of the general hydrophobic nature of these substances (41), the properties of which display complicated concentration and temperature dependencies, as illustrated in Fig. 19 for the influence of ethanol concentration and temperature on the N-state stability of RNase (42). It is seen that at low temperatures and low concentrations ethanol acts as a typical salting-out compound that turns to salting-in behavior at higher temperatures and high concentrations. The destabilizing effect per mol additive increases with the carbon chain length of the additive, evidence for the involvement of hydrophobic interactions in the N –> D process (36). The powerful destabilizing effect of butanol is shown in Fig. 17. Similar salting-in effects are observed for tetraalkylammonium halides, which do not behave as typical electrolytes,

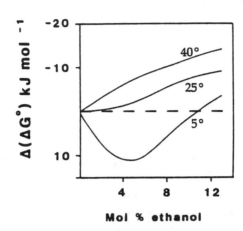

Fig. 19. The influence of ethanol concentration and temperature on the stability of ribonuclease, relative to its stability in aqueous solution at pH 7. $\Delta(\Delta G^0) > 0$ refers to stabilization of the native state (42).

but as hydrophobic alkyl derivatives. The common description of such effects as additive "binding" is erroneous, since the hydrophobic interaction between protein and additive results from the *repulsion* of apolar residues or patches on the protein surface from water rather than from an inherent attraction (van der Waals type) between such domains (*see* Chapter 4). The interactions between proteins and long-chain alkyl derivatives find application in SDS gel electrophoresis. Here the situation is further complicated by micelle formation of the surfactant additive. The exact mode of interaction between SDS and the protein is not yet clear.

3.5. Surface Denaturation

Since proteins are amphiphilic polyelectrolytes, they exhibit some degree of surface activity, i.e., they adsorb to interfaces. Hence, proteins act as emulsifying/dispersing agents, as in the stabilization of fat in blood or in milk, or the stabilization of air bubbles in ice cream. When the surface forces are strong and coupled with a low N-state stability, sorption induces surface denaturation, as, for instance, in the precipitation of blood proteins on contact with some plastic materials.

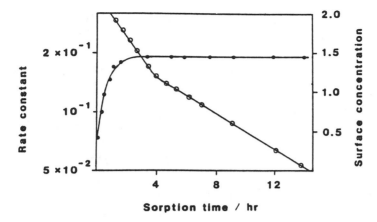

Fig. 20. Comparison of surface concentration-time dependence (● right-hand ordinate) with rate of change of surface pressure (○, left-hand ordinate) for lysozyme. Note that the change in slope occurs when surface concentration reaches its steady-state value *(43,44)*.

Three types of properties are of importance in relation to technological applications of protein sorption:

1. The rate of sorption/denaturation;
2. Sorption isotherms, i.e., the quantity of protein in the interfacial film; and
3. The mechanical properties of the sorbed film.

Quantitative, detailed information on the surface properties of proteins exists for β-casein, BSA, and lysozyme *(43–45)*. The initial sorption of proteins to a clean air/water interface is rapid and diffusion-controlled. This is followed by a slower reorganization/ denaturation step that takes place with an apparent desorption, which is, however, not accompanied by a decrease in the surface pressure. The adsorbed protein therefore unfolds and occupies the surface more economically; i.e., fewer protein molecules in the surface layer will cover the available area. Eventually, more protein will be sorbed to the already existing surface layer. The existence of the two processes is illustrated in Fig. 20. The slower process has a half-life of hours and depends on the protein concentration. Intermolecular effects are important (e.g., aggregation), and the process determines the rheological properties of the sorbed layer.

The initial fast process is associated with the irreversible formation of a monolayer of concentration 2–3 mg/m^2. Eventually, the thickness increases to 5–6 nm. Although the film pressure is determined by the properties of the monolayer, films can build up to >10 nm thickness. The secondary sorption is reversible, indicating that the protein is in a native state.

The actual structure of the sorbed protein depends on the stability of the N state. Lysozyme, which is particularly stable, can exist in the native state at an air/water interface, but at an oil/water interface, even this stable molecule is denatured (compare the effects of alcohols on protein stability). The BSA is partially native, but no definite conformational transition can be observed at a particular packing density during sorption. β-Casein, which is a highly surface active molecule, is extended in the interfacial film; an increase in the concentration leads to densely packed films. The mechanical properties of interfacial protein films are of considerable importance to their role as stabilizers of disperse systems. The dilatational modulus, which measures the response of the film to a compression/expansion cycle, has been measured for β-casein (3–5), BSA (60–400), and lysozyme (200–600 mN/m^2). A comparison of protein gels and interfacial films suggests that the sorbed layer has the properties of a thin gel layer.

Sorption can provide problems in separation procedures and processes involving membranes, i.e., where large surface areas are coupled with high shear rates (35). Protein sorption at interfaces has important medical implications and industrial applications, and is currently developing into a subject of fundamental study by colloid and protein chemists (46).

4. Multistate Denaturation

In many cases, the simple two-state model of protein stability, as discussed in Section 2, is only a crude approximation to the actual course of denaturation, especially when intermediate states have in fact been identified, or even isolated. Various schemes can be written for multistate denaturation, e.g.:

$$N \rightarrow \underset{\downarrow}{I} \rightarrow D_1 \quad \text{or} \quad N \rightarrow D_1 \rightarrow D_2 \qquad (12)$$
$$D_2$$

One or the other of the steps might be fast or slow, reversible or irreversible (e.g., aggregation). A slow process I → D has been identified with a *cis–trans* rearrangement in proline and studied in detail for the unfolding of RNase *(47)*. This protein contains four proline residues at positions 42, 93, 114, and 117, respectively. Of these, only pro-93 and pro-114 are involved in the slow process that accompanies refolding; the other two pro residues are in the *trans* configuration both in the N and the D states of the protein. On the other hand, RNase in the D state contains 70% *cis*-proline-93 and 95% *cis*-proline-114. The refolding kinetics of the denatured protein depends on the *trans–cis* isomerization of these two residues, which are 100% in the *cis* configuration in the N state.

Stable intermediates are often produced by the stepwise dissociation of a multisubunit protein and have been implicated in the phenomenon of cold inactivation. In such cases, the multimer reversibly dissociates into smaller multimers or even single subunits that, under physiological conditions, maintain their native conformations, but lose their biological activity *(47,48)*. Examples of such behavior include phosphofructokinase, glucose-6-phosphate dehydrogenase, ATPase, pyruvate carboxylase, and carbamyl phosphate synthetase.

The various stages of denaturation have been studied in detail for β-lactoglobulin, which normally exists as an α,β-dimer at room temperature and $3 < pH < 6$. The dimer dissociates at $pH > 6$, and this is followed by a conformational transition that, in turn, leads to nonspecific, irreversible aggregation *(16)*. The various processes can be summarized as follows:

pH	3–6	6–9	9–12
Temperature	20°C	0°C	
State	N_2	$N_2 \rightarrow 2R$	$2R \rightarrow S \rightarrow S_n$

where N is the native state (αβ-dimer), R is the dissociated monomer, S is the denatured monomer, and S_n the aggregated state.

Another way of studying the details of unfolding is by the removal of short peptide sequences. For instance, by removing the C-terminal residues 121–124 from RNase, the stability (thermodynamic) is reduced by 70%, but the enzyme activity, by 99.5%. If this modified protein is then denatured, the regain of activity is faster than the rate of refolding, suggesting the existence of inter-mediate states. Urea gradient electrophoresis can be used to detect the existence of intermediate species (49).

DSC is a particularly suitable technique for the study of sequential denaturation of complex proteins. Figure 21 shows the DSC power-time curve corresponding to the heat denaturation of the fibrinogen dimer (50). The sequential unfolding of specific domains can be clearly identified, and thermodynamic informa-tion about their relative stabilities can be obtained by deconvoluting the global DSC profile.

5. Renaturation

It is often assumed that proteins know how to fold spontane-ously. Although this may well be true, some believe that the pro-tein can do with a little help, which takes different forms. Combinations of treatments, e.g., urea + dilution + dialysis; pH + cold + pH, are comnonly used. Most such treatments are based on the principle of salting-in, followed by reorganization, and then salting-out. The kinetics of renaturation depend on the exposure time to the denaturant. If denaturation is performed with urea that is then immediately removed, refolding is very fast. If the dena-tured protein is left in the urea solution for any length of time, 60% is renatured rapidly, followed by a slow renaturation of the re-mainder. The greatest success is achieved by allowing the protein to sample a range of different environments and interactions. Un-der in vivo conditions, specialized proteins, e.g., heat shock pro-teins (hsp 70), help proteins to refold after suffering heat shock.

With the advent of recombinant DNA technology, the ability to renature proteins has assumed great importance. Most proteins produced by such methods are deposited within the host cell in the form of insoluble "inclusion bodies." By systematic variation of the parameters influencing the folding, formation of disulfide bonds, and association of the constituent subunits, renaturation procedures can be designed that permit acceptable levels of native recombinant proteins to be obtained. Thus, Büchner and Rudolph have reported the production of recombinant antibody fragments

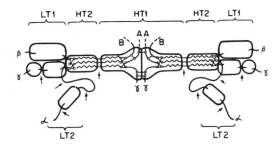

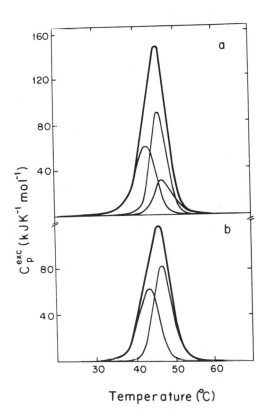

Fig. 21. Subunit structure of the fibrinogen dimer (*see also* chapter 4, Fig. 27). Wavy lines represent thermostable parts of the molecule. The global DSC profiles of the (a) heavy and (b) light fragments have been deconvoluted to obtain the thermal transition enthalpies of the various subunits or groups of subunits. Reproduced, with permission, from ref. 50.

in 40% yields that are identical with native murine F_{ab} in all physico-chemical parameters tested (51). The renaturation procedure included treatments with oxidized and reduced glutathione, aspara-gine, and several buffer systems at several temperatures and for different periods of time. In the eukaryotic cell, helper proteins (pro-line isomerase, thiol disulfide isomerase) assist newly synthesized proteins to gain the native structures (52). An interesting example is protein disulfide isomerase (PDI), which has been isolated fram cytokines, hormones, immunoglobulins, toxins, collagens, and so on. It is a dimer of molecular weight 110,000. It contains 27% asx and glx residues, is soluble, heat stable, and not glycosylated. It forms intra- and intermolecular disulfide bonds and catalyzes disul-fide exchange. Under in vitro conditions, these helpers have to be replaced by carefully chosen chemical and/or physical environments.

In large proteins, e.g., tPA (527 residues), renaturation is sub-ject to several competing reactions, some of which result in incor-rectly folded or aggregated states. By keeping the D-protein concentration low, the reaction D —> I —> N is allowed to reach completion, after which more denatured protein is gradually added. Some renaturation reactions are very slow, with half-times of the order of 300 h. Renaturation is made more complicated by the fact that, even if some steps are reversible, others are not, and whereas some steps are fast and exhibit first-order kinetics, others are slow with second-order kinetics (53).

Detailed studies have been performed of the renaturation kine-tics of ribonuclease by pH jump methods. The enzyme was first denatured at pH 1.7, followed by a pH jump to pH 6, and the refolding was monitored (54). Three steps were observed:

$$D \xrightarrow{\text{fast}} I_1 \xrightarrow{\text{fast}} I_2 \xrightarrow{\text{slow}} N$$

| Folded to inactive sec. structure | Folded, active site, unstable, incorrect pro isomers | Pro isomerization |

The overall folding is not blocked by incorrect pro isomers, i.e., isomerization can follow folding. The same is true for multi-domain proteins. Thus, the refolding of aspartate kinase dehydrogenase is not blocked by proline cis–trans isomerization. Individual domains are reconstructed rapidly, followed by a slow rearrangement of domains with respect to each other to yield the active state.

6. Characterization of Protein D States

The native state of a protein is unique and can be exactly characterized in great detail. For small globular proteins, the N state possesses a high degree of intramolecular order, which, in the case of more complex proteins, extends to the details of the intermolecular interactions and the architecture of the multisubunit structure. The D state, on the other hand, is not as easy to define, since it approaches the random coil configuration more commonly found in synthetic macromolecules dissolved in "good" solvents. (The use of the word "good" in this context is in terms of the Flory-Huggins theory of polymer solutions and refers to solutions for which the χ-parameter does not exceed 0.5.)

The description of the D state is made more difficult by the fact that it is definitely *not* a truly random coil. In fact, denaturation takes place in stages and, depending on the method used to denature a protein, several different D states can be identified. For lysozyme and α-lactalbumin, four intermediate states have been identified, of increasing degree of unfolding, and depending on whether heat, LiCl, LiClO$_4$, GuHCl, or urea is employed as denaturant (55,56). Figure 22 illustrates the effects of chemical denaturing agents on the structural integrity of α-lactalbumin, as monitored by ORD. In each case, the typical N–D curves are obtained (*see* Fig. 1), but of different degrees of cooperativity (steepness) and with different limiting values. According to Fig. 22, GuHCl and urea give rise to the same, maximally unfolded state, and LiClO$_4$ has the least effect, with LiCl occupying an intermediate position. With the aid of sequential denaturation treatments, it can be shown that state IV can be reached by two different routes, as shown in Fig. 23, but that conversions I —> II or I —> III do not take place. State II, which corresponds to the thermally denatured protein, has been observed for lysozyme and RNase, but not for α-lactalbumin. All the evidence suggests that different reagents cause unfolding of different domains within the protein, and that GuHCl and urea produce the most complete unfolding.

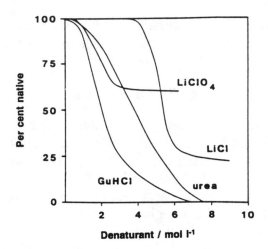

Fig. 22. The degree of denaturation of α-lactalbumin (monitored by ORD) induced by different chemical agents.

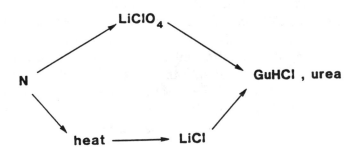

Fig. 23. Intermediate states in the denaturation α-lactalbumin or lysozyme induced by different treatments. For details, *see text*.

Bibliography

1. Mozhaev V. V. and Martinek K. (1990) *Adv. Drug Delivery Rev.* **4**, 387–419.
2. Pain R. H. (1979) in *Characterization of Protein Conformation and Function* (Franks F., ed.) Symposium Press, London, p. 19.
3. Tanford C. (1970) *Adv. Protein Chem.* **24**, 1–95.
4. Pfeil W. and Privalov P. L. (1979) in *Biochemical Thermodynamics* (Jones M. N., ed.) Elsevier, Amsterdam, p. 75.
5. Privalov P. L. and Khechinashvili N. N. (1974) *J. Mol. Biol.* **86**, 4291–4303.
6. Brandts J. F. (1964) *J. Amer. Chem. Soc.* **86**, 665–684.

7. Privalov P. L. (1979) *Adv. Protein Chem.* **33,** 167–241.
8. Baldwin R. L. (1986) *Proc. Natl. Acad. Sci. USA* **83,** 8069–8072.
9. Wolanczyk J. P. (1988) *Cryo-Letters* **9,** 218,219.
10. Schellman J. A. and Hawkes R. B. (1980) in *Protein Folding* (Jaenicke R., ed.), Elsevier/North Holland Biomedical Press, Amsterdam, p. 331.
11. Franks F., Hatley R. H. M., and Friedman H. L. (1988) *Biophys. Chem.* **31,** 307–315.
12. Franks F. (1985) *Biophysics and Biochemistry at Low Temperatures,* Cambridge University Press, Cambridge.
13. Storey K. B. (1990) *Phil. Trans. Roy. Soc. B* **326,** 635–650.
14. Privalov P. L., Griko Yu. V., Venyaminov S. Y., and Kutyshenko V. P. (1986) *J. Mol. Biol.* **190,** 487–498.
15. Hatley R. H. M. and Franks F. (1989) *Eur. J. Biochem.* **184,** 237–240.
16. Douzou P. (1977) *Cryobiochemistry,* Academic, London.
17. Fink A. L. and Painter B. (1987) *Biochemistry* **26,** 1665–1671.
18. Chen B. and Schellman J. A. (1989) *Biochemistry* **28,** 685–691.
19. Franks F. and Hatley R. H. M. (1985) *Cryo-Letters* **6,** 171–180.
20. Hatley R. H. M. and Franks F. (1986) *Cryo-Letters* **7,** 226–233.
21. Griko Yu. V., Privalov P. L., Sturtevant J. M., and Venyaminov S. Yu. (1988) *Proc. Natl. Acad. Sci. USA* **85,** 3343–3345.
22. Griko Yu. V., Venyaminov S. Yu., and Privalov P. L. (1989) *FEBS Lett.* **244,** 276–278.
23. Privalov P. L. (1989) *Rev. Biophys. Biophys. Chem.* **18,** 47–70.
24. Privalov P. L. (1990) *Crit. Rev. Biochem. Mol. Biol.* **25,** 281–305.
25. Hatley R. H. M. (1991) Ph.D. Thesis, Council for National Academic Awards.
26. Franks F. and Hatley R. H. M. (1992) *Adv. Low Temperature Biol.,* **1,** 142–179.
27. Franks F. and Hatley R. H. M. (1991) *Pure Appl. Chem.* **63,** 1367–1380.
28. Gerlsma S. Y. (1968) *J. Biol. Chem.* **243,** 957–961.
29. Arakawa T. and Timasheff S. N. (1982) *Biochemistry* **21,** 6536–6544.
30. Zipp A. and Kauzmann W. (1973) *Biochemistry* **12,** 4217–4228.
31. Hawley S. A. and Mitchell R. M. (1975) *Biochemistry* **14,** 3257–3264.
32. Charm S. E. and Wong B. L. (1970) *Biotech. Bioeng.* **12,** 1103–1109.
33. Charm S. E. and Wong B. L. (1970) *Science* **170,** 466–468.
34. Charm S. E. and Wong B. L. (1975) *Biorheology* **12,** 275–278.
35. Katzenstein G. E., Vrona S. A., Wechsler R. J., Steadman, B. J. Lewis R. V., and Middaugh C. R. (1986) *PNAS* **83,** 4268–4272.
36. Franks F. and Eagland D. (1975) *Crit. Rev. Biochem.* **3,** 165–219.
37. Pfeil W. and Privalov P. L. (1976) *Biophys. Chem.* **4,** 41–50.
38. Filimonov V. V., Pfeil W., Tsalkova T. N., and Privalov P. L. (1978) *Biophys. Chem.* **8,** 117–122.
39. von Hippel P. H. and Wong K. Y. (1965) *J. Biol. Chem.* **240,** 3909–3923.
40. Robson B. (1982) in *Water Biophysics* (Franks F. and Mathias S. F., eds.) John Wiley & Sons, Chichester, pp. 62–66.
41. Franks F. and Desnoyers J. E. (1985) *Water Sci. Rev.* **1,** 171–232.
42. Brandts J. F. and Hunt L. (1967) *J. Amer. Chem. Soc.* **89,** 4826–4838.

43. Graham D. E. and Phillips M. C. (1979) *J. Colloid Interface Sci.* **70,** 403–414.
44. Graham D. E. and Phillips M. C. (1979) *J. Colloid Interface Sci.* **70,** 415–439.
45. Graham D. E. and Phillips M. C. (1980) *J. Colloid Interface Sci.* **76,** 227–250.
46. Symp. Protein and Polyelectrolyte Sorption (1988) *J. Colloid Interface Sci.* **111,** No. 2.
47. Brandts J. F., Halvorson H. R., and Brennan M. (1975) *Biochemistry* **14,** 4953–4963.
48. Bock P. E. and Frieden C. (1978) *Trends Biochem. Sci.* **May,** 100–103.
49. Creighton T. E. (1979) *J. Mol. Biol.* **129,** 235–264.
50. Privalov P. L. (1982) *Adv. Protein Chem.* **35,** 1–104.
51. Büchner J. and Rudolph R. (1991) *Bio/Technology* **9,** 157–162.
52. Pelham H. R. B. (1989) *Ann. Rev. Cell Biol.* **5,** 1–23.
53. Sochava I. V., Belopolskaya, T. V., and Smirnova, O. I. (1985) *Biophys. Chem.* **22,** 323–336.
54. Grafl R. and Schmid F. X. (1986) *J. Mol. Biol.* **191,** 281–294.
55. Kugimiya M. and Bigelow C. C. (1973) *Can. J. Biochem.* **51,** 581–585.
56. Sharma R. N. and Bigelow C. C. (1974) *J. Mol. Biol.* **88,** 247–257.

Twelve

Protein Hydration

Felix Franks

1. Historical

All biological activity critically depends on a correctly folded protein state, and folding takes place on the ribosome in an aqueous medium. Even in vitro, crystalline proteins contain almost 40% water *(1)*. Hydration is therefore likely to play a major role in the maintenance of the native state. The physical properties of water, too, are sensitive to the same factors that influence protein stability, so that some connection is likely. Nevertheless, the realization that hydration might be an important factor in biological and technological function only dates from the early 1970s. Figure 1 summarizes the chronological development of our present understanding of two aspects of enzyme function: specificity and catalytic activity *(2)*. It is now accepted that protein hydration interactions are of crucial importance in the maintenance of higher order structures and in rendering proteins useful as technological macromolecules. As will presently be shown, however, there is as yet little understanding about the details of such interactions and their role in determining the functional attributes of proteins.

From: *Protein Biotechnology* • F. Franks, ed. © 1993 The Humana Press Inc.

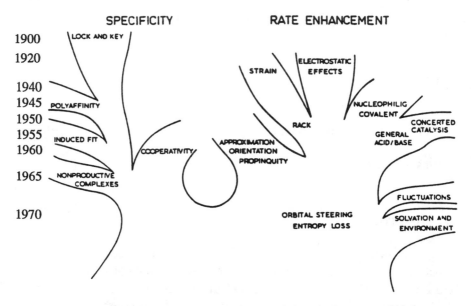

Fig. 1. Chronology of our understanding of the mechanisms that govern enzyme catalysis (ref. 2). Note that a consideration of solvation as a contributing factor dates only from 1970.

2. Classification of Protein Hydration Interactions

In a discussion of protein hydration, three distinct types of solute–water interactions can be distinguished. They are ion hydration (side chains of asp, asn, glu, gln, lys), hydrogen bonding between polar groups and water, and so-called hydrophobic hydration (ala, val, leu, ile, phe), which describes the response of water to chemically inert residues. In principle, the second type should be characteristic of *all* amino acid residues because, quite apart from polar side chains (e.g., —OH in ser, thr, tyr), every peptide can act as proton donor and acceptor. Some amino acids behave, on balance, as typically hydrophobic molecules (3), whereas others more closely resemble polyhydroxy compounds in their hydration properties. The complexity of this particular type of hydration effects has recently been discussed in detail (4).

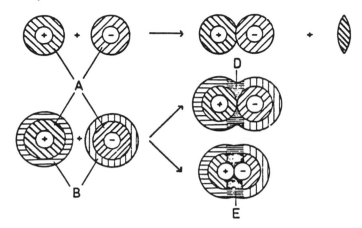

Fig. 2. Schematic representation of ion–solvent interaction in solution: (A) Primary hydration shells. (B) Secondary hydration shells. (D) Outer cosphere overlap (solvent-separated ion pair). (E) Primary cosphere overlap (contact ion pair) (from Franks F., in *Water* [1983] Royal Society of Chemistry Paperbacks, London).

2.1. Ion Hydration

Conceptually, the interaction of water with charged groups is the simplest of the three types of hydration effects. It is of an electrostatic nature (ion–dipole interaction) and has a long range (proportional to r^{-3}, where r is the distance of separation). It is therefore strong compared to most other types of intermolecular interactions. Phenomenologically, ion–water interactions in solution can be represented as shown in Fig. 2. Many quantitative models assign a spherical, monomolecular hydration sphere to the ion, beyond which water is then assumed to be unperturbed (5). A model of this type can be elaborated by including two or more layers of water molecules. Such subtle modifications can sometimes be detected by spectroscopic techniques.

As regards the geometrical details of an ion hydration shell in solutions, neutron diffraction has provided the most detailed information, although mainly for monatomic ions (6), polyatomic ions of the type COO^- or NH_3^+ being too complex for the deconvolutions of the scattering intensity curves. Figure 3 illustrates the hydration

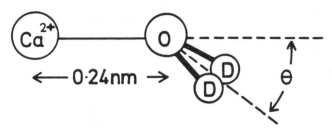

Fig. 3. Hydration geometry of alkali and alkaline earth metals (except Li⁺ and Mg²⁺), showing the tilt of the water molecules (six in the primary hydration shell). The angle θ decreases with decreasing concentration (*see* ref. 6).

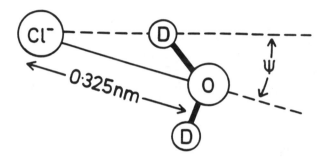

Fig. 4. Hydration geometry of halide anions, showing that the water molecules are not symmetrically disposed with respect to the Cl⁻ — O axis. The angle of tilt ψ is a function of the concentration (*see* ref. 6).

geometry of the alkali metal ions (except Li⁺) and Ca²⁺ in solutions of their chlorides. In all cases, the time-averaged primary hydration shell consists of six water molecules oriented with respect to the cation as shown. The organization of water molecules beyond the primary hydration sphere is still uncertain, but the spatial incompatibility of the primary hydration sphere with the tetrahedral disposition of water molecules in the bulk liquid demands a region of structural mismatch that is believed to have important implications for ion–ion and ion–molecule interactions in aqueous solutions (7). Figure 2 is therefore an oversimplification, because it does not include this region of structural mismatch.

The general hydration geometry of halide anions, as obtained from neutron diffraction, is shown in Fig. 4. Here again, six water molecules form the primary hydration sphere, with one of the—OH bonds pointing toward the anion. Many thermodynamic models of ion hydration have been proposed; all but the most recent ones suffer from the limitation that the solvent is treated as a continuum, possessing only macroscopic physical properties (dielectric permittivity, viscosity) (8). Recent developments in the statistical thermodynamics of electrolyte solutions show promise of more realistic representations of ion–water interactions (5,9). The requirement for more realistic treatments of ion hydration is made particularly urgent by the lack of an explanation for specific ion (lyotropic) effects, discussed in more detail in chapters 5 and 11.

The nature and properties of the hydrated proton in solution are of particular relevance to discussions of protein stability. It has been stated that changes in pH cause less perturbations to proteins than any other destabilizing influence (10). In the solid and gaseous states and in some nonaqueous environments, the species H_3O^+ is well characterized, and other species are also distinguishable and can be prepared. Although the existence of $H_9O_4^+$ clusters has been claimed for liquid water, the evidence is circumstantial. The problems involved in arriving at meaningful descriptions of the hydrated proton have been discussed in depth by Ratcliffe and Irish (11).

2.2. Hydrophilic Hydration

This term describes the *direct* interaction, by hydrogen bonding, between a functional group on a solute molecule and water, the water molecule being able to act either as proton donor or acceptor. This type of interaction is expected to figure largely in the hydration of the peptide bond, with both the >CO and the >NH groups able to participate in hydrogen bonding. The influence of this type of solvation on tertiary structure is graphically illustrated by polyproline, in which the solvent alone determines the respective stabilities of the polyproline-I and II structures.

Among amino acid side chains that might hydrogen bond to water molecules, ser, thr, and tyr possess —OH groups, and his, arg, and trp can interact via the >NH group. The contributions of such side chain hydration interactions to protein conformational stability appear to be much less important than those made by main chain (peptide) hydration. Indeed, hydrophilic hydration may only

Fig. 5. Water cavity capable of accommodating an inert molecule. The particular structure shown consists of a hydrogen-bonded framework of 12 water molecules, eight of which (shaded circles) are nearest neighbors to the apolar molecule.

play a minor role in the maintenance of higher order structures of proteins, as distinct from carbohydrates, the solution configurational properties of which are determined largely by hydrophilic hydration *(4)*. In this respect, carbohydrates differ markedly from peptides, the conformations of which are very dependent on a fine balance between different types of hydration interactions.

2.3. Hydrophobic Hydration and Hydrophobic Interactions

In the vicinity of apolar groups, water molecules reorganize spatially and orientationally so as to be able to accommodate such groups with which they cannot interact directly *(3,12)*. Slight distortions in the normal, predominantly tetrahedral geometry of the water molecules can give rise to a variety of differently shaped cavities, a simple example of which is shown in Fig. 5. Such structures can actually exist in the crystalline state, and they bear a resemblance to ice, in the sense that each oxygen atom is hydrogen bonded to four other oxygen atoms, giving rise to three-dimensional water networks with cavities of various sizes and shapes, with the solute (guest) molecules occupying these cages *(13)*.

As to the importance of the hydrophobic hydration phenomenon to protein function, the main point to note is that the configurational freedom of water molecules, i.e., the number of allowed mutual orientations, is reduced in the proximity of the apolar residue, because a water—OH vector must not be directed toward the center of the cavity. This reduction in configurational freedom leads to the well-known result that when an apolar molecule or residue R is transferred from the gas phase or from an apolar solvent medium to water, this transfer occurs with a loss of entropy. In other words, $T\Delta S_t < 0$, where the subscript t describes the transfer

$$R\,(\text{apolar solvent}) \rightarrow R\,(\text{aqueous medium}) \qquad (1)$$

It is also found that the transfer is thermodynamically unfavorable, i.e., $\Delta G_t > 0$. The above results imply that the reluctance of apolar molecules to mix with water derives not from a net repulsive energy of interaction (which would be expressed through ΔH_t 0), as is usually the case for low solubility. In other words, the low solubility of hydrocarbons in water is the direct result of the entropy lost by the water because of cage formation, rather than a lack of van der Waals attraction between hydrocarbon and water. Thus, for hydrophobic hydration, another important result is that I ΔH_t I < Tl ΔS_t I, whatever the sign of ΔH_t (3). Yet another symptom of hydrophobic hydration is that $\Delta C_p \gg 0$, consistent with the cage model, since a large specific heat change is indicative of the creation of structure or order in the liquid that is sensitive to thermal disruption.

The amino acid residues that are subject to hydrophobic hydration include ala, val, leu, ile, and phe; possibly sulfur-containing amino acids cys and met should also be included, as well as the imino acid pro. The above discussion demonstrates that apolar residue/water contacts are thermodynamically (configurationally) unfavorable, and it is to be expected, therefore, that the apolar residues will be found predominantly in the interior of a folded protein structure, if the stability of the protein in its native environment is to be maximized. Indeed, it is believed that hydrophobic hydration (or rather, its converse, the hydrophobic interaction) provides much of the driving force for folding (14–17).

Since the transfer of an apolar residue R from hydrocarbon to water, as depicted in Eq. (1), is seen to be thermodynamically unfavorable, then the converse, i.e., the association of R residues in water, should be accompanied by a negative free energy change.

At the simplest level, two hydrocarbon molecules or two alkyl residues, each with its hydration cage, would gain in stability by their association, because this would "release" water molecules from the cages, which could then relax into their more stable, unperturbed configurational states. The process

$$2R(\text{hydrated}) \rightarrow R_2(\text{hydrated}) + \text{water} \tag{2}$$

would therefore be expected to take place spontaneously. Once again, it is emphasized that the driving force for such an association does *not* derive from an attraction (e.g., by van der Waals forces) between alkyl groups, but from an extrusion of alkyl groups by water for configurational reasons. The process is said to be entropy-driven, in the sense that $T\Delta S_\phi > 0$ and $T|\Delta S_\phi| > |\Delta H_\phi|$, where the subscript ϕ describes the association in Eq. (2). Thus, what *appears* to be an attraction between alkyl residues (negative free energy) is actually the sum of two repulsions, and the term hydrophobic interaction, which is commonly used to describe the process, is really a misnomer.

It has been powerfully argued (18) that a lack of appreciation of the configurational origin of the hydrophobic interaction can lead to misleading conclusions and some confusion about protein structure. Thus, the free energy associated with the process in Eq. (2) can easily be fitted by simple one-parameter models, e.g., the molecular surface area of the alkyl residues. ΔG_t is then expressed as the work done to create the cavity in water and described in terms of a surface tension. Such simplistic fitting devices may be able to model ΔG, but they fail to model ΔH and break down completely for the second T and P derivatives of ΔG (heat capacity and compressibility). However, the reader is warned that the protein literature contains many accounts purporting to describe the role of hydrophobic effects in the stabilization of protein tertiary structures in terms of van der Waals interactions. Indeed, such interpretations are even now still being advanced from time to time (19). For up-to-date thinking and discussions of hydrophobic effects, the reader is referred to the 1982 Faraday Symposium (no. 17, The Hydrophobic Interaction) of the Royal Society of Chemistry.

Before leaving the subject of the "simple" hydrophobic interaction as represented by Eq. (2), we stress that the association of many R residues, according to the scheme

$$n\text{R(aqueous)} \rightarrow \text{R}_n\text{(aqueous)} + \text{H}_2\text{O} \qquad (3)$$

where $n \gg 2$, as for instance in the formation of a surfactant micelle, cannot be correctly described by a stepwise aggregation starting from Eq. (2). The thermodynamics describing the association of pairs or small clusters of hydrated alkyl residues are *qualitatively* different from those describing multiple aggregation (20). This observation is of importance in calculations of protein stability, but is not even mentioned in most monographs dealing with the subject.

The simple model described by Eqs. (1) and (2) is in any case an oversimplification of the real processes that occur in water when it is perturbed by apolar residues. For instance, the cavity-forming effects produced by the introduction of the apolar group can hardly be confined to the primary hydration shell; the configurational perturbation is apparently felt by water molecules as far distant as 5 nm from the center of the primary cavity (21). This makes hydrophobic hydration a long-range interaction, and calculations of hydrophobicity should take this into account, but rarely do so (20,22).

Equation (2) can be taken as the elementary unit step in calculations of protein folding and stability in the sense that in the D state most of the apolar residues are believed to be fully exposed to the solvent, but that the folding process reduces the solvent accessibility of these residues. As more apolar residues become involved in the folding process, so eventually the volume of apolar residues in the center of a globular protein increases and a considerable number of such residues will become completely inaccessible to the solvent. The interior of a protein (like that of a surfactant micelle) is often compared to a liquid hydrocarbon environment. This type of argument has led to the simple concept of hydrophobicity that is expressed as the free energy change ΔG_ϕ associated with the transfer of an apolar group from a hydrocarbon (typically octane) to an aqueous environment.

Table 1
Order of Decreasing Hydrophobicity Estimated
as the Free Energy of Transfer (kJ/mol) for a Transfer
from a Hydrophilic to a Hydrophobic Phase[a]

Residue	Water → ethanol	Transfer free energy Protein folding	Water → vapor
Ile	21	2.9	9.0
Phe	21	2.1	−3.2
Val	12.5	2.5	8.4
Leu	14.4	2.1	9.6
Trp	27.2	1.3	−24.7
Met	10.5	1.7	−6.3
Ala	4.2	1.3	8.0
Gly	0	1.3	10.1
Cys	0	3.8	−5.2
Tyr	18.8	−1.7	−25.6
Pro	6.3	−1.3	−
Thr	2.1	−0.8	−20.6
Ser	−2.1	−0.4	−21.3
His	4.2	−0.4	−43.3
Glu	−	−2.9	−43.0
Asn	−6.3	−2.1	−40.5
Gln	−4.2	−2.9	−39.4
Asp	−	−2.5	−45.8
Lys	−	−7.5	−40.0
Arg	−	−5.8	−83.6

[a]Data from ref. 23.

$$R(octane) \rightarrow R(aqueous) \tag{4}$$

A problem arises when ΔG_ϕ is to be calculated for amino acids, peptides, and proteins, because the contribution of the peptide group(s) has to be estimated. Different methods have been proposed, and they have been discussed by Eisenberg et al. (23). Hydrophobicity scales, derived on the basis of different models, are summarized in Table 1. Here the values in the second column (24)

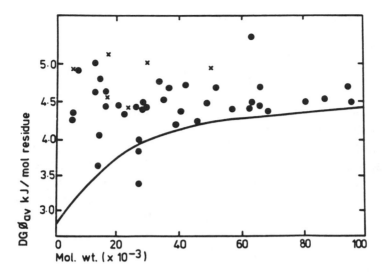

Fig. 6. Average hydrophobicities of globular proteins as a function of their molecular weights; crosses refer to associating proteins. The curve shows the lower limit of ΔG_ϕ compatible with a globular conformation. Approximately 50% of globular proteins have ΔG_ϕ = 4.6 kJ/mol (for details, *see* ref. 25).

refer to the transfer of the residue from ethanol to water, and the other columns all refer to other modifications of Eq. (4) in the sense that the transfer is not necessarily from liquid hydrocarbon.

Bigelow, who calculated the average hydrophobicities of 150 globular proteins, concluded that for a globular protein to be stable, it must possess a certain minimum average ΔG_ϕ per residue (25). This is illustrated in Fig. 6. Of the 150 proteins examined, 50% were found to have ΔG_ϕ (average) > 4.6 kJ/mol. The curve in Fig. 6 denotes the estimated lower limit of stability for a globular protein.

3. Medium Effect on Protein Stability

It is now possible to attempt an analysis of the various interactions that might contribute to the stability of the N state relative to the D state, and to speculate on the effects of temperature or additives on these various interactions. As discussed earlier, $\Delta G(N \rightarrow D)$ for most single subunit proteins lies in the range of

20–50 kJ/mol. This is equivalent to only three hydrogen bonds in energy, whereas it is known from the secondary structures that the N state of a typical small globular protein (say of 100 residues) may contain approx 150 hydrogen bonds. It must be concluded that $\Delta G(N \rightarrow D)$ is the sum of at least two contributions, one of which stabilizes the N state, and one that destabilizes it.

Table 2 is one attempt at a free energy balance sheet for a small globular protein (17). The first term is the experimentally determined $\Delta G(N \rightarrow D)$, which is assumed to be compounded of the contributions that follow. ΔG_{titr} expresses the transfer of ionized side chain groups from the protein interior to the bulk aqueous solvent during protein unfolding. An estimate can be obtained from the potentiometric titration curve. ΔG_{es} accounts for the net change in charge–charge interactions (repulsions) on the periphery of the protein. ΔG_{ϕ} is large, but its actual magnitude depends critically on the assumed range of the hydrophobic hydration effect. The model based on the simple water cage and Eq. (1) gives $\Delta G_{\phi} \cong 500$ kJ/mol, but by allowing for an increase in the range of the interaction, ΔG_{ϕ} can assume values of 2500 kJ/mol or larger.

The net polar interactions are treated as the sum of three contributions, of which the last one arises from polar groups in the interior of the protein, that, for steric reasons, are unable to participate in hydrogen bonds. The configurational entropy term originates from the gain in configurational freedom of a residue once it is transposed to the solvent during unfolding. On the other hand, for each dissociated intrapeptide hydrogen bond, several water molecules become hydrogen bonded to >CO and >NH groups, so that the loss of the secondary structure contains a negative contribution to the entropy change that originates from this peptide hydration process.

There may well be other contributions to $\Delta G(N \rightarrow D)$, the magnitudes of which are not easy to estimate. Three possible effects are included in Table 2. The total van der Waals energies between amino acid residues in the compact folded protein are likely to differ from those between the residues and the surrounding water molecules in the D state, although it would at present be pointless to attempt to put a value on this contribution to the stability. The same applies to the interaction between polar and apolar groups in close

Table 2

Tentative Balance Sheet of Contributions to the Free Energy Change Accompanying the Unfolding of Lysozyme and Ribonuclease (17)[a]

Term		Comments	Estimated magnitude, kJ/mol
$\Delta G_{N \to D}$	Free energy change on unfolding	Small ~ 3–5 hydrogen bonds	20–50
ΔG_{titr}	pK_a changes on unfolding		–20?
ΔG_{es}	Changes in surface charge–charge interactions	Estimated at neutral pH	0?
ΔG_{ϕ}	Hydrophobic free energy change	Surface exposure model, allow for possible long range[b]	500
			100–2500
ΔH_{ψ}	Several hydrogen bonding effects		
	Difference in strength of water–polar group and polar–polar group— hydrogen bond distortion	Average of several quantum mechanical values	300
		Using water–water energy surface	–400 to –800
	"Unsaturated" polar groups		–300
$n_c T\Delta S$ config.	Configurational entropy change per residue	Value of between 8–20 J/K (residue mol)$^{-1}$ used	–300 to –800
$n_w T\Delta S$ release	Water forms polar interactions with peptide groups	Assumes 25% of value for water freezing (underestimate?)	500
Other terms, e.g., ΔH_{VDW}	Changes in van der Waals interactions	Protein denser than water, implying some stabilization? Possibly included in G_{ϕ}	?
$\Delta H_{\psi-\phi}$	Polar–apolar interactions within protein		?
$T\Delta S_{vibr}$	Vibrational entropy changes	Potentially large and solvent-related	?

[a]n_c and n_w are the respective number of mols of residues and water involved in the unit process (17). [b]Evidence is growing that the simple proportionality between buried accessible area and free energy change is probably not valid, and that the hydrophobic interaction is of a long-range nature.

proximity, although the indications are that such interactions are of a net repulsive nature (i.e., destabilizing), at least in aqueous solution (26,27).

The final contribution in Table 2 arises from changes in the vibrational and librational modes experienced by individual covalent bonds in the amino acid residues during unfolding. In principle (but hardly in practice), such information is obtainable from infrared and Raman spectra. At the present state of our knowledge, it is only possible to speculate that such effect is certainly solvent-related and may be large.

Lest it be concluded that attempts at calculations of $\Delta G(N \rightarrow D)$ from a summation of the individual contributions have not been very successful, it must be stressed that other, empirical or semiempirical approaches, e.g., those based on so-called solvent accessibilities of individual residues in the folded and unfolded states of proteins, may have produced numbers that fit $\Delta G(N \rightarrow D)$ for individual proteins (28,29), but the theoretical basis for such calculations is weak. In addition, all such treatments are based on the simplistic (and in our view, incorrect) model of the hydrophobic interaction as a van der Waals type of effect.

Essentially, the crystal structure forms the starting point and the fraction of the total residue surface area of the protein that is "accessible" to water molecules (assumed to be spherical and closely packed) are estimated. A simple proportionality between the accessible area and $\Delta G(N \rightarrow D)$ is assumed. Clearly the results are extremely sensitive to the value chosen for the van der Waals radius (r^*) of the water molecule, because cavities in the surface topology of a protein might be accessible to solvent molecules with $r^* = 0.14$ nm, but not to those with $r^* = 0.18$ nm (29).

The weaknesses of unrealistic treatments of protein stability are usually revealed, if the temperature dependence of ΔG is considered. The representation of hydrophobic effects in terms of a van der Waals type of (potential energy) interaction cannot account for the skewed parabolic nature of $\Delta G(T)$, which has already been referred to. Its origin must be related to changes in the balance of the various contributions, such as those shown in Table 2, to the total stability. In Table 3, we attempt to present, in a semiquantitative manner, the temperature coefficients of the major contributing interactions.

Table 3
Signs and Magnitudes of Individual Contributions to ΔG
at Physiological Temperatures (+ Corresponds to Stabilization)
and of Their Temperature Coefficients (+ Corresponds to
an Increasing Contribution with Rising Temperature)[a]

Interaction	Contribution to stability	Temperature coefficient
Hydrophobic effects	+++	+
Salt bridges	++	+
Configurational free energy	−	+
Intrapeptide H-bonds	+	−
Water–peptide H-bonds	−	−
Van der Waals	+	−
Polyelectrolyte effect		?

[a]If the hydrophobic interaction is indeed the major stabilizing factor, then at low temperatures, its magnitude becomes insufficient to balance the growing strength of water–peptide and other polar interactions that promote unfolding (instability). The nature of the cold-unfolded state is therefore likely to differ from that of the heat-unfolded state, as witnessed by the absence of aggregation of cold denatured proteins; *see also* Chapter 11.

4. Hydration Sites in Protein Crystals

High-resolution diffraction methods (X-ray and neutron) have made possible the assignment of spatial coordinates to water molecules in protein crystals *(1)*. Such "structural" water molecules, which constitute only a small proportion of the total number of water molecules in the crystal, can be regarded as part of the secondary/tertiary peptide structure. Five typical locations for structural water molecules can be distinguished:

1. As metal ligand, e.g., in carboxypeptidase and carbonic anhydrase, the Zn atom is coordinated to three his residues and one water molecule that is displaced by the substrate during the formation of the enzyme–substrate complex;
2. Between residues at different locations within the same peptide chain. Such water bridges can consist of one or more molecules. A three-molecule water bridge linking >CO and —NH groups in papain is shown in Fig. 7 *(31);*

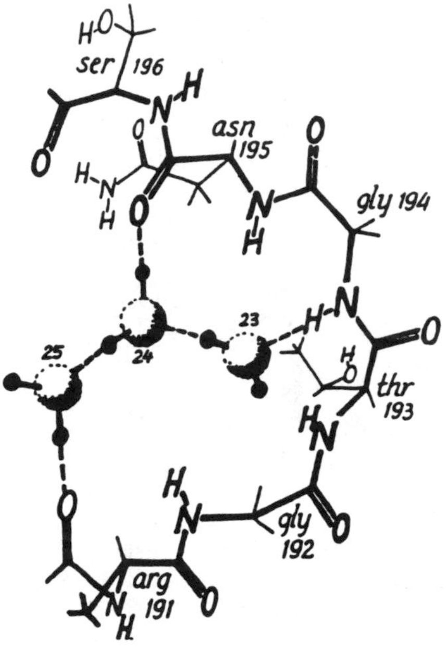

Fig. 7. Three-water-molecule bridge stabilizing the papain main chain by linking arg-191, gly-194, and asn-195 *(see* refs. *31,32)* (from Berendsen H. J. C. [1975], in *Water—A Comprehensive Treatise* vol. 5 [Franks F., ed.] Plenum, New York).

3. Between ionogenic groups in close proximity. Thus, in papain two water molecules are found in a hydrophilic cleft that contains two lys and one glu residues. The water molecules are linked to four more water molecules, which together constitute a chain to the protein periphery and the bulk solvent. Such a water chain provides an efficient means for proton transfer *(32);*

4. Between main chain >CO and—NH groups and polar side chains, e.g., in papain, the —OH groups of tyr-48 and tyr-82 are hydrogen bonded via water molecules to the >CO group of ala-104;

5. The active site of an enzyme is sometimes extensively hydrated. In carbonic anhydrase C, nine water molecules have been identified in the active site, four of which are linked to

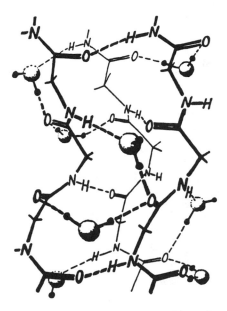

Fig. 8. Speculative stabilization of the three-stranded collagen structure by hydration interactions; water molecules are shown as spheres (from ref. 33).

amino acid residues, but the remainder are only linked to other water molecules; water bridges are also operative in linking different peptide chains in a quaternary structure. Figure 8 shows one suggested hydration structure for collagen (33).

Since collagen cannot be crystallized, diffraction experiments can only be performed on aligned fibrils, and the diffractogram must be interpreted in terms of an assumed structural model of the molecule, thus making the assignment of coordinates to labile water molecules somewhat uncertain. For this reason, attractive though it is, the Ramachandran hydration model is by no means unique or generally accepted. Indeed, there is no agreement about the locations of water molecules in the collagen myofibril (34). Water can also form complex hydration bridges between different protein molecules in a crystal, as shown for α-chymotrypsin in Fig. 9. Here again, several of the water molecules participating in the bridge structure are not linked directly to the peptide, but only to other water molecules (35). The symmetry of the water molecule positions is remarkable.

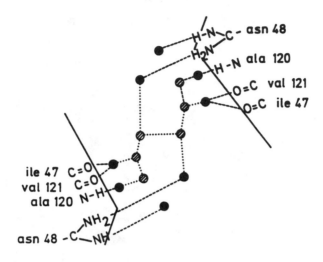

Fig. 9. Intersubunit hydration bridges in α-chymotrypsin *(35)*. The symmetry relating the two protein molecules is retained by the water network. Water molecules marked in black constitute the primary hydration shell; those shaded are linked only to other water molecules (secondary hydration).

The advent of neutron diffraction has made possible the extensive refinement of previous X-ray data. Figure 10 shows a cross-section through the coenzyme B_{12} hydrate structure. The water is distributed over two regions—a pocket and a channel, the latter region being the more disordered *(36)*. It is seen that both the water molecule geometry (valence angles) and the hydrogen bond lengths and angles display a remarkable variety. In the particular model used, O—O bonds of <0.33 nm and H—O bonds of <0.24 nm were accepted as hydrogen bonds. Water is seen to be very adaptable in filling space, and adjusting to the distances of separation and orientations of the polar sites on the macromolecule. A similar hydration study on the small protein crambin has also revealed a rich variety of water "structures," dominated by pentagons *(37)*, reminiscent of the well-known gas hydrate (clathrate) structures.

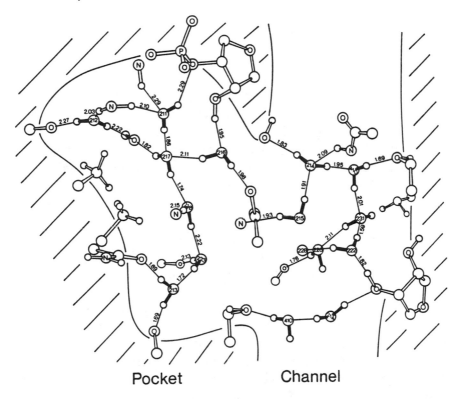

Fig. 10. Water networks in the coenzyme B_{12} hydrate structure: the filled bonds represent water —OH bonds. Hydrogen bond distances are given in angstroms. Reproduced, with permission, from ref. *36*.

5. Dynamic Aspects of Water in Protein Solutions: Protein Hydration

Protein–water interactions are frequently described in terms of the effect of the macromolecule on the diffusional motions of the water molecules. Of all the techniques employed in studies of protein hydration, nuclear magnetic resonance (NMR) in its various modes has received the most attention *(38,39)*. In particular, nuclear magnetic relaxation rates of ^1H, ^2H, and ^{17}O, as well as line splittings, have been used to determine the rotational and translational diffusion of water. Various interpretations of the experimental data (decay of nuclear magnetization) in terms of diffusional parameters have been proposed, but there is as yet no universal agreement (40). The systems to be investigated are often heterogeneous and anisotropic because, for experimental reasons, high protein concentrations must be employed. The interpretation of experimental data is further complicated by water exchange between two or more sites; such exchange can be fast or slow. Proton exchange must also be a possibility, unless ^{17}O spectra are being considered. Furthermore, water can also perform anisotropic diffusion parallel to the protein surface.

Qualitatively, three types of motion that differ by orders of magnitude can generally be distinguished with time scales (τ). A small fraction of the molecules has $\tau > 10^{-7}$ s. Their motions are governed by those of the macromolecule, i.e., on the time scale of a molecular rotation of the protein; such water molecules appear to be "bound" to a specific site on the protein. The fraction of water molecules with $\tau < 10^{-11}$ s is almost unaffected by the protein. For intermediate diffusion times, with τ of the order of 10^{-9} s, the water molecules are believed to perform slow, anisotropic tumbling motions, characteristic of water in small pores or narrow capillaries or in the proximity of the protein surface. Table 4 provides a summary of typical water diffusion rates in proteinaceous systems.

6. Protein Hydration and Bound Water: Some Popular Misconceptions

The protein literature contains numerous references to "bound" water (see, for instance, refs. 41,42), unfreezable water, or different "states" of water (43). In extreme cases, it is even claimed that the water in living cells is structured in some peculiar way that differs from water in its normal state, and that this difference is intimately related to cellular functions, such as the transport of ions across cell membranes (44). It has also been argued that, taking into account the total surface areas of organelles, membranes, and cytoskeletal components, most of the cell water can be regarded as a sorbed layer, not many molecules thick, on the surface of the

Table 4

Typical Tumbling Times (τ) or Diffusion Times of Water
in Proteinaceous Systems Under a Variety of Conditions[a]

System	Experimental conditions	τ, s
Lysozyme	<0.3 g water/g protein	10^{-9}
	>0.3 g water/g protein	Approaches 10^{-11}
Apotransferrin	10–70 mol water/mol protein	10^{-7}
	>70 mol water/mol protein	10^{-9}
Frog gastrocnemious muscle	250 K (unfrozen water)	2×10^{-9}
Collagen	0.2–0.5 g water/g protein	10^{-7}
Corneal stroma	0.8 g water/g protein, 145 K	10^{-9}
Striated muscle		2×10^{-8}

[a] In all cases, tumbling is anisotropic; see ref. (38).

cell matrix components, most of which are proteinaceous. The water is thus said to be bound to the various protein components, and to have properties that differ from those of "ordinary" or "free" water.

Water binding is seldom defined either in terms of specific sites of binding, energetics (presumably by hydrogen bonds) of binding, or lifetimes of the bound states. This is not the place to discuss the pros and cons of such views, but the reader should beware of hypotheses that offer simple explanations for very complex phenomena. Too frequently the evidence for such hypotheses is based on unwarranted assumptions and/or incorrect data handling or ambiguous interpretations of experimental results. This has been particularly true in the interpretations of NMR results.

The question is often asked: How many water molecules constitute the hydration shell of a protein? The literature contains many such estimates of protein hydration expressed as g water/g protein. However, the measured hydration number depends on the experimental technique used for the measurement. By and large, the degrees of hydration calculated from calorimetric and NMR measurements of frozen solutions agree well, probably because both techniques really measure the unfrozen water. Differential scanning calorimetry (DSC) results are expressed as $C_p(T)$. At low water contents, e.g. <0.3 g water/g protein in the case of tropocollagen

(45), no absorption of heat is apparent at 0°C, and only at 0.465 g water/g protein is the normal melting behavior of ice observed. On the basis of a gly-pro-hypro model for tropocollagen, this corresponds to 2.4 mol water/residue, in good agreement with the value 2.7 obtained from NMR measurements *(46)*. The same degree of correspondence is also achieved with several globular proteins, and the results have been accounted for in terms of a protein model in which only polar groups on the exterior surface are hydrated. It must, however, be noted that all such earlier interpretations of hydration shells could not take into account the nonequilibrium nature of freeze-concentrated solutions or the significance of the glass transition, which sets a limit to freeze concentration; *see* Chapter 14 for a full discussion. The expression of hydration numbers as mol water/mol residue is therefore likely to be meaningless, because the residual water is trapped randomly within the freeze-concentrated amorphous solid, rather than being associated with specific polar residues.

Infrared spectroscopy is of particular value in the study of hydration at low moisture levels, but some results can also be utilized to describe the hydration in aqueous solution. McCabe and Fisher *(47)* were able to distinguish between the absorption caused by water excluded from the hydration sphere, the water affected by the peptide, and that caused by the protein itself. In this context, there is general agreement that —NH_2 groups are more extensively hydrated than —NH_3^+ groups.

On a molecularly less detailed level, circular dichroism (CD) has been used to estimate the solvation contribution to the observed spectra of polypeptides *(48)*. The assumption is made that the amide groups are the primary solvation sites, and the hydration estimate is based on the calculated molecular asymmetry introduced by placing water molecules at these sites.

Protein hydration can be expressed in terms of structure, energetics, and hydrodynamics, but no quantitative relationship exists between hydrodynamic and thermodynamic approaches to water binding. The former approach is very model-sensitive, and difficulties arise in attempts to define the boundary of the hydrodynamic particle in the (continuum) solvent. Thermodynamic criteria, such as water activity (a_w), are frequently used to distinguish (allegedly) between bound or free (or trapped) water. Their application becomes progressively more uncertain as the assumed pockets of trapped water decrease in size. Questions also arise regarding the probable nonequilibrium nature of such systems, since a_w is defined in terms of equilibrium thermodynamic quantities (49).

The assignment of notional hydration numbers, whether derived from spectroscopic or thermodynamic measurements, can sometimes lead to absurd results that illustrate the inherent weaknesses of such procedures. A striking example is provided by calculations of hydration numbers of a series of glycols, based on cryoscopic measurements (50). By postulating simple hydration equilibria, the author was driven to the conclusion that the degree of water binding *increases* with increasing temperature, to the extent that at the boiling point of the solution, the solute is quite extensively hydrated. This and other similar treatments show up the futility of approaches based on stoichiometric hydration equilibria.

In summary, it is unwise to express protein hydration in solution in terms of g water/g protein, as though there were a sharp cutoff between water of hydration and bulk water. (In technological applications, where protein preparations are dried to a given moisture content, such descriptions are admissible for purposes of quality control.) The perturbation of water by the protein decays as a function of distance, the decay function depending on the particular hydration site on the polypeptide chain. Thus, for ionogenic side chains, the interaction is strong and has a long range. For apolar sites, hydrophobic hydration is weak, but probably of long range (12,21). For polar sites (—OH, > CO,—NH), the interaction approximately equals in strength that between two water molecules and is likely to be short-ranged. The problem of expressing hydration in aqueous solution in a meaningful manner resembles that of describing a water sorption isotherm, in which the sorbent is highly heterogeneous and contains an irregular array of different sorption sites with different sorption energies and is likely to orient the sorbed water molecules in different ways.

7. Proteins at Low Moisture Content:
Sequential Hydration

If a protein in solution is carefully concentrated under conditions that do not disrupt the N state (i.e., moderate temperature), then in some favorable cases, water can be almost completely removed and an almost anhydrous protein obtained. If water vapor is then readmitted, the interactions in the system can be monitored by any one of several physical or biochemical techniques, and the sites of interactions characterized as a function of the moisture content. Table 5 summarizes the techniques and the type of information that can be obtained (51).

A popular method for studying water-sensitive or water-soluble polymers (and this includes proteins) is by the measurement of water vapor sorption isotherms. Usually the sorption and desorption isotherms do not coincide. Sorption hysteresis at low moisture levels is symptomatic of thermodynamic metastability. The hysteresis loops are experimentally reproducible, and suggest conformational changes and/or slow water migration on the surface of the protein to sites of high sorption energy; *absorption* (plasticization) must also be considered a possibility) if the system has undergone a glass/rubber transition.

Despite all these complications, moisture sorption isotherms are often subjected to thermodynamic analyses based on the Langmuir, BET, or some other sorption theory. This makes possible the calculation of a notional monolayer coverage, the sorption energy, and other details of the sorption process. It should be remembered, however, that the assumptions on which such *equilibrium* sorption theories are based cannot possibly apply to a nonequilibrium system that is subject to sorption hysteresis. Furthermore, under the experimental conditions employed, the measured relative humidity, p/p_o, is not to be equated to the water activity a_w. Indeed, a_w has no meaning where true thermodynamic sorption/desorption equilibrium does not exist, a fact that is still not widely appreciated, especially by food technologists who habitually express microbiological safety and textural quality in terms of a measured relative humidity that is mistakenly referred to as a_w.

Table 5
Available Experimental Techniques
for the Study of Protein–Water Interactions

	State of sample		Type of Information			Information about	
	Solution	Solid	Time-average	Dynamic	Structural	Water	Protein
Diffraction	+	+	+	(+)	+	+	+
Spectroscopy	+		+		(+)	(+)	+
Thermodynamics	+	+	+			+	+
Sorption		+	+			+	+
Relaxation (nmr, esr)	+	+		+		+	+
Hydrodynamics	+			+			+
Kinetics (e.g., H$^+$ exchange)	+	+		+			+

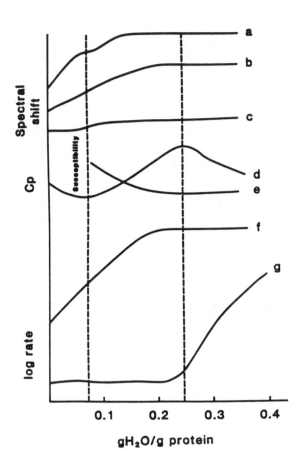

Fig. 11. Effect of sequential hydration on various experimental parameters of the lysozyme/substrate complex (for details, *see text* and ref. *51*).

Among thermodynamic properties, the heat capacity is the
most sensitive indicator of overall hydration interactions, although
various spectroscopic techniques provide information about interactions between water and specific chemical groups on the
polypeptide. Figure 11 shows the changes in various experimental parameters as a function of the moisture content, expressed
as g water/g protein, that occur when water vapor is admitted to
previously dried lysozyme *(51)*. The curves labeled a, b, and c are,
respectively, the infrared intensities of the carboxylate band (1580
cm⁻¹), the amide I band (1660 cm⁻¹), and the water OD stretching

Table 6

Events Associated with the Sequential Hydration (Water Vapor Uptake)
of the Lysozyme/Substrate Complex, as Monitored by the Parameters
Shown in Fig. 11

Degree of hydration gH$_2$O/g protein	Site of interaction	Event
0.1	Ionic side chains glu, asp, lys	Acids reach "normal" pK$_a$ values
	Main chain > CO	Minor conformational transition
	Main chain > NH	Side chain mobility
0.2	peptide groups "saturated"	>CO, >NH stretching frequencies attain "normal" values
	Side chain —OH	
0.3	water clusters, bridges	τ H$_2$O assumes normal value (~ 3 ps). Detectable enzyme activity

frequency (2570 cm^{-1}). Curve d is the apparent specific heat, e is the diamagnetic susceptibility, f is the log rate peptide proton exchange, and g is the log enzyme activity. Curve g coincides exactly with the rotational diffusion rate of the protein, as monitored by an electron spin probe.

From the moisture dependence of the various parameters in Fig. 11, a picture can be constructed that describes the events that take place on the surface of the protein as moisture is admitted. This is shown in Table 6. Where the various properties level off near 0.4 g water/g lysozyme, water is observed to take up its "normal" physical properties. The indications are that the first traces of moisture produce a "minor" conformational reordering of the protein resulting from the hydration and ionization of carboxylate groups. As the degree of hydration increases, so the acidic and basic groups become saturated and water molecules diffuse from there to other sites, specifically the peptide backbone —NH and >CO groups. At this stage, a low level of enzyme activity is first observed that then increases with hydration, but only reaches its "normal" value at 9 g water/g protein. The decrease in C_p at hydration levels above 0.25 is yet to be explained; it may be related to hydrophobic hydration phenomena or the water bridging of several hydration sites.

Controlled desiccation also enhances the thermal stability of globular proteins; the following water contents (per g protein) will stabilize the proteins against heating at 100°C: myoglobin, 8%; hemoglobin, 11%; chymotrypsinogen, 11%; collagen, 28%; lysozyme, 15%; and β-lactoglobulin, 22%. Similar effects of desiccation are observed in the inactivation of enzymes. Thus, the heat inactivation of glucose-6-phosphate dehydrogenase (also glucose dehydrogenase and leucine dehydrogenase) for which the inactivation rate constant at 50°C (in dilute solution) is 0.12/min is reduced to 0.015/min at 145°C and R. H. = 11%.

With the more recent realization of the important part played by glassy states in the stabilization of proteins, the above results are hardly surprising. It is to be expected that, below the glass transition, the protein would be stable toward inactivating influences, such as those discussed in Chapter 11.

Bibliography

1. Finney J. L. (1979) in *Water—A Comprehensive Treatise*, vol. 6 (Franks F., ed.) Plenum, New York, pp. 47–122.
2. Rupley J. A. (1979) in *Characterisation of Protein Conformation and Function* (Franks F., ed.) Symposium Press, London, pp. 54–61.
3. Franks F. (1975) in ref. 1, **4**, 1–94.
4. Franks F. and Grigera J. R. (1990) *Water Sci. Rev.* **5**, 187–289.
5. Friedman H. L. and Krishnan C.V. (1973) in ref. 1, **3**, 1–118.
6. Enderby J. E. and Neilson G. W. (1979) in ref. 1, **6**, 1–45.
7. Frank H. S. and Wen W. Y. (1957) *Disc. Faraday Soc.* **24**, 133.
8. Conway B. E. (1981) *Ionic Hydration in Chemistry and Biophysics*, Elsevier, Amsterdam.
9. Friedman H. L. (1978) *Faraday Disc. Chem. Soc.* **64**, 1.
10. Privalov P. L. and Khechinashvili N. N. (1974) *J. Mol. Biol.* **86**, 4291–4303.
11. Ratcliffe C. I. and Irish D. E. (1988) *Water Sci. Rev.* **3**, 1–78.
12. Franks F. and Desnoyers J. E. (1985) *Water Sci. Rev.* **1**, 171–232.
13. Davidson D. W. (1973) in ref. 1, **2**, 115–234.
14. Eagland D. (1975) in ref. 1, **4**, 305–518.
15. Franks F. and Eagland D. (1975) *Crit. Rev. Biochem.* **3**, 165.
16. Pain R. H. (1979) in ref. 2, 19–36.

17. Finney J. L. (1982) in *Biophysics of Water* (Franks F. and Mathias S. F., eds.) John Wiley & Son, Chichester, pp. 55–57.
18. Chan D. Y., Mithcell D. J., Ninham B. W., and Pailthorpe B. A. (1979) in ref. *1*, *6*, 239–278.
19. Privalov P. L. and Gill S. J. (1989) *Adv. Protein Chem.* **39**, 19–234.
20. Tanford C. (1973) *The Hydrophobic Effect.* John Wiley & Sons, New York.
21. Okazaki S., Nakanishi K., and Touhara H. (1983) *J. Chem. Phys.* **78**, 454.
22. Dec S. F. and Gill S. J. (1985) *J. Solution Chem.* **14**, 417.
23. Eisenberg D., Weiss R. M., Terwilliger T. C., and Wilcox W. (1982) *Faraday Symp. Chem. Soc.* **17**, 109.
24. Nozaki Y. and Tanford C. (1971) *J. Biol. Chem.* **246**, 2211.
25. Bigelow C. C. (1967) *J. Theor. Biol.* **16**, 187.
26. Nord L., Tucker E. E., and Christian S. D. (1983) *J. Solution Chem.* **12**, 889.
27. Franks F. and Pedley M. D. (1983) *J. Chem. Soc. Faraday Trans I* **79**, 2249.
28. Lee B. and Richards F. M. (1971) *J. Mol. Biol.* **55**, 379.
29. Shrake A. and Rupley J. A. (1973) *J. Mol. Biol.* **79**, 351.
30. Savage H. F. J. (1986) *Water Sci. Rev.* **2**, 67–148.
31. Drenth J., Jansonius J. N., Koekoek R., and Wolthers B. G. (1971) *Adv. Protein Chem.* **25**, 79.
32. Berendsen H. J. C. (1975) in ref. *1*, *5*, 293–330.
33. Ramachandran G. N. and Chandrasekharan R. (1968) Biopolymers **6**, 1649.
34. Chapman G. E., Danyluk S. S., and McLauchlan K. A. (1971) *Proc. Roy. Soc.* **B 178**, 465.
35. Birktoft J. J. and Blow D. M. (1972) *J. Mol. Biol.* **68**, 187.
36. Vovelle F., Goodfellow J. M., Savage H. F. J., Barnes P., and Finney J. L. (1985) *Eur. Biophys. J.* **11**, 225–237.
37. Teeter M. M. and Kossiakoff A. A. (1984) in *Neutrons in Biology* (Schoenborn B. P., ed.), Plenum, New York, p. 335.
38. Packer K. W. (1977) *Phil. Trans. Roy. Soc. Ser. B.* **278**, 59–86.
39. Derbyshire W. (1982) in ref. *1*, *7*, 339–430.
40. Bryant R. G. and Halle B. (1982) in ref. *17*, 389–393.
41. Kuntz I. D. and Kauzmann W. (1974) *Adv. Protein Sci.* **28**, 239–255.
42. Cooke R. and Kuntz I. D. (1974) *Ann. Rev. Biophys. Bioeng.* **3**, 95–126.
43. Clegg J. S. (1982) in ref. *17*, 365–383.
44. Ling G. N. and Negendank W. (1980) *Persp. Biol. Med.* **24**, 215.
45. Privalov P. L. and Mrevlishvili G. M. (1967) *Biofizika* **12**, 22.
46. Kuntz I. D. (1971) *J. Amer. Chem. Soc.* **93**, 516.
47. McCabe V. C. and Fisher H. F. (1970) *J. Phys. Chem.* **74**, 2990.
48. Pysh E. S. (1974) *Biopolymers* **13**, 1557.
49. Franks F. (1991) *Trends in Food Sci. Technol.* **2**, 68–72.
50. Ross H. K. (1954) *Ind. Eng. Chem.* **46**, 601–610.
51. Rupley J. A., Gratton E., and Careri G. (1983) *Trends Biochem. Sci.* **8**, 18–22.

Thirteen

Recombinant Protein Technology

Linda A. Fothergill-Gilmore

1. Introduction

Every cell has a remarkably complex array of processes that ensure that its genetic information is maintained and reproduced, and also ensure that the genes are manifested in the enzymes and other proteins required for the cell to function and survive. Thus, for example, an assembly of enzymes brings about the very accurate replication of DNA once during each cell cycle. Another group of enzymes and other factors catalyzes the transcription of DNA into RNA, which in turn is translated into proteins on the ribosomes. Superficially, it would seem that all these processes are far too complex to be manipulated at will in the laboratory. However, it was realized in the 1970s that it is possible to harness viruses, plasmids, and bacteria to do the manipulations on behalf of the experimental molecular biologist. A wide range of powerful and deceptively straightforward techniques has subsequently been developed. These techniques are the basis for the production of recombinant proteins.

1.1. What Are Recombinant Proteins?

Genetic engineering and recombinant DNA technology are terms used to refer to the collection of procedures for the experimental manipulation of genetic material. At the heart of these

From: *Protein Biotechnology* • F. Franks, ed.   ©1993 The Humana Press Inc.

techniques is the process of gene cloning. Proteins that are produced as the products of genetic engineering can be referred to as recombinant proteins.

1.2. Why Produce Recombinant Proteins?

In detail, there are probably as many reasons for producing recombinant proteins as there are scientists doing the experiments. However, it is probably possible to group the aims into four broad categories:

1. To obtain large quantities of a protein;
2. To study site-directed mutant proteins;
3. To produce proteins for biotechnology; and
4. To manipulate metabolism in vivo.

1.2.1. Large Quantities of a Protein

The expression of recombinant proteins under the control of strong promoters gives the possibility of obtaining very large quantities of proteins. For example, the expression of yeast phosphoglycerate kinase from a multicopy plasmid can attain levels of up to approx 80% of the total cell protein *(1)*. Thus, even a small laboratory-scale fermentation in a couple of shaking flasks can yield several grams of protein. This not only yields large quantities, but can also greatly simplify the purification procedures. These advantages are of particular importance for the production of clinically useful proteins, such as the clotting factor VIII. Highly expressed recombinant proteins are also very suitable for study by biophysical techniques, such as X-ray crystallography or nuclear magnetic resonance spectroscopy.

1.2.2. Site-Directed Mutant Proteins

The recently developed methods for the production of proteins with specifically engineered mutations (*see* Section 4.) have opened the door to an exciting and powerful new approach for studying protein structure and function. In most cases, there is a requirement for the expression of the mutant form of the protein, so that it can be purified and subjected to further characterization and study.

1.2.3. Biotechnology

Enzymes are used in many industrial processes, such as for the production of metabolites, antibiotics, and so forth. Recombinant protein technology can be used not only to yield large quantities of these enzymes, but also to produce enzymes with improved properties for industrial processes. Thus, some industrial enzymes have been engineered by site-directed mutagenesis to have enhanced stability to high temperatures or to certain solvents.

1.2.4. Manipulation of Metabolism

There are many situations where it may be advantageous to be able to manipulate the metabolism of an organism. Examples might include the engineering of a strain of bacteria with an improved yield of an antibiotic or a strain of yeast capable of producing an unusually high concentration of ethanol. In medicine, there are many metabolically based diseases that might respond to approaches that involve an alteration of the metabolism of the patient. Similarly, in agriculture, there is much interest in the possibilities of engineering new strains of plants or animals with improved properties. An example would be the production of fruits and vegetables with increased resistance to frost damage. All of these situations require the use of recombinant techniques to produce proteins with novel properties. In these cases, the recombinant protein is produced *in situ* where it is required by the relevant animal or plant.

2. Gene Cloning

It is outside the scope of this chapter to present a survey of the techniques of gene cloning. Nevertheless, some of the procedures are fundamental to the production of recombinant proteins, and it is probably helpful to mention them briefly here. For further information, the reader is referred to an excellent introductory book in this field: *Gene Cloning: An Introduction* by T. A. Brown (2). Several manuals of experimental protocols are available, and among these, two are generally found to be particularly useful: *Guide to Molecular Cloning Techniques* by Berger and Kimmel (3), and *Molecular Cloning* by Sambrook, Fritsch, and Maniatis (4).

2.1. Manipulation of DNA

Crucial to gene cloning is the ability to manipulate DNA. Chromosomes, viruses, and plasmids must be purified, usually by a combination of centrifugation and solvent extraction. Genes must be removed from chromosomes and inserted into cloning vectors, such as viruses or plasmids. The ends of DNA fragments frequently must be altered to tailor them for specific purposes. Most of the technology of DNA manipulation is enzyme based.

The enzymes used to manipulate DNA can be grouped into four main categories:

1. *Nucleases* cut, shorten, or degrade nucleic acid molecules;
2. *Ligases* join DNA molecules together;
3. *Polymerases* make copies of nucleic acid molecules; and
4. *Modifying enzymes* remove or add chemical groups.

Table 1 gives some examples of each of these categories of enzymes, together with an indication of their activities.

2.2. Cloning Vectors

The central component of a gene cloning experiment is the vector or vehicle (sometimes referred to as a molecular taxi!), which transports the gene of interest into a host cell and is responsible for its replication. The vector must be capable of entering a host cell, and once inside, the cell must be able to undergo replication to produce multiple copies of itself (Fig. 1). There are three types of DNA molecules that satisfy these requirements: plasmids, bacteriophages, and phagemids.

Plasmids are small, circular, double-stranded DNA molecules that occur naturally within bacteria, yeast, and certain other organisms. Plasmids can replicate independently of the host cell chromosome, and can thus be present in several tens of copies per cell. A number of naturally occurring plasmids have been modified for convenient use in the laboratory, such that their presence in a cell can be selected for. Also they have within them multiple cloning sites that are artificial stretches of DNA designed to contain a number of different restriction sites. Examples of frequently used plasmids are pBR322 and pUC18.

Table 1
Enzymes Used To Manipulate DNA

Nucleases	
Exonuclease Bal31	Removes nucleotides from both strands of double-stranded DNA
Exonuclease III	Removes nucleotides from the end of just one strand of double-stranded DNA
Restriction endonucleases	Recognize specific sequences of four to six nucleotides in double-stranded DNA, and cleave both strands once within the recognition sequence
Ligases	
DNA ligase	Forms phosphodiester bonds to link two pieces of DNA together. Will repair single-stranded breaks in double-stranded DNA, or can join two restriction fragments
Polymerases	
DNA polymerase I	Synthesizes a new strand of DNA complementary to an existing DNA template. Also has nuclease activity that provides an editing function
Klenow fragment	Derived from DNA pol I and has only the polymerase activity
Reverse transcriptase	Synthesizes DNA from an RNA template
DNA-modifying enzymes	
Alkaline phosphatase	Removes the phospho group at the 5' terminus of a DNA molecule
Polynucleotide kinase	Adds phospho groups to free 5' termini
Terminal deoxynucleotidyl transferase	Adds one or more nucleotides to the 3' terminus of a DNA molecule

Bacteriophages are viruses that infect bacteria. During infection, the bacteriophage DNA molecule is injected into the host cell, where it undergoes replication. Some bacteriophages have the great practical advantage that they occur as both single-stranded and double-stranded versions at different stages of their infectious cycle. The double-stranded form is used for the insertion of restriction

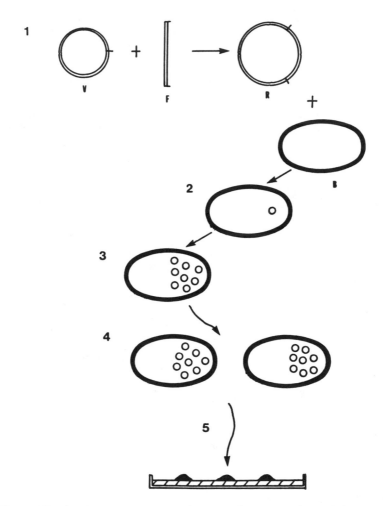

Fig. 1. The basic steps in gene cloning (drawing adapted from Brown, [2]). 1. Construction of a recombinant DNA molecule from a vector (V) and a fragment of DNA (F). 2. Introduction of the recombinant DNA molecule (R) into a host bacterial cell (B). 3. Amplification of the recombinant DNA molecule using the DNA replication machinery of the host. 4. Division of the host cell with approximately equal distribution of the recombinant DNA molecules between the two daughter cells. 5. Numerous cell divisions yield clones of cells that can be grown as bacterial colonies on solid medium, such as agar, in a Petri dish.

fragments, whereas the single-stranded form is convenient for use as a template for DNA sequencing and for site-directed mutagenesis. As in the case of plasmids, naturally occurring bacteriophages have been modified to include multiple cloning sites and selection systems. Commonly used bacteriophages include the filamentous phage M13 (*see* Fig. 2) and the phage lambda, which is often used for the construction of DNA libraries (*see* Section 2.3.).

As their name implies, phagemids are DNA molecules derived from a combination of plasmids and bacteriophages. These molecules are suitable for a wide range of applications, such as cloning, sequencing, mutagenesis, and expression (*see* Section 3.), without the need for cumbersome and time-consuming subcloning steps. Phagemids pEMBL8 *(5)*, pVT-L, and pVT-U *(6)* are examples of versatile and convenient vectors (Fig. 3).

2.3. DNA Libraries

An essential early step in gene cloning is to isolate the gene of interest separated from all the other DNA of the organism. For most organisms, this is a daunting prospect, because there are many thousands of genes spread over many millions of bases of DNA. The construction of a DNA library is a procedure that involves the insertion of manageably small pieces of DNA into a convenient bacteriophage. Restriction fragments derived from a digestion of genomic DNA are frequently used for the construction of genomic libraries, as are fragments generated by mechanical shearing. Alternatively, cDNA (complementary DNA), which has been copied from messenger RNA by reverse transcriptase, may be used to produce cDNA libraries. Various derivatives of phage lambda (which have different selection systems and which are able to accept a range of sizes of inserted DNA) are in common use. The recombinant phage particles are transfected into a bacterial host and these bacteria are then spread out on agar plates such that individual clones of cells are allowed to grow. The phages present in the different clones are probed for the gene of interest, and the appropriate clone can thus be isolated in a relatively pure form. These procedures are represented schematically in Fig 4.

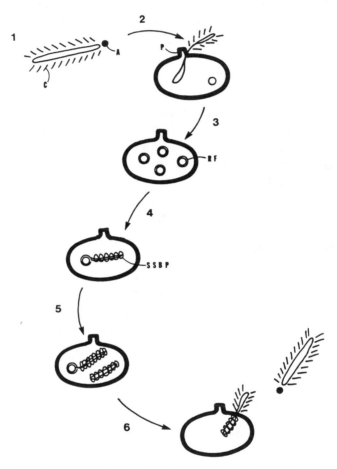

Fig. 2. The infectious cycle of bacteriophage M13. 1. A particle of bacte-riophage M13 consists of a covalently closed piece of single-stranded DNA approx 7 kb in length that is covered with many molecules of coat protein (C). A single molecule of A protein (A) is at one end. The DNA is always in the same orientation and is by convention termed the plus (+) strand. It can accept approx 1000 extra bases of DNA in the intergenic space at the end away from the A protein. 2. The A protein mediates binding of the M13 particle to the pilus (P) of an *E. coli* cell. The DNA enters the cell, whereas the coat protein is retained at the cell membrane for reuse. 3. The phage DNA then undergoes replication until approx 100 copies of the double-stranded or replicative form (RF) have been synthesized. 4. The phage genes are expressed during this phase of the infectious cycle and eventually a critical

3. Expression of Recombinant Proteins

Probably the most difficult and least understood aspect in the field of recombinant protein technology is the actual expression of the protein. The experimenter is faced with an almost bewildering choice of different expression systems that involve various combinations of plasmids, phagemids, and host organisms. There is the additional decision of whether it may be advantageous to have the recombinant protein secreted from the host cell. Having made all these decisions, it is necessary to choose which type of promoter to use. Should the protein be expressed constitutively or after an appropriate induction signal? Should the strongest possible promoter be used? Moreover, it must be decided whether it might be an advantage to express the protein as a fusion with a completely different, but well-expressed protein.

The state of the art is such that there are no hard and fast answers to any of these questions. This is clearly a very unsatisfactory situation, but at the moment, it is wise advice to adopt the empirical approach of trying a variety of systems until one that works well is found. It is to be hoped that current progress toward understanding the processes of gene expression will enable the experimental approach in the future to be rather more rational.

3.1. Which System?

Probably the logical first decision concerns the choice of host organism. The three most commonly used systems are bacteria (e.g., *E. coli*), yeast (e.g., *S. cerevisiae*), and mammalian cells. Other possibilities include filamentous fungi, insect cells with baculovirus vectors, frog oocytes, transgenic animals and plants, and in vitro systems. The choice of the system is influenced primarily by the type of protein to be expressed. Some of the factors involved are summarized in Table 2.

---

concentration (approx 100 copies/cell) of one of the gene products—the single-strand binding protein (SSBP)—builds up. These proteins bind to newly synthesized DNA and cause a switch to rolling circle replication, which has single-stranded DNA as a product. 5. Many copies (1000) of the (+) strand of M13 are thus synthesized, which (6) then extrude from the *E. coli* cell, picking up coat protein and protein A in the process.

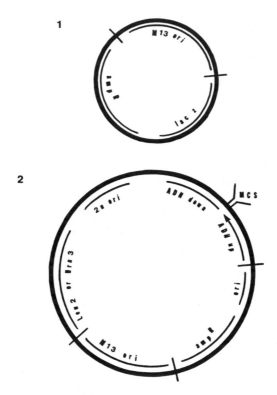

Fig. 3. Some properties of phagemids pEMBL8, pVT-L, and pVT-U. 1.
Phagemid pEMBL8 (approx 4 kb) will accept much larger inserts than M13
(approx 4 kb instead of 1 kb), but needs helper phages because a substantial
portion of the M13 genome has been deleted. The phagemid contains the
M13 origin of replication, a selection system for propagation in *E. coli* (ampicillin
resistance), and a portion of the *lac z* gene to give a convenient color assay
for recombinant molecules. The helper phages provide single-strand binding protein and coat protein as well as other functions. 2. Phagemids pVT-L
and pVT-U can be used as plasmids, phages, or as shuttle vectors. They can
be propagated as plasmids in *E. coli* to yield the double-stranded form or as
phages to give the single-stranded form. In the latter case, a helper phage is
required. These vectors can also be used as multicopy plasmids in yeast,
which is especially useful for expression. The phagemid has ampicillin resistance for selection in *E. coli*, and either the Leu2 or Ura3 genes for selection
in yeast. It also has three different origins of replication depending on the
system to be used. Expression in yeast is under the control of the upstream and
downstream portions of a yeast alcohol dehydrogenase (ADH) gene. There
is a multiple cloning site (MCS) for convenient insertion of DNA fragments.

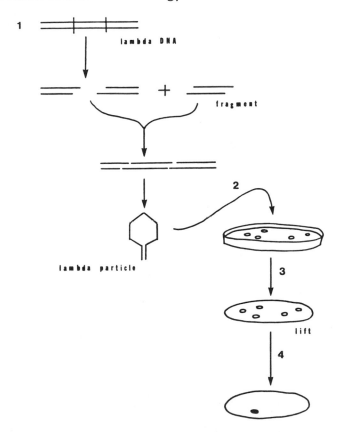

1 lambda DNA

fragment

lambda particle

2

3 lift

4

Fig. 4. Construction and probing of DNA libraries. 1. The genome of phage lambda is a linear double-stranded molecule that has been modified to contain a convenient multiple cloning site and various other functions. Recombinant molecules containing inserts of DNA fragments are readily prepared, and can be packaged into phage particles for transfection into *E. coli* cells. 2. A suspension of phage particles is placed on a lawn of bacteria growing on solid medium. The phages cause lysis of the cells they have transfected, and this can be seen as small circles of clear regions or plaques in the bacterial lawn. 3. A piece of filter membrane usually made of nitrocellulose or nylon is placed on top of the plaques to lift off the phage particles that have been released into the plaques. The membrane is then treated with alkali to lyse the phages and to denature the DNA. 4. The membrane "lift" is exposed to a suitable radioactive probe designed to bind specifically to the DNA of interest. In this example, one of the plaques has given a positive result. Fresh phage particles containing this DNA can be recovered by picking them from the corresponding plaque in the dish used to make the "lift."

Table 2
Choice of Expression Systems

Bacteria	Convenient to grow in large fermenters
	May obtain very high yields (0.1–5 g/L)
	Best for bacterial proteins
	Proteins will have N-terminal methionine
	Will not do postsynthetic modification reactions, such as glycosylation
	Eukaryotic proteins frequently form insoluble inclusion bodies that are relatively easy to purify, but difficult to renature
	Best to use bacterial secretion signals
	Will not cope with introns
Yeast	Convenient to grow in large fermenters
	Good yields (0.01–1 g/L)
	Best for yeast proteins, but may be good for other eukaryotic proteins
	Postsynthetic modification reactions occur, but usually differ in detail from mammalian reactions
	Will not cope with introns from higher organisms
	May successfully use secretion signals from other organisms
	Proteolysis can be a significant problem
Mammalian	Difficult and expensive to grow in large-scale cultures
	Relatively low yields (0.001–0.1 g/L)
	May be the only successful system for some mammalian proteins
	Postsynthetic modification reactions occur normally
	Introns spliced normally
	Mammalian secretion signals recognized successfully

3.2. Secretion

Whether or not to try to achieve secretion of a protein will depend largely upon the protein itself. Thus, if it normally occurs in an intracellular location, then it will probably not be an advantage to secrete it. The secretion process itself is complex and involves various modifications to the structure of a protein, such as unfolding and renaturation. It is possible that some intracellular proteins will not survive the secretion process in a native form. On the other hand, if the protein is usually secreted, then it is probably best to use a secretion system for the recombinant version. In this way, the recombinant protein will undergo all the usual postsynthetic modification reactions that are part of the secretion process.

Another consideration about whether or not to use secretion concerns the purification of the recombinant protein. If the protein is retained within the cell, then it must be purified from the very complex mixture of cellular components. If the protein is secreted, then the purification may be a relatively simple separation from the constituents of a defined growth medium. This is usually particularly straightforward for the bacterial and yeast expression/secretion systems.

3.3. Promoters

It is essential to match the promoter to be used with the type of host cell expression system. One must use bacterial promoters with bacterial hosts, yeast promoters with yeast hosts, and so on. In principle, any promoter should be satisfactory to control the expression of a gene, but in practice, each host system has a short list of favorite promoters. These have been chosen to provide two main functions: (a) constitutive/inducible expression and (b) strong/weak expression. The main consideration is the potential toxicity to the host cell of the expression of the recombinant protein. If toxicity is anticipated, then it is probably best to grow the host cells to high density in the absence of expression and then induce the expression. It may also prove necessary to accept a relatively low level of expression. However, if it is unlikely that the recombinant protein will be toxic, then constitutive expression controlled by a strong promoter will probably give the highest yields.

Unfortunately, the factors that control the levels of expression are as yet so ill understood that it is not feasible to predict which promoter will turn out to be the most effective in a particular system. The only sensible approach is to plan at the outset to try a number of different promoters in a number of different host cell strains.

3.4. Fusion Proteins

Some proteins are already well known to be highly expressed under a particular set of conditions. One way to avoid the trial-and-error approach advocated in the previous section is to attach the new recombinant protein onto an existing well-expressed protein, such that a fusion protein is generated. For example, fusions with the enzyme β-galactosidase have proven to be very successful in an *E. coli* expression system. There are two important considerations to be kept in mind. First, the joining of the DNA to specify the fusion protein must be done so as to maintain the correct reading frame of the downstream gene. Second, it is necessary to have a strategy to separate the desired recombinant protein from its carrier. This process usually involves some type of specific peptide bond cleavage. For example, recombinant proinsulin can be recovered from its fusion with β-galactosidase by treatment with CNBr to cleave the methionine peptide bond linking the two proteins (7).

4. Protein Engineering

In the early 1980s, Michael Smith from the University of Vancouver devised a relatively simple and straightforward procedure to introduce specific base change mutations into cloned DNA (8). The method is known as site-directed mutagenesis (SDM) or oligonucleotide-directed mutagenesis. This method and extensions of the method have had an extraordinary impact on a wide range of research projects in biology, because they have enabled experimenters to introduce any desired base change, insertion, or deletion at will into specific sites in DNA. Site-directed mutagenesis can be used to study the structure and function of DNA (e.g., repressor binding sites), RNA (e.g., control of translation) and proteins (e.g., enzyme mechanism). It is the use of site-directed mutagenesis to study and exploit proteins that is relevant here.

4.1. What Is Protein Engineering?

Protein engineering can be defined as the design and construction of proteins by recombinant DNA techniques. Engineered proteins are sometimes referred to as factitious proteins. The potential of protein engineering is really only limited by the imagination of

Table 3
Classes of Protein Engineering Experiment

Protein structure and function
    Signal sequences
        Targeting to organelles
        Secretion
    Enzyme properties
        Mechanism
        Substrate binding
        Transition state stabilization
        Allostery and cooperativity
    Polypeptide folding and stability
        Folding pathways
        Energetic contributions
        Subunit interactions
Construction of industrially useful proteins
    Increased thermal and solvent stability
    Alteration of substrate specificity
    Novel enzyme reactions
Introduction of reporter groups
    Spectroscopy—esr, nmr, fluorescence
    X-ray crystallography
Design of novel drug carriers
Design and production of clinically useful proteins

---

the experimenter and frequently by the lack of knowledge of the detailed tertiary structure of a protein. The reasons for doing protein engineering are thus very varied, but can be grouped under five main headings as described in the next section.

4.2. Types of Protein Engineering

The five broad categories of protein engineering project are summarized in Table 3.

4.3. Outlines of Strategies

The procedure for oligonucleotide-directed mutagenesis as devised by Zoller and Smith (8) is summarized in Fig. 5. The gene to be mutated is inserted into phage M13, and the single-stranded form is isolated to act as template. An oligonucleotide (usually about

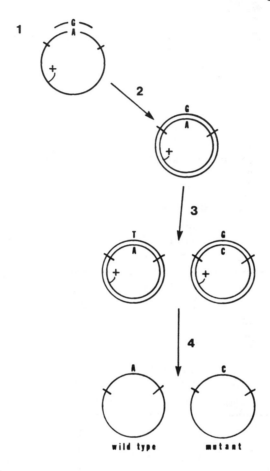

Fig. 5. Outline of the Zoller and Smith method of oligonucleotide-directed mutagenesis. 1. An oligonucleotide specifying the desired mutation is annealed to a recombinant phage M13 DNA molecule containing the gene to be mutated. 2. The Klenow fragment of DNA polymerase is added to synthesize a second complementary strand, and DNA ligase is used to complete the covalently closed second circle of DNA. The (+) strand contains the wild-type sequence (nucleotide A in this example), whereas the complementary strand has the mutation (nucleotide G). 3. This double-stranded molecule is then used to transform E. coli cells. 4. The normal M13 infectious cycle takes place with two types of replicative forms. In one case, the (+) strand is wild type, and in the other, the (+) strand is mutant. 5. Eventually, the mutant and wild-type, single-stranded phage particles are produced.

20 bases) is synthesized to be complementary to the site of mutation and to specify the desired mutation. The oligonucleotide is then annealed to the gene (there will be a mismatch at the mutation), and DNA polymerase and ligase added to yield the double-stranded form of the recombinant phage. This double-stranded molecule is transformed into *E. coli* as if it were a plasmid. The recombinant phage then undergoes the normal rounds of replication, and will ultimately yield single-stranded phage particles that should theoretically be in equal proportions of mutant and wild type. In practice, the efficiency of mutagenesis in this system yields <10% mutant phage particles, probably because of the mismatch repair mechanisms present in the *E. coli*.

Several different approaches have been adopted in attempts to improve the efficiency. A chemical approach was devised by Eckstein and colleagues (9), and this is summarized in Fig. 6. This procedure is extraordinarily efficient, and yields of 95% mutant phage particles are routine. The reason for the dramatic increase in efficiency is owing to the fact that the double-stranded recombinant phages used to transform the *E. coli* contain mutations in both strands. There is thus no mismatch for the host defense mechanisms to recognize and correct. A completely different approach to improve efficiency was developed by Kunkel (10). This procedure exploits strains of *E. coli* that will preferentially destroy the strand of the recombinant phage bearing the wild-type gene. The method is outlined in Fig. 7.

4.4. Types of Possible Mutation

In the early days of oligonucleotide-directed mutagenesis, it was expected that only single-base changes would be possible. It was considered that more extensive mismatches would not anneal well enough. Moreover, in the early to mid-1980s, automated DNA synthesizers capable of producing long oligonucleotides were not as widely available as they are now. Subsequently, it has been found that quite extensive mutations can be specified by a single oligonucleotide, providing that at least ten perfectly complementary bases flank the positions of mutation. An indication of the scope of oligonucleotide-directed mutagenesis is given

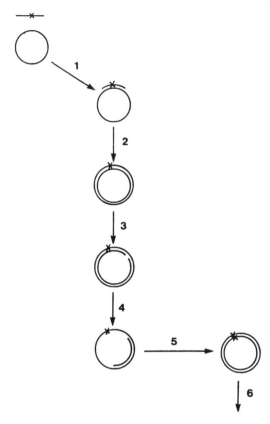

Fig. 6. A chemical approach to improving mutagenesis efficiency: the Eckstein method. 1. An oligonucleotide specifying the desired mutation is annealed to a recombinant M13 DNA molecule containing the gene to be mutated. 2. The complementary strand of DNA is synthesized with Klenow fragment and DNA ligase. In this method, an analog of dCTP is used in which one of the oxygen atoms on the α phosphorus is replaced by a sulfur atom. 3. The replicative form is treated with the restriction enzyme *Nci*I. This enzyme will bind to its recognition sequence despite the presence of a sulfur atom, but only the normal strand with oxygen atoms in the phosphodiester bond is cleaved. 4. The nicked molecule is treated with exonuclease III, which removes nucleotides specifically from the nicked strand. The other strand is left intact. 5. This partially digested molecule is used as the substrate for DNA polymerase and DNA ligase to synthesize a complete double-stranded molecule. This new molecule now has both strands mutant, and there is no mismatch. 6. The mutant M13 molecule is used to transform *E. coli* cells, and mutant phage particles are recovered with very high efficiency.

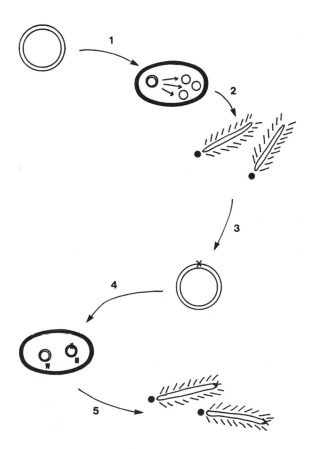

Fig. 7. A genetic approach to improving mutagenesis efficiency: the Kunkel method. 1. A recombinant M13 molecule containing the gene to be mutated is transformed into a special strain of *E. coli* cells. This strain contains two mutations: *dut⁻* and *ung⁻*. These mutations specify (a) a deficiency in dUTPase, giving an increased intracellular dUTP concentration, and (b) a deficiency in uracil glycosylase, thus allowing an incorporation of uracil residues into newly synthesized DNA instead of thymidine. 2. Particles of M13 are produced in which 20–30 residues of uracil have been incorporated into the genome. 3. This uracil-containing M13 DNA is used for site-directed mutagenesis, such that the new mutant strand contains no uracil. 4. The double-stranded molecule is then used to transform a strain of *E. coli* that is *ung⁺*. In this case, the wild-type (W), uracil-containing molecules will be destroyed, but the mutant (M) all-thymidine molecules will be processed (5) to yield mutant M13 particles with very high efficiency.

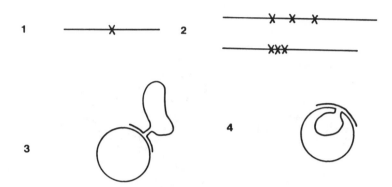

Fig. 8. Possible types of mutation engineered by oligonucleotide-directed mutagenesis. 1. Single base changes can be specified by an oligonucleotide of 20 bases with the mismatch placed near the center. 2. Multiple base changes are possible, provided that approx 10 bases of perfectly complementary sequence occur on either side of the mutations. The base changes can be contiguous (e.g., to specify a complete codon change) or can be spaced out (e.g., to specify three different codons). 3. Long oligonucleotides can be used to specify insertions, provided that approx 10 bases of perfectly complementary sequence occur on either side of the insertion. 4. In a similar way, deletions can be specified with oligonucleotides that have about 10 bases of perfectly complementary sequence on either side of the deletion.

in Fig. 8. It can be seen that there is in effect almost no limit to the types of mutation that can be engineered. Moreover, the procedures have now been so well streamlined that, once a system is set up, it is possible to generate the mutant gene, sequence it, and express the mutant protein all within a couple of weeks. It is easy to appreciate why these new methods have had such a major impact on biological research.

Bibliography

1. Wilson C. A. B., Hardman N., Fothergill-Gilmore L. A., Gamblin S. J., and Watson H. C. (1987) *Biochem. J.* **241,** 609–614.
2. Brown T. A. (1990) *Gene Cloning: An Introduction*, 2nd ed., Chapman and Hall, London.
3. Berger S. L. and Kimmel A. R. (1987) *Guide to Molecular Cloning Techniques*, Academic, New York.

4. Sambrook J., Fritsch E. F., and Maniatis T. (1989) *Molecular Cloning,* 2nd ed., Cold Spring Harbor Laboratory Press, Plainview, NY.
5. Levinson A., Silver D., and Seed B. (1984) *J. Mol. Appl. Genet.* **2,** 507–517.
6. Vernet T., Dignard D., and Thomas D. Y. (1987) *Gene* **52,** 225–233.
7. Goeddel D. V., Kleid D. G., Bolivar F., Heneker H. L., Yansura D. G., Crea R., Hirose T., Kraszewski A., Itakura K., and Riggs A. D. (1979) *Proc. Natl. Acad. Sci. USA* **76,** 106–110.
8. Zoller M. and Smith M. (1982) *Nucleic Acids Res.* **10,** 6487–6500.
9. Taylor J., Ott J., and Eckstein F. (1985) *Nucleic Acids Res.* **13,** 8764–8785.
10. Kunkel T. (1985) *Proc. Natl. Acad. Sci. USA* **82,** 488–492.

Fourteen

Storage Stabilization of Proteins

Felix Franks

1. Need for Stabilization

Proteins in solution are subject to more or less rapid deterioration; the causes are not clearly established. Some deterioration processes were discussed in Chapter 11. The view is held by some that proteins can never be purified to the extent that all traces of proteolytic enzymes are removed, thus making a gradual deterioration in a liquid medium unavoidable. For a proteinaceous preparation to be turned into a viable commercial product with a reasonable shelf life capable of being shipped and stored, it must be subjected to some form of treatment that will substantially retard all possible inactivation processes. Standard stabilization methods include:

Chemical additives [glycerol, $(NH_4)_2SO_4$];
Undercooling (subzero temperatures, unfrozen);
Chemical modification/immobilization;
Sequence alteration/"protein engineering";
Freeze/thaw, lyophilization;
Water-soluble glasses.

The preferred method must depend on the use to which the product will be put. Thus, for general biochemical reagent use, chemical additives offer a convenient and simple choice, although salts and/or glycerol may have to be removed by the end user. In the case of industrial enzymes, heat stability is important and can

From: *Protein Biotechnology* • F. Franks, ed. © 1993 The Humana Press Inc.

often be achieved by immobilization, chemical modification, or, more rarely, by purposeful protein engineering. For the laboratory storage of small quantities of labile, but valuable intermediates or chromatographic fractions, undercooling provides an effective method (*see* Section 3.2.3. *below*). For therapeutic products, especially those that are administered by injection, the choice is strictly limited, mainly for safety (regulatory) reasons. The current preferred method for parenteral preparations is to convert the product into the solid state by freeze-drying, largely because of the perceived benefits of freezing as a means of stabilization. In some cases, products are stored frozen and shipped on dry ice, despite the considerable inconvenience and cost involved. Recently developed techniques, based on the evaporation of water (rather than freezing plus the sublimation of ice) and on "capturing" the active material in a water-soluble glass, have drastically simplified the drying process, while achieving superior recoveries of products with a high degree of shelf stability.

For some commercial applications, proteins can be chemically modified, provided that such modifications do not adversely affect the function that is to be preserved. This is particularly true where enzymes are used to promote biotransformations. Crosslinking the enzyme molecule, attachment to the packing material in a reactor, or modification of certain amino acid residues may extend the useful life of the enzyme.

This chapter deals primarily with proteins that are destined for therapeutic and diagnostic use. The processor is then much more restricted in the choice of stabilization treatments, with chemical modification being out of the question. Although liquid state stabilization would often be the method of choice, this is rarely possible in practice. In this chapter, therefore, emphasis is placed on lyophilization, which, despite its unfavorable economics, has established itself as the most widespread stabilization method for the manufacture of stable pharmaceutical products.

2. Chemical Additives

The efficacy of high concentrations of additives rests on two principles: salting-out and/or the colligative freezing point depression. The effects of salting-out have already been described in previous chapters; several enzymes are supplied as suspensions in 3M ammonium sulfate, in which state they have adequate, although not remarkable, shelf lives at ambient temperatures.

More labile products, e.g., protein kinases or restriction enzymes, have to be maintained at low temperatures, but are not freeze/thaw stable. They are supplied in 50% aqueous glycerol solution, which remains unfrozen at all normal subzero temperatures. The solutions are shipped and stored at –20°C. Even then, lifetimes can be quite short; some products have to be freshly prepared before use. The presence of glycerol may also be undesirable, because it can adversely affect the specificity of enzymes (1). Although the glycerol is being removed by dialysis at low temperature, the enzyme may suffer partial inactivation.

3. Stabilization with the Aid of Low Temperatures (Unfrozen)

At present, the stabilization methods that enjoy the greatest popularity involve freezing, followed either by deep-frozen storage or by drying from the frozen state (lyophilization). The rationale is that, since freezing is performed at low temperatures, such techniques therefore profit from the obvious benefits of low temperatures, e.g., the reduction of chemical deterioration rates and the inhibition of microbiological attack. Actually, this type of reasoning fails to recognize that exposure to low temperature (chill) and freezing are two completely different processes and have little in common. Although chill conditions do indeed retard kinetic rate processes of all types, freezing (separation of pure water in the form of ice) is invariably damaging and can be lethal, unless performed under very carefully controlled conditions.

3.1. The Physical Properties of Undercooled Water

Figure 1 shows the subzero temperature behavior of water in schematic form. Under equilibrium conditions, water freezes at its "normal" freezing point (T_f) of 0°C. Below this temperature, ice is the stable phase. In practice, unless seeded with crystalline ice, water does not freeze at 0°C. Instead, it will undercool, i.e., remain liquid at subzero temperatures. The degree of undercooling that is actu-

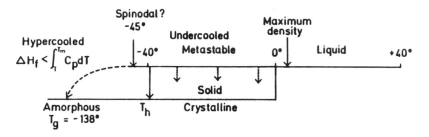

Fig. 1. Forms of stable and metastable (unstable) water that can exist at subzero temperatures.

ally achieved depends on the volume of water, its purity, the cooling rate, and other factors to be discussed later. Undercooled water is unstable, and once frozen, the undercooled liquid state cannot be regained directly by heating; the ice must first be melted at 0°C. Bulk samples of highly purified water can usually be undercooled to –20°C. The mechanism of undercooling and freezing is further described in Section 4.

All physical properties of water become increasingly sensitive to temperature with decreasing temperature. The eccentric properties of undercooled water have been described by Angell (2). Some of these changes are illustrated in Fig. 2. Ionization equilibria are particularly affected: thus, pK_w increases with decreasing temperature according to:

$$\ln K_w = - (34{,}865/T) + 939.8563 + 0.22645T – 161.94 \ln T \qquad (1)$$

At –20°C, pK_w = 16; i.e., the neutral pH = 8. Similarly, pK_a values of all acids are more or less temperature sensitive, with the temperature dependence of pK_a exhibiting a complex shape (*see* Fig. 3). This factor should be taken into account in the choice of pH buffers.

The density of undercooled water decreases steeply with decreasing temperature, giving rise to a large negative coefficient of expansion. If extrapolated, the density curves of water and ice intersect at –45°C, which is also the temperature at which most physical properties of water appear to diverge to infinity (2). This

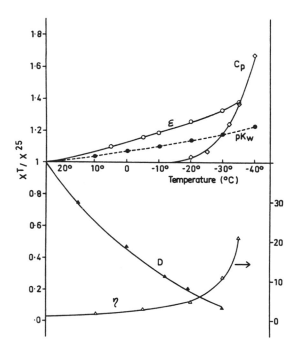

Fig. 2. Temperature dependence of some physical properties of under-cooled water (X^T) compared to the property at 25°C (X^{25}): pK_w, dielectric constant (ε), isobaric specific heat (C_p), and self-diffusion coefficient (D)—left ordinate; viscosity (η)—right ordinate. Reproduced from Franks F. (1982) in *Biophysics of Water* (Franks F. and Mathias S. F., eds.), John Wiley & Sons, Chichester, p. 279.

behavior, which is thought by some to be a spinodal-type phenomenon, is a mystery that is still baffling scientists.

The specific heat of water increases rapidly with the degree of undercooling below –20°C; this leads to a decrease in the latent heat of crystallization (*not* the latent heat of fusion) of ice with decreasing temperature. Eventually, the so-called hypercooled state is reached, where the enthalpic driving force for ice crystallization reaches zero (Fig. 1).

The diffusive behavior of water exhibits a highly anomalous, non-Arrhenius behavior. Thus, undercooled water at –90°C has a viscosity comparable to that of glycerol at room temperature.

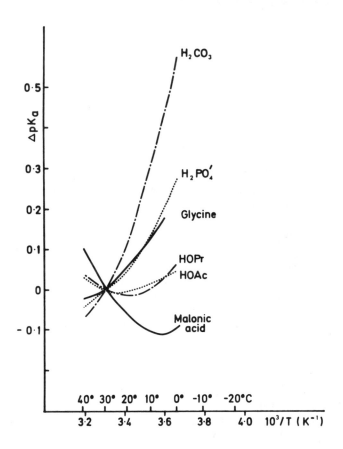

Fig. 3. The effect of temperature on pK_a of some acids used as pH buffers, normalized to 30°C.

3.2. Biochemical Processes Under Chill Conditions

Conditions on the low-temperature side of the maximum protein stability are being referred to here, as described by the $\Delta G(T)$ profile in Fig. 6 of Chapter 11. The aqueous system containing the protein remains unfrozen, and thus, of constant composition and concentration. If the temperature drops below the equilibrium freezing point, T_f, then the liquid is said to be undercooled and in a thermodynamically unstable state (*see* Fig. 1), because under these conditions, ice is the stable state. However, both in vivo and in vitro, such conditions can be stabilized by kinetic means.

3.2.1. Uncoupling of Reaction Sequences

One of the most dramatic biochemical consequences of low-temperature exposure is the uncoupling of reaction sequences. Biochemical reaction cascades have evolved, such that the rates of the various constituent reactions are equal and no intermediate species are generated at appreciable concentrations. Consider the following reaction sequence:

$$A \rightarrow I \rightarrow B \rightarrow \qquad (2)$$
$$k_1 \quad k_2$$

where I is a transient intermediate state of very low concentration, and $k_1 = k_2$ at the optimum physiological temperature. However, the activations energies, E^*, of the two processes are not equal. At any temperature other than the optimum, the reactions become uncoupled. Let $k_1 = k_2$ at 25°, $E^*_1 = 50$ and $E^*_2 = 70$ kJ/mol, then at –20°C, $k_1 = 3k_2$ and at –40°C, $k_1 = 6k_2$. With decreasing temperature, the degree of uncoupling increases, resulting in the enrichment of intermediate I, which will not be converted at a sufficiently high rate into the product B, which itself serves as the reactant for further downstream processes.

3.2.2. Reversible Cold Inactivation of Proteins

The general thermodynamic principles of cold denaturation have been discussed in Chapter 11 on conformational stability. Cold dissociation (and/or unfolding) appears to be a general phenomenon that has been described for many different proteins, e.g., phosphofructokinase, lactate dehydrogenase, tubulin, and virus coat proteins (3). Unlike heat denaturation, cold inactivation appears to be fully reversible (even at very high protein concentrations), and does not lead to permanent damage or loss of activity (4). The detailed shape of the stability profile depends on the peptide sequence and on environmental factors. Cold inactivation and destabilization play an important role in natural cold adaptation/resistance mechanisms of plants, microorganisms, and insects (5), which all rely on the synthesis of low-molecular weight carbohydrates for cold survival. For example, when pyruvate kinase becomes blocked during cold exposure, the glycolytic pathway will be diverted to the synthesis of glycerol. Similarly, at even lower temperatures, phosphofructokinase is inactivated, leading to the conversion of hexose phosphates to sorbitol (6). The carbohydrate protectants fulfill a dual role: They stabilize proteins in solution by a general lyotropic effect (salting-out), but they also reduce the water content and, hence, the degree of freeze concentration of the cytoplasm. It has also been suggested that, in extreme cases, they may even facilitate the vitrification of the whole cell. It is not known, however, whether cold inactivation/destabilization also takes place in vitro during the low-temperature storage of proteins.

It is noteworthy that the same naturally produced polyhydroxy cryoprotectants (PHC) also find application as excipients in freeze-drying. Presumably, they fulfill a dual role: While a protein is in solution, it is stabilized against denaturation, and in the freeze-concentrated state, the PHCs can replace water as "solvent" and help to vitrify the amorphous mixture, as described in detail in Section 5.4.

3.2.3. Undercooling as a Protein Stabilization Process

As will be described below in more detail, the formation and separation of water as a pure phase (ice) from an aqueous solution relies on the generation of crystal nuclei, which are capable of growth into macroscopic crystals. The nucleation rate depends on several factors, including the solution volume (7). The nucleation probability at any given subzero temperature can be reduced by several orders of magnitude if the aqueous sample is dispersed in the form of microscopic droplets (μm diameter) in an immiscible phase (8). This principle forms the basis of undercooling as a viable method for stabilizing proteins. The organic phase must be chosen such that it is not active as a catalyst for ice nucleation; nor must it permit the aqueous droplets to coalesce. The aqueous phase should be readily recoverable after the emulsion is returned to ambient temperature. Protein stabilization by undercooling has been described in detail (9), and the effectiveness of the technique is demonstrated in Fig. 4. Several grades of dispersion fluids are obtainable commercially under the trade name U-COOL™.

Undercooling is particularly suitable for the in-house stabilization of small samples of valuable, but highly labile products, e.g., protein kinases, or fractions collected from a chromatographic

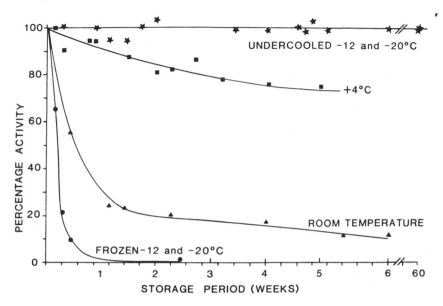

Fig. 4. Maintenance of enzyme activity by LDH stored under different conditions; LDH concentration: 10 μg mL^{-1} in phosphate buffer pH 7. Reproduced from ref. 9.

column during a purification procedure. Complete retention of enzyme activity for several years in dilute solutions can regularly be achieved with storage at –20°C in the undercooled state. The technique does not lend itself well to use with large volumes of solutions, because an organic:aqueous phase ratio of 4:1 is recommended for optimum results. Also, the dispersions have to be shipped at temperatures not exceeding 4°C, which they can tolerate for limited periods.

Benefits of undercooled storage include:

1. The process is simple and cost-effective, especially when compared to freeze-drying;
2. Protecting additives are not usually required;
3. Proteins can be processed and supplied over a very wide range of concentrations;
4. Aliquots can be taken from a sample without affecting the activity of the remainder;

5. Products are stable and not sensitive to medium-term temperature fluctuations, provided that the temperature never exceeds ca. 4°C;
6. The storage temperature is not critical, provided that it does not approach the ice nucleation temperature (in the region of −40°C);
7. Products are not susceptible to moisture pick-up; and
8. The process is essentially the same for all proteins.

4. Physics of Freezing

The chief difference between undercooling and freezing is that the latter process, which describes the separation of ice as a pure solid water phase, is of necessity accompanied by a concentration increase of all water-soluble species in the system (freeze concentration).

4.1. Ice Nucleation

As was previously mentioned, the liquid/solid transition of water requires the nucleation of the solid phase in the undercooled system. Similarly, for any solutes to crystallize from solution under eutectic conditions, a nucleation event must precede the growth of crystals; otherwise, the solution becomes supersaturated.

Nucleation is a stochastic process, caused by random density fluctuations owing to Brownian diffusion of molecules, to produce a transient embryo that resembles the crystal in geometry and has a sufficiently long lifetime for further condensation of molecules to occur *(7)*. The probability of nucleation is therefore a function of the temperature and the volume (the total number of molecules available to form such a "critical" nucleus).

In the absence of catalysts, nucleation is said to be *homogeneous* (spontaneous). For ice, the nucleation rate increases rapidly (by orders of magnitude) just above −40°C, which is referred to as the homogeneous nucleation temperature, T_h. Nucleation is a first-order kinetic rate process, but *not* of the Arrhenius type, i.e., the nucleation rate $J(T)$ *increases* with *decreasing* temperature according to the relationship *(10)*:

$$J(T) = A \exp (B\theta) \tag{3}$$

where $\theta = [(\Delta T)^2 T^3]^{-1}$ and ΔT is the degree of undercooling $(T_f - T)$.

In the presence of particulate matter able to act as heterogeneous catalyst (dust, pyrogens, cell debris, membrane fragments, and so on), nucleation takes place at a *higher* temperature. The *heterogenous* nucleation temperature, T_{het}, depends on the catalytic efficiency of the catalyst particle, its radius of curvature, and its degree of wetting by ice and water (7). Otherwise, the theory resembles that for homogeneous nucleation. In practice, unless extreme precautions are taken, nucleation in bulk aqueous phases is always heterogeneous.

In recent years, biogenic ice nucleation has become a prominent field of study. It was discovered that certain microorganisms, e.g., *Ps. syringae* and *Erwinia herbicola*, reduce the ability of water to undercool. The sites of nucleation activity are protein structures; so-called ice-minus mutants have been successfully prepared (11). At the other end of the nucleation scale, the Antarctic fish serum antifreeze glycoproteins (and other antifreeze proteins) inhibit ice nucleation, permitting blood to undercool to some extent ($\Delta T \approx$ 2°C), and thus causing the Antarctic fish to remain unfrozen at temperatures below the normal freezing point of blood (12).

4.1.1. Nucleation of Ice in Aqueous Solutions

Solutes are able to depress T_h and T_{het}, the depression depending on the concentration, as shown in the form of a Clapeyron-Clausius representation in Fig. 5. For instance, in a 40% solution of ethylene glycol, $T_h = 90°C$. There appears to be an almost linear correlation between the equilibrium freezing point depression and the depression of T_h (13). The reason is unclear, since the two phenomena are completely unrelated.

4.2. Ice Crystal Growth, Morphology, and Size Distribution

The ice morphology and crystal size distribution depend in a complex manner on the viscosity, cooling rate, temperature, nucleation density, and the heterogeneity of the substrate (14). The Avrami equation relates the fraction crystallized, f, at time t to the nucleation rate, J, and the linear crystal growth rate, u:

$$f = (4\pi J u^3 t^4)/3 \tag{4}$$

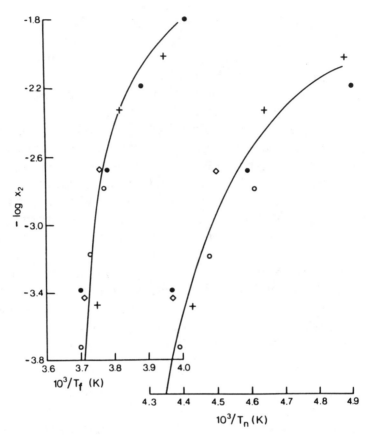

Fig. 5. Melting point and nucleation temperature data for aqueous solutions as functions of solute concentration, plotted according to the Clapeyron-Clausius equation: ●, ethane diol; ○, glucose; ◇ urea; + NaCl; data from ref. 13. Reproduced from ref. 14.

TTT (time/temperature/transformation) plots are of complex appearance (*see* Fig. 6 for the growth of ice in LiCl solutions) because of the different kinetics that govern nucleation and crystal growth: The lower limb corresponds to rapid nucleation, the rate being limited by crystal growth; the upper limb reflects rapid crystal growth, with the rate limited by nucleation *(15)*. It is noteworthy that, over the 20-degree interval spanned by the *TTT* plot, the diffusion coefficient of water increases by 10^5 (viscosity decreases by 10^5).

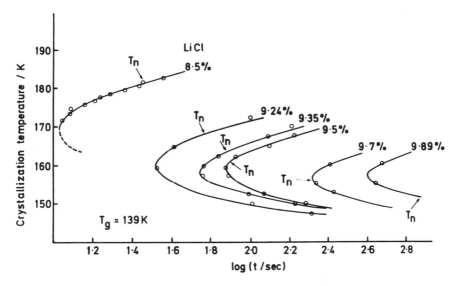

Fig. 6. Time-temperature-transformation curves of ice crystallization in concentrated LiCl solutions. Reproduced from ref. *10.*

The cooling rate for a liquid drop is given by:

$$dT/dt = \sigma\,[(\partial^2 T/\partial r^2) + (2/r) \cdot (\partial T/\partial r)] + \Delta H/dC \qquad (5)$$

where σ is the thermal diffusivity and is given by k/Cd. The quantities d, C, and k are the mean values of the density, specific heat, and thermal conductivity, respectively; r is the position in the sample ($0<r<R_2$), R_2 is the radius of the sample, and ΔH is the latent heat of crystallization (Fig. 7A). The first term on the right-hand side in the above equation describes the heat removal rate, and the second term describes the warming rate owing to the liberation of latent heat during freezing at R_1. Note that nucleation may take place anywhere within the sample, although in practice it often occurs at the surface ($r = R_2$). For water and ice, k is small and C is large; hence, it is difficult to achieve fast cooling. Figure 7B shows the dependence of crystal size and cooling rate on r, when freezing occurs within the drop.

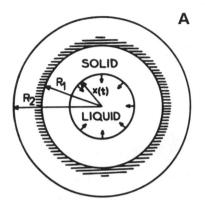
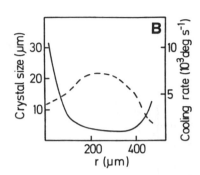

Fig. 7. (A) Significant parameters for the calculation of the cooling rate as a function of droplet dimensions. (B) Crystal radii (......) and cooling rates (———) as functions of location within a drop of 0.55 mm radius. Reproduced from ref. *14*.

dT/dt and the degree of undercooling also determine the ice crystal habit; examples are shown in Fig. 8 *(16)*. Slow cooling produces large dendritic crystals; at higher rates, spherulites of decreasing dimensions are formed. In rapidly quenched solutions, small spherulitic crystals are formed, and freezing may be incomplete (partial vitrification). With ultrarapid quenching of small volumes, such as are used in electron microscopy, freezing can be completely inhibited, and the undercooled liquid is then vitrified *(17)*.

If, after rapid freezing, the temperature is allowed to rise, two processes can occur: (1) crystallization of previously unfrozen water (or solutes), also referred to as devitrification, and (2) maturation: a disproportionation whereby large crystals grow at the expense of small crystals. The rate of maturation depends on the crystal polydispersity and the temperature fluctuations. This process is also referred to as recrystallization and can be utilized to increase the rate of ice sublimation during lyophilization. The results of partial recrystallization are apparent in Fig. 7B; larger crystals are found in the region where the exothermic process of freezing was triggered.

In aqueous solutions, J, u, and the rate of maturation depend on the volume concentration of solute, mainly via its diffusion coefficient (i.e., molecular weight). The rate of crystallization

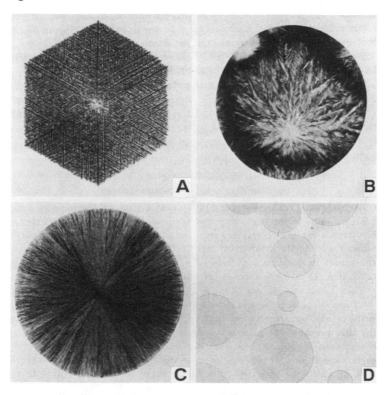

Fig. 8. Principal ice morphologies encountered in aqueous solutions subjected to different cooling rates: (A) dendrites, based on hexagonal patterns; (B) transition forms, irregular dendrites; (C) spherulites; (D) "evanescent" spherulties. Reproduced, with permission, from ref. *16.*

depends on the local concentration of the slowest diffusing component in a mixture *(18);* hence, polymers are most effective in retarding the crystallization of ice.

5. Chemistry and Biochemistry of Freezing

All the chemical processes, frequently of an undesirable nature, that occur during freezing are secondary effects of concentration changes that accompany the removal of water from the solution phase as ice.

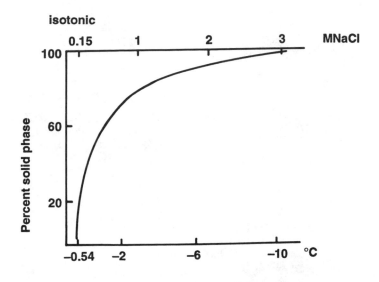

Fig. 9. Freeze concentration of a 0.154M NaCl (isotonic saline) solution. For details, *see text*. Reproduced from ref. *10*.

5.1. Freeze Concentration

The course of freeze concentration is determined by the initial concentrations of solutes (water activity) and the cryoscopic freezing-point depression. In the limit of $c \rightarrow 0$, the slope of the liquidus curve $T_f(c)$ is given by the cryoscopic constant (1.86 deg/mol). The curvature depends on the nonideality of the solution (solute–solute interactions). Apart from its conventional form, the phase behavior can also be represented by the volume (or mass) of frozen water as a function of temperature and/or solute concentration in the residual liquid phase (*see* Fig. 9). The liquidus curve terminates at the eutectic point T_e, provided that the solutes actually crystallize at this temperature. This point is important when pH buffer mixtures are used. The eutectic data for phosphate buffers are summarized in Table 1. It is seen that at the ternary eutectic temperature, the mol ratio $NaH_2P/Na_2HP = 57$. Any mixture that contains the salts in another ratio suffers pH shifts *if* one or the other of the salts crystallizes as a binary eutectic during cooling. The mol ratio for a pH 7 buffer is 0.72. The eutectic ratio is therefore

Table 1
Eutectic Data of Phosphate Solutions:
(1) Monosodium (Potassium) Salt and (2) Disodium (Potassium) Salt

	T_e°	$C_e, 1, M$	$C_e, 2, M$	$C_e, 1/C_{e'}, 2$
NaH_2PO_4	−9.7	3.42		
Na_2HPO_4	−0.5		0.11	
$NaH_2PO_4 - Na_2HPO_4$	−9.9	3.42	0.06	57 $(0.72)^a$
KH_2PO_4	−2.7	0.92		
K_2HPO_4	−13.7		2.85	
$KH_2PO_4 - K_2HPO_4$	−16.7	1.30	2.70	0.48 $(0.72)^a$

^{a}Corresponds to the pH 7 buffer ratio.

not appropriate for pH buffering of proteins. The mol ratio for the potassium salt ternary eutectic is 0.48, with $T_e = -16.7°C$, i.e., much closer to the mol ratio corresponding to pH 7 and of a better freeze stability than the Na salt. In the presence of other neutral salts, quaternary eutectics (three salts + ice) can be formed with different composition ratios. Even if the protein is not damaged by excessive pH shifts during freezing, the buffer salt concentration will increase several-fold, and this is likely to have deleterious effects.

5.2. Crystallization and Supersaturation

As distinct from predictions based on the equilibrium-phase behavior, the *actual* crystallization kinetics of buffers and other solution components during freezing depend on the sample size, the cooling rate, and the particular solutes, their initial concentrations, and their nucleation rates in the mixture. Under freezing conditions, salt crystal nucleation and growth are very slow compared to ice crystal growth. With regard to the phosphates, it is found in practice that KH_2PO_4 and Na_2HPO_4 crystallize readily, K_2HPO_4 crystallizes slowly, and Na_2HPO_4 does not crystallize at all at subzero temperatures *(19)*. However, as Fig. 10 shows, complete crystallization of salts from a freezing solution is not common for salt concentrations <0.5M. This is even true for NaCl. Similar results are obtained with mannitol, a commonly used freeze-drying additive. Furthermore, no ternary eutectic phase separation appears to take place in mixed

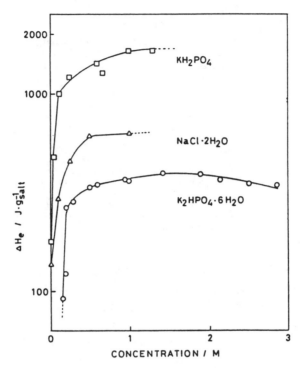

Fig. 10. Dependence of the heat of eutectic melting (degree of precipitation) of salts on the initial solution concentration. Reproduced from ref. *19*.

phosphate solutions at any initial concentrations *(19)*. It is therefore common for freezing solutions to become supersaturated either wholly or partly. Supersaturation can sometimes be prevented by annealing the frozen mixture, i.e., by temperature cycling below the freezing point. It is frequently observed that salt crystallization that is inhibited wholly or partially during freezing will take place when the frozen mixture is heated. It should, however, be noted that the aim in freezing processes is frequently to achieve supersaturation of some solutes. This is in fact the very basis of action of cryoprotectants.

5.3. Secondary Effects of Freeze Concentration

Reaction rates in partially frozen systems show a complex temperature dependence (Arrhenius equation becomes inoperative), because the decreasing temperature is more than counterbalanced

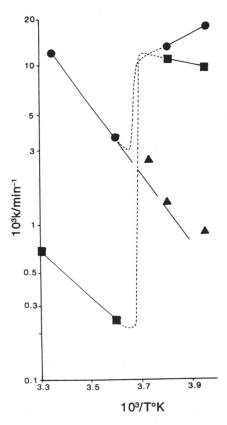

Fig. 11. The effect of temperature on the oxidation rate of ascorbic acid, illustrating the rate enhancement due to freezing (broken lines); circles: concentrated solution; squares: dilute solution; triangles: unfrozen (undercooled) solution. Reproduced, with permission, from ref. 20.

by a rapidly increasing concentration; see Fig. 11 for the oxidation of ascorbate (20). Note also that, for undercooled solutions, the Arrhenius relationship is observed, i.e., the rate continues to decrease with decreasing temperature, with no break at the freezing point. Large rate enhancements during freezing are common, especially for enzyme-catalyzed reactions (21).

Freeze denaturation/aggregation/insolubilization of proteins is the result of the combined effects of freeze-concentration of the protein itself and of salts, possibly coupled with pH shifts caused

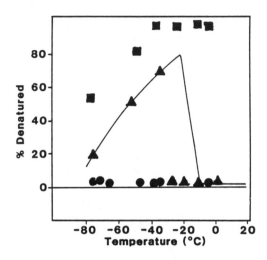

Fig. 12. The effects of additives on the freeze denaturation of chymo–trypsinogen: 0.1*M* NaCl (▲), 0.1*M* sucrose (●) and 0.1*M* urea (■). Data from ref. 22.

by freeze perturbation of the buffer (*see* Fig. 12). Denaturation can be minimized or completely prevented by protectants (salting out?); usually polyhydroxy compounds (glycerol, sorbitol, inositol, sugars) are used. The same substances also fulfill the role of in vivo freeze protectants (*6*). As shown in Fig. 13, the rate of denaturation is a complex function of temperature (*22,23*), but *see also* Fig. 11. Freeze denaturation rates and the limiting degrees of denaturation exhibit maxima at intermediate subzero temperatures for reasons that are discussed more fully in Section 5.5.

5.4. Freezing in Cellular Systems

The freezing process in biological systems composed of cells or tissues is accompanied by complex water and solute fluxes in and out of the cell (*10*), because freezing always begins in the extra-cellular space. Intracellular freezing is invariably lethal. Cryo-protectants are used (1) to lower T_f and reduce the degree of freeze concentration, and (2) to maintain the volume of the cell and prevent cell shrinkage while water is removed as a result of the

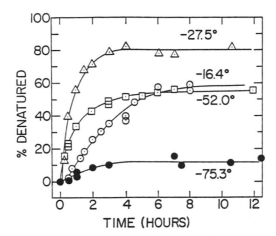

Fig. 13. The rate of chymotrypsinogen denaturation in frozen solutions at four temperatures. Solutions contain only protein (0.25 wt%) and HCl (pH 1.78 at 25°C). Reproduced, with permission, from ref. 22.

osmotic imbalance across the cell membrane. Cryoprotectants must therefore be able to penetrate the membrane. Glycerol or dimethyl sulfoxide is commonly used. The coupled flux rates of water (out) and cryoprotectant (in) depend on the cooling rate and the permeability of the plasma membrane, as well as other factors (24), a discussion of which is beyond the scope of this book. The interested reader is referred to texts on cryobiology.

5.5. Supersaturation and Vitrification: The Role of Excipients

The foregoing discussion has emphasized that a dividing line must be drawn between processes that *may* take place (thermodynamics) and those that actually do take place within a measurable period of time (kinetics). It was shown that, in practice, kinetics appear to control crystallization, especially from concentrated aqueous solutions. In such systems, supersaturation is therefore common. This phenomenon is utilized in vivo by freeze-tolerant organisms to manage their water contents and in vitro for the protection of labile substances that are subjected to freezing. It is a

common practice to add substances (excipients) that do not themselves crystallize readily from aqueous solutions at low temperatures and that can inhibit the crystallization of other species.*

Excipients in common use include monomeric and oligomeric
PHCs and some amino acids (e.g., glycine, alanine, and histidine).
Such substances also tend to retard the crystallization of salts from
solution. As freezing proceeds, the solution viscosity increases with
the increasing solute concentration; the temperature dependence
of the viscosity at a given concentration approximately follows the
Arrhenius equation. However, as the viscosity rises to very high
values of the order of 10^8 Pa s and above, its temperature dependence becomes much greater and, eventually, the mixture is said
to undergo a glass transition, where the viscosity increases by several orders of magnitude over a narrow temperature interval, during which the physical state of the homogeneous mixture changes
from a fluid, through a syrup, and a viscoelastic "rubber" to a brittle
solid ("glass"). The viscosity/temperature relationship of such an
amorphous material is shown diagrammatically in Fig. 14. In a frozen aqueous PHC solution, this transition, which is characterized
by a so-called glass temperature, T_g, typically occurs when the solute concentration in the mixture has reached 80% by weight (25).
The viscosity at T_g is then of the order of 10^{14} Pa s. A glass is thus to
be regarded as an undercooled liquid (noncrystalline) with a very
high viscosity. This viscosity is equivalent to a flow rate of the
order of 1–10 μm/yr. Freeze concentration therefore comes practi-
cally to a halt at T_g, and the amorphous product is characterized by
its glass temperature and composition. Figure 15 shows a typical
state diagram, in this case for the binary water/sucrose system (25).
This state diagram is not to be confused with a phase diagram,
because, unlike the freezing point curve, the $T_g(c)$ curve is not a
phase-coexistence curve, but represents the locus of all points hav-

---

*The word "excipient" has crept into freeze-drying jargon, but its mean-
ing is not exactly clear. According to some authors, an excipient is any substance in the solution other than the active material; this would classify salts
and pH buffers as excipients. We here refer to excipients as substances that are
added to the aqueous solution in order to protect and stabilize the active material against freezing/drying damage.

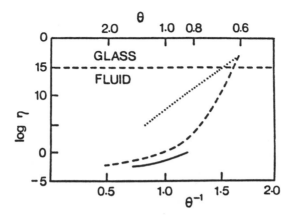

Fig. 14. The relationship between log (viscosity) and θ^{-1}, the reciprocal reduced temperature ($\theta = T/T_f$) for silica (. . . .), a typical organic glass former (----), and water(———). An approximate extrapolation suggests that for the vitrification of water $\theta^{-1} \approx 2$, i.e., $T_g \approx 136$ K. Reproduced from ref. 10.

ing the same viscosity. According to the equilibrium-phase diagram, sucrose should crystallize at its eutectic temperature, −14°C. However, the nucleation of sucrose at this temperature is a most improbable event. Instead, the solution continues to freeze and becomes progressively more supersaturated (thermodynamically unstable, but kinetically very stable). Thus, the point at which the liquidus curve intersects the glass curve, referred to in the literature as (T_g',c_g'), is *not* an invariant point like the eutectic point T_e in the thermodynamic sense. In practice, it is an important point, because freezing has then become so slow that it can no longer be detected on a measurable time scale. Also, the physical properties of a glass are such as to inhibit chemical and biological reactions. The maximally freeze-concentrated system is therefore mechanically and chemically stable.

However, on the way to maximum freeze concentration and the vitreous state, the aqueous mixture has to traverse the viscoelastic "rubber" state, where it becomes vulnerable to rapid deterioration. It has already been mentioned that, as the viscosity approaches the glass value, so rate processes become more temperature-sensitive than is predicted by the Arrhenius equation. Several kinetic models have been proposed to describe physical and chemical rate

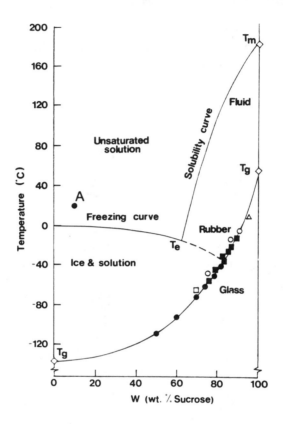

Fig. 15. Solid–liquid state diagram for the sucrose–water system, show-ing the glass transition T_g/concentration profile, the equilibrium solubility curve, running between the melting point and the eutectic temperature, T_e, and the region of metastability (supersaturation) beyond T_e (broken line). Symbols represent experimental T_g data taken from different sources. Point A represents a typical dilute solution that is to be dried to a stable solid product.

processes in "rubbers" (26). In the field of protein stabilization, the equation commonly used is the so-called WLF equation, accord-ing to which kinetic rate constants k are given by:

$$\ln k = -C_1 (T - T_g)/[C_2 + (T - T_g)] \tag{6}$$

where C_1 and C_2 are constants. This equation differs from the more common Arrhenius equation in that the reference temperature is

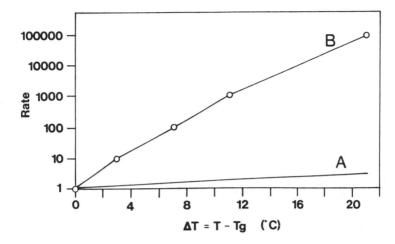

Fig. 16. Relative dependence of kinetic rate processes on $(T–T_g)$ for A: typical Arrhenius fluids and solids (with an assumed activation energy of 80 kJ/mol) and B: rubbery materials in which rates are governed by WLF kinetics. For explanation, *see text*. Reproduced from ref. 32.

T_g, whereas in the Arrhenius formalism, the reference temperature is zero Kelvin. Figure 16 dramatically illustrates the difference in the relative rates for a process with a nominal unit rate at T_g. For the Arrhenius representation, an activation energy of 80 kJ/mol has been chosen.

The consideration of glass transition parameters and possible non-Arrhenius kinetics becomes important where products are subjected to accelerated storage testing. As can readily be deduced from an inspection of Fig. 16, extrapolations of kinetic data by means of the Arrhenius equation will provide highly erroneous results when a glass transition falls within the temperature range employed in the storage tests. For example, if some deterioration process has a nominal unit rate at 50°C, and the glass transition of the system on test is 30°C, then the relative rate at 30°C is only 10^{-5}, i.e., orders of magnitude less than would be predicted by the Arrhenius equation. As a general rule, the WLF kinetic equation should be applied for temperatures up to 50°C above T_g.

5.6. Very Concentrated Mixtures: Water as Plasticizer

Reference to Fig. 15 shows that T_g is a monotonic function of c, rising steeply as c approaches 100%. At this point, T_g refers to the dry product. In the neighborhood of $c \longrightarrow 100\%$, it is more useful to express T_g in terms of the residual moisture content, w ($=100-c$). In the language of polymer chemistry, water here acts as a plasticizer, softening the glassy material (26). To achieve a good storage stability at ambient temperature, it is therefore necessary to reduce w to the extent that T_g of the final product lies above the storage temperature.

Applying these considerations to the practice of freeze-drying and referring to Fig. 15, it is seen that the freezing process can remove water until the concentration of the initially dilute solution, A, has reached 82% ($w = 18\%$). The ice so formed can be removed by sublimation. The residual moisture must be removed by diffusion through, and desorption from, the product until an acceptable moisture level for ambient temperature storage has been achieved (ca. 2%). Emphasis in the foregoing sections has been placed on the thermal and mechanical properties of excipient solutions, rather than on proteins, because in most pharmaceutical formulations, the protein is a very minor component, seldom exceeding a few percent of the total solids.

6. Process Analysis of Freeze-Drying

Of the various techniques available for the removal of water, freeze-drying is by far the most complex (27). It is surprising, therefore, that lyophilization rarely even merits mention, let alone a detailed process analysis, in chemical engineering texts. Scheme 1 summarizes the various physical processes, already discussed, that occur during the freezing of an aqueous solution and the subsequent removal of water, whereas Scheme 2 outlines the experimental variables that are under the control of the operator and the manner in which they affect the quality of the dried product. It is here assumed that the process is applied to labile biologicals, where

FREEZING

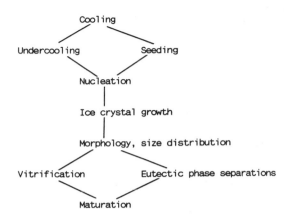

DRYING

Ice sublimation (Product Temperature, Chamber pressure)

Desorption (Temperature profile, Residual moisture)

STORAGE   Residual moisture/temperature control

Scheme 1. Physical principles of freeze drying.

FORMULATION: Individual solute concentrations, mass ratios of solutes, vial dimensions, fill volume

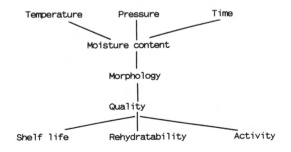

Scheme 2. Experimental variables.

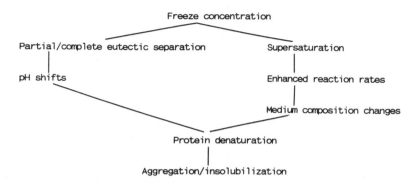

Scheme 3. Chemistry and biochemistry of freeze concentration.

the aim is to recover and preserve some form of biological activity. Scheme 3 summarizes the chemical and biochemical consequences of freezing, also described above in Section 5.3.

6.1. Freezing and Ice Sublimation

The products are supplied either dosed by volume, usually in vials, or in bulk solution in liter quantities. Depending on the production batch, freezing is carried out in the lyophilizer, or the solutions are prefrozen, often by immersion in liquid nitrogen, and then loaded onto the precooled lyophilizer shelves.

If complete eutectic crystallization of all nonaqueous material takes place during cooling, then ice can be removed by simple sublimation, leaving the pure, crystalline product; this is rare and does not occur in protein-containing systems. Problems can arise where a salt crystallizes in the form of a hydrate that might subsequently be dehydrated by the protein (e.g., citrate, acetate).

More commonly, the frozen solid is a mixture of ice crystals embedded in a freeze-concentrated amorphous phase (glass). Because at subzero temperatures ice is the stable water phase, the vapor pressure of ice is lower than that of the unfrozen water held in the solute matrix. What matters, however, is the *rate* of sublimation: Ice sublimes faster than unfrozen water in the amorphous product.

Table 2
Sublimation Rates and Pressures for Isolated Ice Crystals
of 1 mm Diameter at Several Temperatures

Temp. °C	Time	Pa
−10	3 s	2.61×10^2
−30 (sucrose)	20 s	3.81×10^1
−60	11.1 min	1.08
−100 (glycerol)	5.8 d	1.32×10^{-3}
−130	26.3 yr	1.15×10^{-8}

^{a}Excipients with corresponding T_g' values are shown in parentheses.

While freezing goes to completion, the condenser is precooled, and the ice is then sublimed by applying the vacuum. The sublimation rate can be calculated from the following quantities: T_g', fill volume, vial dimensions, total solid content, and pressure. Details of such a calculation are given in Section 6.5. The actual pressures and temperatures to be used during the primary drying stage need to be carefully determined and controlled. For instance, sublimation rates can be increased by employing gas (nitrogen) bleeds and/or raising the temperature of the refrigerated shelves to compensate for the absorption of the latent heat of sublimation by the product. Good practice (rarely observed) demands that, at the end of the sublimation period, the shelf temperature should be lowered back to T_g'. The process details, as affected by heat and mass transfer, have been lucidly explained by Pikal (28). Typical sublimation rates for isolated ice crystals are shown in Table 2. They graphically illustrate the importance of choosing formulations with T_g' values as high as possible. A general rule is that ice sublimation rates decrease by 50% for every three-degree decrease in temperature. In a "real" product, i.e., one consisting of active components, excipients, salts, and so forth, the ice sublimation rates are substantially lower than those quoted in Table 2 because the product matrix imposes a resistance to the migration of water vapor from the ice crystal surface. This resistance is a function of the product density (total solid content) and the fill depth (28).

6.2. Secondary Drying

The composition of the freeze-concentrated phase, which contains all uncrystallized material, and especially its water content, is of primary importance in freeze-drying. Note that the composition of the freeze-concentrated phase does *not* depend on the initial solution concentration, but the *degree* of freeze concentration and the total volume of ice formed do. Any water that remains in the amorphous solid acts as plasticizer. It is *not* unfreezable and it is *not* bound. It is prevented from freezing in a measurable time by the high viscosity of the amorphous solid *(29)*. The secondary drying cycle consists of the removal of the remaining moisture from the product matrix, which should now be porous, after the removal of the ice crystals. It is this porosity (large surface area) that will make the dried product easily rehydratable. If the temperature at any time exceeds T_g by more than a few degrees (*see* Fig. 16), the viscosity will rapidly drop, the solid will collapse, the surface area will decrease, and the final product will not be readily rehydratable. Equally important, all damaging processes (protein denaturation and aggregation, microbiological attack, biochemical reactions) that had been inhibited in the glassy state can now take place in the so-called "rubber," which is viscoelastic. The product should preferably be dried from T_g' to its final state by ramping the temperature to track the $T_g(c)$ profile shown in Fig. 15.

6.3. Choice of Formulation

Where freedom exists in the choice of a suitable formulation, the excipient(s) should be selected for their T_g' and c_g' values, and their glass transition temperatures in the dry state. Ideally, the salt content of the product should be reduced to a minimum, as should the content of any other low-molecular weight components. Table 3 provides a summary of the important glass parameters for a series of PHCs. Consideration must also be given to the possibility of crystallization of some constituents from the product during drying or storage. Where this is deemed to be desirable, then such crystallization should be encouraged by annealing the frozen mixture over a temperature range between T_g' and the onset of ice melting. On the other hand, if crystallization is to be avoided, then

Table 3
Glass/Rubber Transitions of Anhydrous and Freeze-Concentrated PHCs
and Concentrations (Wt Percent) of the Freeze Concentrates

	T_g °C	T_g' °C	Wt %
Glycerol	–78	–95	
Xylitol	–39	–47	
Ribose	–10	–47	
Xylose	9	–48	
Glucitol (sorbitol)	≈0	–44	88
Glucose	39	–43	83
Mannose	30	–41	
Tagatose	40.5	–41	
Galactose	110 (?)	–41	84
Fructose	16	–42	85
Maltose	≈100	–30	81.5
Cellobiose	77		
Trehalose		–30	80
Sucrose	65	–32	83
Glucose:fructose (1:1)	20	–32	
Maltotriose	76 (?)	–24	

[a]Data from ref. 29 and C. van den Berg, personal communication.

excipients with high T_g values should be used, and the product
should be dried to a low residual moisture content.

6.4. Storage Stability

The shelf life of the dried product depends on (1) the storage
temperature and (2) the residual moisture content of the *amorphous*
phase. For safe storage and insensitivity to temperature fluctuations, a low residual moisture is a prime condition. In practice,
quality is traded off against time required for drying and residual moisture content. Ideally, a freeze-dried proteinaceous product should have a final T_g >40°C and should therefore be stable at
room temperature. This is rarely achieved in practice. Most producers recommend that freeze-dried products should be stored at
–20°C, a sign of partial failure or lack of confidence in the process.
Both the ice sublimation and the subsequent water desorption

Table 4
Calculated Freeze-Drying Conditions for Two Formulations
of a Labile Cytotoxic Product, Based on the Determination
of the Glass Transition Parameters, as Described in the Text

		Product A		Product B
Active ingredient (mg)		2		5
Stabilizer (mg)		51		51
Fill vol (mL)		10		10
Vial diameter (mm)	46		26	26
Fill depth (mm)	6		13	13
T_g' (°C)		–33		–36
Moisture content of freeze concentrate (g/g)		0.49		0.49
Sublimation pressure (mbar)		0.28		0.20
Sublimation rate (g/vial/h)	1.47		0.27	0.167
Total mass of ice (g)		9.7		9.7
Primary drying time (h)	7		37	58
Secondary drying time	12		16	18

stages are frequently performed suboptimally, allowing partial melt-back of ice and/or partial collapse of the amorphous phase to occur. Alternatively, the composition of the product may be such that shelf stability is unattainable (e.g., high salt content). If at any time during storage the product reverts to the "rubbery" state, by temperature abuse or by moisture migration from the stopper, then crystallization of salts and/or excipients may occur, thus further reducing the stability of the remaining amorphous (protein) phase.

6.5. Examples

An effective freeze-drying process can be developed *ab initio*, based on a knowledge of the formulation details, the thermal parameters already discussed in detail, and the vial dimensions. Consider a product consisting of 2 mg active component and 51 mg stabilizer in 10 mL solution, in a vial of 46 mm diameter. Recommended freeze-drying parameters are shown in column 2 of Table 4. The manufacturer reasons that, by reducing the vial diameter,

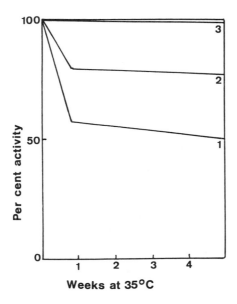

Fig. 17. Typical activity/time curves for a diagnostic enzyme, freeze-dried under different conditions and subjected to storage at 35°C; for details, *see text*.

the production batch fed to the lyophilizer can be increased, and he has a vial of 26 mm diameter available. Column 3 in Table 4 illustrates that the primary drying time, corresponding to the increase in the product fill depth, would need to be increased from 7 to 37 h, casting doubts on the economic feasibility of the vial replacement. The manufacturer also wishes to market a new product in which the active substance is to be increased to 5 mg/vial. For this product, $T_g' = -36°C$. As column 4 shows, the ice sublimation at this temperature and in the narrower vial will take 58 h. The table also demonstrates that economic and marketing decisions can now be performed on a sound technical basis.

An example of process improvement is shown in Fig. 17, which refers to the large-scale lyophilization (500 L) of a purified enzyme extract, obtained from 4 t of a plant raw material. The lowest curve shows the stability profile at 35°C of the dried product, which had been subjected to an indifferent freezing protocol, and arrived at by *ad hoc* experimentation. The intermediate curve shows a similar profile for a product, where freezing and ice sublimation had been performed according to the principles described earlier in this chapter; the immediate losses could then be reduced from 40 to 20%. The top trace shows the stability for the same product after optimal primary, as well as secondary drying, with temperature ramping parallel to the $T_g(c)$ curve.

6.6. Aide Memoire for Effective Freeze-Drying

Given below is a checklist containing the factors that need to be considered for the optimization of a product formulation designed for freeze-drying, and for the correct adjustment of process conditions/variables to achieve maximum product yield and stability.

Formulation
   Choice of buffer: pH(T) profile, aim at precipitation or super-saturation?
   Choice of excipient: Aim at high T_g', high c_g', aim at crystallization during process, or supersaturation (glass)?
   Mass ratio protein:excipient:salt(s)
   Total solid content <10%
   Vial fill depth, ideally of the order of 5 mm
Process
   Cooling rate: quenching or freezing *in situ*?
   Thermal cycling (annealing) of frozen product?
   Hold temperature (frozen) must be <T_g'.
   Performance of equipment: temperature uniformity across shelves
   Primary drying cycle: Product temperature must not exceed T_g', especially toward the completion of the ice sublimation stage
   Knowledge of c_g' (typically 80% by wt)
   Pressure control; gas bleed
   Locus of $T_g(c)$ curve, especially for c >90%
   Desorption conditions: Temperature should be ramped, close to $T_g(c)$ curve
   T_g and moisture content of "dry" product determine safe storage conditions

7. Experimental Determination
of Thermal Parameters

It has been shown that knowledge of the various thermal transitions that invariably occur in complex mixtures during cooling, warming, and drying determines the correct choice of formulation and process protocol. Since all these transitions are associated with enthalpy and/or specific heat changes, calorimetry is indicated as the direct means of monitoring such processes. By far the most effective and rapid diagnostic technique for the determination of eutectic phenomena, crystallization and melting processes, and glass/rubber transitions is differential scanning calorimetry (DSC), which monitors changes in specific heat as a function of temperature *(30)*. In commercial freeze-driers, the "state" of the product is usually monitored *in situ* by its electrical resistance. This is convenient; some transitions, e.g., crystallization and melting, can be adequately identified, but the electrical resistance is not always sensitive to glass transitions, especially in systems of low moisture contents. There are also other technical problems related to the physical contact of the probe with the product during the drying stages.

Most commercial DSC instruments are designed for studies of materials, such as polymers and metals, where the sample size presents no practical problems. With valuable protein-based products, such instruments generally lack the necessary sensitivity required for the measurement of glass transitions. Where instruments do possess the required sensitivity range, they are not designed for measurements at subzero temperatures; inadvertent freezing may actually damage the instrument. A modified DSC instrument, based on the Perkin-Elmer DSC-2, has been described, by means of which all the operations described in this chapter can be performed *(31)*.

Figure 18A shows a typical DSC power-time trace recorded during the heating of a freeze-concentrated mixture. T_g' and the ice-melting endotherm are clearly distinguishable. Figure 18B shows the trace for the same product in the final dried state (2% residual moisture). A $T_g(c)$ profile can be constructed by determining T_g at different moisture contents. It defines the temperature/moisture conditions that determine the shelf life of the dried product, as shown in Fig. 15.

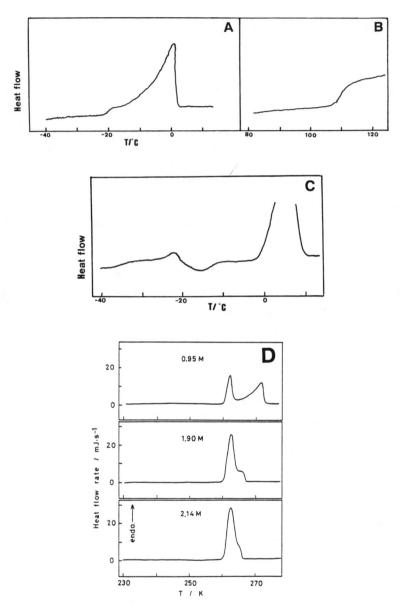

Fig. 18. DSC power–time curves obtained during the heating of (A) a freeze concentrated Ficoll™ solution, (B) freeze-dried Ficoll, (C) freeze-concentrated mannitol solution (note the crystallization exotherm), and (D) freeze-concentrated K_2HPO_4 solutions of increasing concentration, clearly showing the eutectic melting endotherms.

Other transitions that influence product stability include eutectic-phase separations during cooling or drying and crystallization processes during storage. They can all be usefully monitored by DSC methods. Examples are shown in Fig. 18 A–D; *see also* ref. 7.

A practical example of the use of DSC as a quality-control tool is shown in Fig. 19. A product containing a blood coagulating factor is routinely prepared by adding a constant volume of the chromatographic column eluate to a constant volume of a buffer/excipient solution. The solution is frozen to –50°C and lyophilized. Some batches pass quality control, whereas others fail. The DSC traces show that T_g' is by no means identical, the reason being that the column eluate contains variable amounts of protein, so that the mass ratio protein:salt:excipient is not constant from batch to batch. It is, however, this ratio that determines T_g'; thus, the temperature at which ice sublimation is performed must take this variation into account. Setting a fixed process temperature is hazardous, where the product composition from batch to batch is not known to be constant.

8. Solid-State Stabilization Without Freezing: the Permazyme Process

The central problem of dry-state stabilization is how to convert a dilute solution of approx 5% total solids and containing a labile product into a solid solution, containing, say, 2% residual water, without causing the active component to lose its bioactivity. This is shown by the path AE in Fig. 20 (which is a more elaborate version of Fig. 15). Lyophilization has become the most popular method for achieving this end (path ABCDE in Fig. 20), probably because of the (mistaken) assumption that great benefit is to be derived from the low temperature at which it is carried out. As has already been explained, the objective should be to transform a dilute solution into a glass of low residual moisture content as rapidly as possible, consistent with avoiding protein inactivation. The rate of water removal at a given temperature, T, must be balanced against the rates of deleterious processes, physical and chemical, that in a very concentrated solution are governed by $(T - T_g)$ according to the WLF kinetic equation (Eq. [6]). On any industrial scale, freezing is a slow process, so that the freeze concentration of a dilute

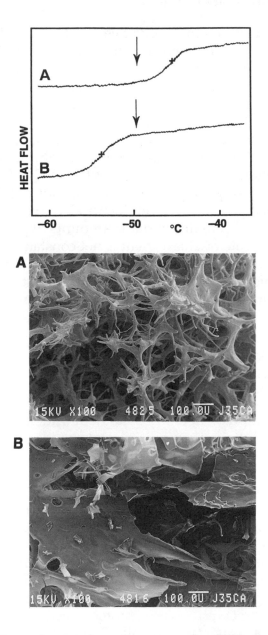

Fig. 19. Electron micrographs and corresponding DSC power–time traces for two freeze-dried blood product samples, frozen to and sublimed at –50°C: (A) product had retained activity (T_g lies above operating temperature); (B) product had lost activity (T_g' lies below operating temperature). For further details, *see text*. Bar = 100 μm. Reproduced from ref. 27.

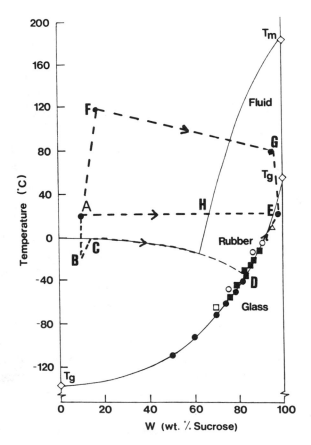

Fig. 20. State diagram for the sucrose/water system as in Fig. 15, but elaborated to show various possible pathways for drying a dilute solution (A) to a glassy solid (E) with a low residual moisture content.

solution takes place over several hours, even up to a day, during which period the protein will have to survive in an increasingly hostile chemical environment.

Water can be removed more rapidly by evaporation, e.g., along path AHE in Fig. 20. Under optimum conditions, the concentration step for products in typical vials containing approx 5 mL solution from, say, 5 to 95% total solids can be performed in 1 h. For bulk volumes (several liters) spray-drying offers an attractive alternative (path AFGE in Fig. 20). The drying period can then be reduced to <1 s. Evaporation offers several technical and marketing advantages over freezing (32), quite apart from the obvious economic benefits.

The freezing pathway ABCD is prescribed by the course of the liquidus curve and cannot be modified. For reasons already discussed, freezing terminates at T_g', point D. The moisture still left in the product must be removed by diffusion and desorption, so that the stable state E may be reached. Evaporation, on the other hand, is much more flexible; a large variety of pressure/temperature pathways can be employed, i.e., the process can be time-optimized. The resultant product tends to be more stable than its lyophilized counterpart. This is owing to the more uniform distribution of the residual moisture. Yet another advantage of evaporation lies in its endothermic nature, whereas freezing is exothermic. It is easier to control temperature by heating rather than by cooling the product during processing. Water, once it has left the product by evaporation, has been removed permanently, whereas in a frozen solution, the water is still present (as ice), so that any inadvertent rise in the temperature will result in a partial melt-back into the product, with damaging consequences. Perhaps the most important advantage of evaporation at ambient temperatures is the much shorter processing period, i.e., the more effective utilization of capital equipment.

Various attempts are on record that describe the stabilization of proteinaceous products by the evaporative removal of water. The disaccharide trehalose has been claimed to act in a unique manner as being able to protect food, biotech, and therapeutic products during conventional air-drying at ambient temperatures, but few long-term storage stability data are available (33). Although trehalose is indeed found as a very effective in vivo protectant against freezing and desiccation in many species, it is by no means universal in this respect, nor does it ever occur alone, but always in conjunction with other PHCs, especially glycerol (5). Indeed, it can only occur in organisms that possess a trehalose metabolism. As an in vitro protectant, trehalose is subject to some practical limitations: It is expensive, is not a permitted excipient for injectable preparations, and is not an approved food additive. Trehalose is also difficult to dry to a low moisture content. Products that have been claimed to be "dried" may well have a residual moisture content of the order of 30% (34).

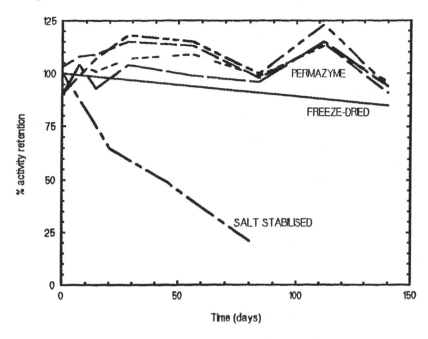

Fig. 21. Comparative stabilities, at 35°C, of lactate dehydrogenase in ammonium sulfate salt suspension (ex Sigma), freeze-dried (ex Sigma) and Permazyme-evaporation-dried (*see text*).

The Permazyme™ process was developed to exploit evaporative drying techniques for the stabilization of labile bioproducts (*35*). It is based on the optimization of pressure/temperature/formulation conditions, so as to minimize the deterioration of the bioproduct while it traverses the rubber state, where WLF kinetics govern most damaging reaction processes, as rapidly as possible. Depending on the processing method adopted (vials or bulk solutions), the dried, shelf-stable products can be presented in the form of transparent films or bulk powders. As expected, the long-term stability of such a product critically depends on the shape of its $T_g(c_g)$ profile. Figure 21 provides a comparison of the stabilities of LDH-based products stabilized by ammonium sulfate, commercial freeze-drying, and Permazyme technology, respectively. The technology has also been applied successfully to stabilize a range of enzymesin the solid, glassy state, e.g., restriction endonucleases and alcohol oxidase, which had previously been considered incapable of being dried and had to be formulated in concentrated glycerol solution, shipped on dry ice, and stored at –20°C.

9. Conclusions

Protein activity can be affected by the process conditions chosen for the stabilization and also by the physical state of the final product. The single most important property that determines the maintenance of biological activity is the glass transition profile of the product. This, in turn, depends on the product formulation.

Numerous claims that formulation and process must be developed by trial-and-error methods are considered to be out of date. Instead, a proper appreciation of the interplay of the physics, chemistry, and engineering aspects of drying processes, including freeze-drying, enables useful principles to be applied and predictions to be made, the observance of which leads to economic processes and stable products.

Bibliography

1. Wilson G. A. and Young F. E. (1975) *J. Mol. Biol.* **97,**123.
2. Angell C. A. (1982) in *Water–A Comprehensive Treatise*, (Franks F., ed.), Plenum, New York, p. 1.
3. Privalov P. L. (1990) *Crit. Rev., Biochem. Mol. Biol.* **25,** 281–306.
4. Franks F. and Hatley R. H. M. (1991) *Pure Appl. Chem.* in press.
5. (1990) *Life at Low Temperatures.* The Royal Society, London.
6. Storey K. B., Baust J. G., and Storey J. M. (1981) *J. Comp. Physiol.* **144,** 183–190.
7. Hobbs P. V. (1974) *Ice Physics.* Clarendon Press, Oxford
8. Michelmore R. W. and Franks F. (1982) *Cryobiology* **19,** 163–171.
9. Hatley R. H. M., Franks F., and Mathias S. F. (1987) *Process Biochem.* **22,** 169–172.
10. Franks F. (1985) *Biochemistry and Biophysics at Low Temperatures.* Cambridge University Press, Cambridge.
11. Warren G. J. and Wolber P. K. (1987) *Cryo-Letters* **8,** 204–215.
12. Tomchaney A. P., Morris J. P., Kang S. H., and Duman J. G. (1982) *Biochemistry* **21,** 716–721–730.
13. Rasmussen D. H. and MacKenzie A. P. (1972) in *Water Structure at the Water–Polymer Interface* (Jellinek H. J. G., ed.), Plenum, New York, p. 126.
14. Franks F. (1982) in *Water, A Comprehensive Treatise*, vol. 7 (Franks F., ed.), Plenum, New York, p. 215.

15. MacFarlane D. R., Kadiyala R. K., and Angell C. A. (1983) *J. Chem. Phys.* **79**, 3921–3927.
16. Luyet B. (1964) *Proc. NY Acad. Sci.* **125**, 502–512.
17. Mayer E. and Brüggeller P. (1983) *J. Phys. Chem.* **87**, 4744–4749.
18. Körber C. (1988) *Quat. Rev. Biophys.* **21**, 229–298.
19. Murase N. and Franks F. (1989) *Biophys. Chem.* **34**, 293–300.
20. Hatley R. H. M., Franks F., and Day H. (1986) *Biophys. Chem.* **24**, 187–192.
21. Fennema O. (1975), in *Water Relations of Foods* (Duckworth R. B., ed.), Academic, London, p. 397.
22. Brandts J. F., Fu J., and Nordin J. H. (1970) in *The Frozen Cell* (Wolstenholme G. E. W. and O'Connor M., eds.) J. & A. Churchill, London, p. 189.
23. Tamiya T., Okahashi N., Sakuma R., Aoyama T., Akahane T., and Matsumoto J. J. (1985) *Cryobiology* **22**, 446–456.
24. Diller K. R. and Lynch M. E. (1983) *Cryo-Letters* **4**, 295–308.
25. Hatley R. H. M., van den Berg C., and Franks F. (1991), *Cryo-Letters* **12**, 113–126.
26. Levine H. and Slade L. (1988) *Water Sci. Rev.* **3**, 79–185.
27. Franks F. (1990) *Cryo-Letters* **11**, 93–110.
28. Pikal M. J. (1990) *BioPharm.* **3**, 18–27.
29. Levine H. and Slade L. (1988) *Cryo-Letters* **9**, 21–63.
30. Hatley R. H. M. (1992) *Dev. Biol. Standardization* **74**, 105–122.
31. Hatley R. H. M., Franks F., and Green M. (1989) *Thermochim. Acta* **156**, 247–257.
32. Franks F. (1991) *BioPharm.* **4**, 38–42, 55.
33. Roser B. J., US Patent 4,891,319, January 2, 1990.
34. Carpenter J. F. and Crowe J. H. (1988) *Cryobiology* **25**, 459–470.
35. Franks F. and Hatley R. H. M., U.S. Patent 5,098,893, March 24, 1992.

Fifteen

Process Purification

James L. Dwyer

1. Introduction

This chapter addresses itself to the isolation and purification of high-value peptides/proteins. There is also a set of process operations within the food and chemical industries that deals with protein purification. Purity requirements and process economics requirements of the latter materials are so different as to require a separate set of unit operations.

Figure 1 is a generic process flow sheet for a recombinant protein process. Although it is true that a process is designed around the unique characteristics of a product molecule, one can nonetheless make some general observations about these processes.

Production process may be classified into four phases:

1. Removal of insolubles from the starting biomass;
2. Product isolation;
3. Product purification; and
4. Polishing and packaging.

The first phase of product separation from starting biomass involves primarily mechanical operations, such as cell disruption, extraction, centrifugation, and filtration.

The second phase yields a low-resolution separation of product molecule from the multitude of other materials that may be present. The object at this stage is to increase the relative mass con-

From: *Protein Biotechnology* • F. Franks, ed.   ©1993 The Humana Press Inc.

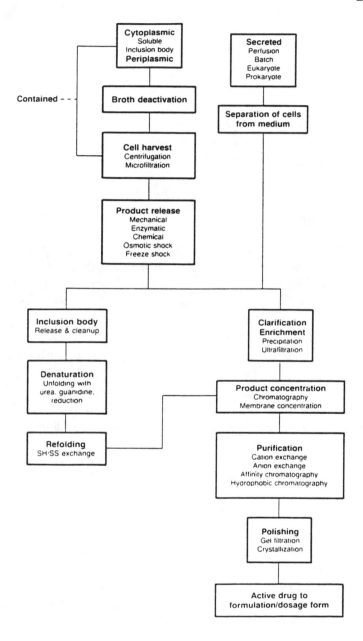

Fig. 1. Representative downstream processing flow sheet for protein purification. This figure originally appeared in *BioPharm* (Volume 3, Number 5, May 1990) and is used with the permission of Aster Publishing Corp.

centration of product and to reduce the total mass of material for downstream processing. The third phase employs processes of high resolution and selectivity to refine the product from molecular species that may be quite similar in physical/chemical characteristics.

The final phase will often employ purification techniques of lower resolution to remove single specific components that either passed through the previous refinement stages or were contributed to the product by the process itself (for example, the leaching of small amounts of ligand from an affinity separation). This phase also employs processes, such as formulation, freeze-drying, sterile filtration, and so on, to prepare the product for packaging and storage.

There are a variety of process unit operations in use, and when designing a process (a linked series of these unit operations), there are a number of considerations in selection. In general, protein purification techniques should be readily scaled, economic, simple, and possess the maximum attainable mass and activity recovery.

2. Source Materials

What constitutes high-value peptides currently in production? One immediately thinks of the products of genetic engineering, but this excludes a large body of high-value, high-purity materials. Foremost among these would be the blood-derived therapeutic products.

In general, peptide and protein products come from the following starting sources:

Blood/plasma and human tissues;
Plant and animal tissues;
Bioreactors: fermentation and cell culture;
Monoclonal antibodies; and
Chemical synthesis (peptides):
  Bulk chemical synthesis;
  Enzymatic synthesis; and
  Solid-phase synthesis.

Levels of expression vary significantly in the single-cell vehicles. Bacteria can express >20% of their mass as a recombinant protein. Yeast typically expresses 3–5%. By contrast, the mammalian cells secrete proteins only at the mg/L concentration.

2.1. Human Source Proteins

Generally speaking, biologics (products derived from human sources, such as blood, placenta, cord-blood, milk) have somewhat unusual purity requirements. Although they cannot be contaminated with foreign (nonhuman) materials, they can frequently contain significant concentrations of other nonimmunogenic blood factors. Human albumin, in fact, is frequently added to stabilize a product.

An increasing concern in blood-derived products is the presence of pathogenic viral materials. At present, we must live with standards built around "best available technology" with respect to detection and elimination of viral materials. In many situations, the infectivity level of virions is assumed to be well below assay sensitivities. In contrast to bacterial elimination processes, we presently lack absolute methods of separating virions or active nucleic acid constituents from a blood-derived product.

2.2. Plant and Animal Tissues

A wide variety of plant and animal tissues are the source of proteins, such as enzymes. These materials must be won from the crude source by a combination of chemical and mechanical processes. Tissues are chopped or ground, and then extracted in a bath that has good solution properties for the product of interest. Products, such as membrane proteins, are strongly associated with the solid structure of the crude source, and avid extractants must be employed.

2.3. Bioreactors

Bioreactors here refer to any growth system for expression. This encompasses fermenters tissue culture vessels and solid-state vehicles used for cell culture. Some of the commonly used cellular sources include:

Bacteria;
Yeast;
Virus—vaccines;
Insect cells;
Mammalian cells;
Hybridomas; and
Plant cells.

With sources, such as recombinant bacteria, one can achieve very high expression levels of the desired product. Generally speaking, high expression levels facilitate the purification task.

2.4. Monoclonal Antibodies

Considered from a protein-purification standpoint, monoclonal antibodies are a mammalian cell culture product. The upstream technology of gene fusion required to produce the antibody is different, but the isolation and purification aspects are by and large the same as those for eukaryote-expressed recombinant proteins.

2.5. Chemical Synthesis

Polypeptides must be synthesized one amino acid at a time, in order to achieve the proper sequence. Economic chemical synthesis is thus not practical for high-mol-wt polypeptides. Low-mol-wt polypeptides are chemically synthesized via economically competitive processes. Synthesis can proceed via sequential addition of appropriate amino acids to a terminal amino acid. Failure to achieve complete reaction in each step leads to a multiplicity of incomplete or improper structures, but for peptides containing only a few amino acid residues, one can use a bulk synthesis.

A variant of this is solid-phase synthesis. Here, the initial amino acid is covalently bound to a solid substrate. This allows efficient contact of reactants and reagents at each step of the process.

A third synthesis route is enzymatic synthesis. Side reaction problems are addressed here, because the various carboxypeptidase enzymes are used to ligate amino acids. These enzymes are specific to the various amino acids. Unreacted material can be removed after each additional step, and can be recycled to yield better synthesis economy. Another benefit of the enzymatic synthesis is its stereospecificity. No "off isomers" will arise in such a synthesis.

3. Cell Harvest Processes

Some of the bioexpression systems produce product proteins that are secreted into the culture broth. Where this is not the case, crude harvest of fermentation products is a two-step separation of

(1) cell paste concentration, and (2) separation of the protein from cell mass. Cell paste concentration is achieved by either centrifugation or filtration, simple sedimentation being inadequate for large-volume processes.

3.1. Centrifugation

Centrifugation methods are used for solids separation throughout the bioprocess train of operations. Large-scale centrifugation is practical, and there are a variety of sizes and types of process centrifuges available to answer specific needs. It can be misleading to take a procedure developed with laboratory centrifuge equipment and expect a direct scale-up to process volumes. This is because it is difficult to scale up the high g performance of a laboratory unit. Large process volumes mean large centrifuge equipment, and mechanical stress builds up as the square of the radius of rotation. One must accommodate bigger centrifuge bowls by reducing rotation rates (and achievable performance).

The equation below is a general throughput equation for a centrifuge. Some of the terms are empirical and are influenced by the particular type of equipment, but it is illustrative of performance factors. Centrifuge throughput:

$$T = [d^2 (p_s - p_e)g/18\mu] \cdot \Omega^2 rV/Sg)$$

where
T = throughput,
d = particle diameter,
p_s, p_e = particle, fluid density,
g = gravitational constant,
Ω = angular velocity,
r = rotation radius,
V = liquid volume in centrifuge,
μ = kinematic viscosity, and
S = thickness of liquid layer in centrifuge.

In this equation, the first set of terms relates to the feed fluid characteristics, whereas the second set of terms addresses the centrifuge characteristics. With respect to the fluid being processed, efficiency is enhanced by:

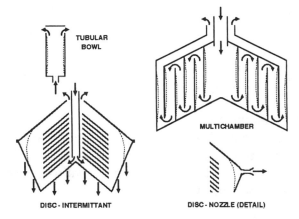

Fig. 2. Principal centrifuge configurations used in protein processing.

1. Large particle size;
2. Large density difference between the liquid and the solids; and
3. Low viscosity.

Equipment performance is increased by:

1. High angular velocity;
2. Large centrifuge bowl radius; and
3. A thin liquid layer in the centrifuge bowl.

There are a number of process centrifuge designs, three of which are generally used for cell processing (Fig. 2) and other protein process operations. They are all continuous-feed devices. In contrast to laboratory practice, a batch loading centrifuge is simply too small to be of use in process operations.

3.1.1. Tubular Bowl

The tubular bowl centrifuge is one of the simplest of designs. The core element is a vertical tube, approx 50 cm long and 10 cm in diameter. The tube is spun about the cylinder axis, and feed is introduced at a bottom center port. The dense solids deposit on the wall of the spinning cylinder as liquid flows up the walls. At the top of the cylinder, the liquid discharge is thrown from the bowl and is caught in a discharge casing. A bowl 10.8 cm in diameter rotating at 15,500 rpm produces approx 14,000g. The tubular bowl centrifuge with its high attainable g forces, provides good clarification and good dewatering of retained solids. It is a simple and reliable type of equipment. The high *g* capability of tubular bowls permits efficient cell harvest of bacterial recombinant cell sources. The drawbacks of the tubular bowl include low capacity, foaming and aeration of the liquid supernatant, and the generation of aerosols. The centrifuge is semicontinuous, and capacity is limited by the buildup of the retained solids in the rather small bowl. Conditions of high fluid shear exist, both at the fluid inlet port and at the top of the cylinder, where the liquid is flung to the wall of the chamber that houses the bowl. It is at this location that foaming and aeration of the liquid become a possible issue.

3.1.2. Multichamber

The multichamber centrifuge increases solids handling capacity by using a maze of nested cylindrical surfaces. The larger radius (≈50 cm) of rotating parts, while providing more solids capacity, restricts rotational speed. Solids recovery is more complex, as is cleaning.

3.1.3. Disk Bowl

The disk bowl is comprised of a set of closely stacked cones on a rotor, surrounded by a biconical bowl that conforms to the conical stack. Fluid is pumped into the top of the rotor and flows down a central tube. It flows inward between the cone stacks and is discharged at an annular port at the top. As the liquid flows between the cone stacks, the solids sediment onto the cone faces. The solids slide out and down the cone faces, and are thrown to the widest perimeter of the bowl.

There are two versions of this type of centrifuge. In one, the solids accumulate in the bowl and are discharged by periodically opening the halves of the bowl. Solids (and some liquids) are flung out to a circular collection basin. Another variant of this design has a set of discharge nozzles placed at the widest perimeter of the bowl, such that solids are continuously discharged. Liquid content of discharged solids is a function of nozzle size, suspended solids content, and operating settings of the centrifuge.

These centrifuges offer continuous performance, a desirable process attribute. Balanced against this, they are more complex and provide lower *g* forces (hence lesser solids harvest efficiency) than a tubular bowl. Solid discharge will generally contain more liquid. With the continuous discharge designs, one must ensure that the separated solids contain no particles or aggregates large enough to clog the discharge nozzles.

Centrifugal separations provide large-scale process capability and are ubiquitous in protein process operations. There are some drawbacks, however. In general, the protein products and source materials tend to foam and entrap air. The result is a low effective settling velocity and denaturation at air–liquid interfaces. Aerosol generation is a real problem with centrifuges processing biohazardous materials. Cleaning tends to be difficult and provides potential hazardous exposure of operators.

3.2. Filtration

Filtration provides an alternative to centrifugation of cell harvesting and/or clarification of protein source materials. Conventional "dead-end" filter methods are of little utility because of the rapid blocking of a filter matrix. Although some mycelial biomass can be harvested by plate-and-frame filter presses or continuous discharge drum filters, bacterial or mammalian cell mass proves intractable. The brewing industry uses body feed filtration methods for clarification and separation of yeast cell mass. A slurry of rigid, porous particles (filter aid—diatomaceous earth) is admixed with the crude product stream, such that an open, free-flowing cake is formed on the filter face as solids accumulate.

Body feed techniques are unsuitable for cell harvest methods. The cells are admixed in a large mass of the filter aid, necessitating another separation process. It is also difficult to achieve a high retention efficiency.

Microporous filters, whose pore size is smaller than the cell (or viral) material to be harvested, provide virtually 100% harvest efficiency. Unfortunately, when run in a "dead-end" mode, these filters have extremely low capacity.

In recent years, the techniques of tangential or crossflow filtration have been developed to provide quite effective cell harvest capability. Figure 3A schematically depicts bacterial (or other solids) retention in conventional "dead-end" flow. As a feed-stream is processed through the filter, solids larger than the pore size are

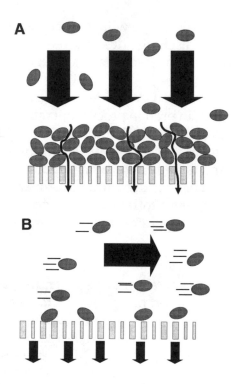

Fig. 3. "Dead-end" filter retention and tangential flow retention. Concentration polarization. (A) Retained material forms a pseudo-membrane and diminishes flow rate. (B) Tangential flow.

retained on the filter surface. Because of the micron size of such particles, the accumulating layer of solids offers a very high flow resistance. Milligram/cm² quantities can block membrane filter flow.

If, however, a vigorous sweeping flow (Fig. 3B) of feed stock is maintained *normal to the filter face* (hence the names crossflow, tangential flow), accumulating solids are swept into suspension, rather than forming an impermeable mat on the filter face. Cell suspensions can be concentrated to a paste containing very little free liquid, and harvest efficiency is virtually 100%. The tangential flow harvest systems have been widely adopted for recombinant cell harvest and for virus harvest (vaccines). The circulating hydraulics actually impose less harsh shear conditions than exist in centrifuge feed/discharge ports, and it is thus common to see higher

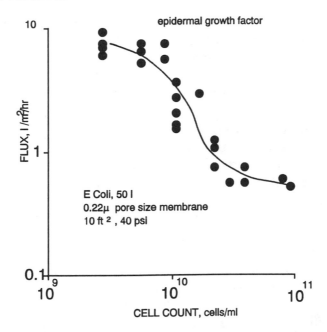

Fig. 4. Cell harvesting of *E. coli* cultures by tangential flow filtration. Reproduced with the permission of MILLIPORE Corp.

viable virion activity in vaccine processing. The membrane systems employed are either hollow fiber membranes or parallel plate membrane stacks, with thin flow channels between parallel membrane faces.

Figure 4 shows tangential flow batch harvest data for an *E. coli* culture. Ferment mass is continuously recirculated across the face of the filter. The flow restriction of the discharge valve creates a modest pressure drop across the membrane, resulting in flux of the clear broth across the membrane and concentration of retained cell mass. As the retained solids content increases, the membrane flux declines (*see* concentration polarization effect in Section 5.). Nonetheless, it is practical to membrane process a broth to >90% cell paste by tangential flow.

Membranes can provide closed, contained multiple-use systems that can be sanitized and sterilized without disassembly. No aerosols are generated during operation. Equipment is simple and relatively low in cost. For small batch operation (<1000 L), they provide an attractive alternative to centrifugation equipment.

When using membranes for cell processing, particularly from microbial fermentations, another factor that may present itself is membrane fouling. This refers to a strong adsorption of species on the face of the membrane. This results in a loss of membrane flux. Unlike polarization, it cannot be alleviated by increased tangential flow. Certain bacterial pigments and some of the antifoams used in fermentation can give rise to fouling.

There are some things one can do to minimize this problem. Reducing concentrations of antifoam agents or switching to other antifoams can help. In many cases, a membrane can be restored by contact with solvents or detergents.

Membrane harvesting of mammalian cell cultures poses some unique problems. These cells are "soft" deformable entities compared to yeast and bacteria. If transmembrane flux is permitted to go above critical limits during crossflow, the transmembrane pressure gradient can become high enough to deform cells at the membrane surface, forcing the cells into the pore structure, and eventually causing cell crushing and rupture.

This problem can be addressed in membrane cell harvest by rigorous control of the transmembrane flux. A low controlled flux will limit the transmembrane pressure and avoid cell damage. Typically, mammalian cell harvest systems use positive volume pumps both for the retentate circulation and for control of the permeate flux. Low-shear peristaltic tubing pumps are commonly employed. Such systems can even be employed to harvest such fragile cells as erythrocytes.

3.3. Cell Disruption

These processes disrupt cells so as to release protein. Cell disruption is accompanied by liquid/solid separations and, in some cases, solubilization of released protein. Even in the case of secreted proteins, there is a need for removal of cells, cell fragments, and solid materials arising from media and metabolites.

In the main, physical/mechanical processes are employed, although some chemical effects are utilized. Table 1 summarizes prevalent methodologies of cell disruption (2).

Release of materials from mammalian cell mass is generally easier than from yeast and bacterial sources. The bacterial cell wall

Table 1
Cell Disruption Methods

Mechanical	Nonmechanical
Liquid shear	Desiccation
Ultrasonic	Freeze-drying
Agitation	Air-drying
Liquid extrusion	Solvent-drying
Solid shear	Lysis
Grinding	Physical
Pressure	Osmotic shock
	Pressure release
	Freeze/thaw
	Chemical
	Detergents
	Glycine
	Enzymatic
	Antibiotics
	Lysozyme, others
	Phage

is durable and able to withstand strong osmotic shock. When isolating material from animal tissue source, it may be necessary to isolate the specific organ tissue from surrounding cell mass and fat.

The liquid shear systems more frequently employed are homogenizer nozzle systems. Material is forced through orifice nozzles at pressures of several hundred kg/cm^2, causing cell disruption. By optimizing conditions, it may be practical to achieve sufficient cell disruption in a single pass. A variant nozzle system is the ultrasonic nozzle. The ultrasonic energy may be generated by the fluid flow itself or be input to the nozzle by piezoelectric drivers. The mechanical processes of cell disruption may be combined with the physical and chemical procedures to optimize product release.

Other methods of mechanical disruption include colloid mills where the biomass is fed through a rotor-stator system. High rotation speeds (several thousand rpm) and close clearances provide the necessary shear. Bead mills admix the cell mass with hard beads, which by their tumbling action serve to bread cell walls.

The nonmechanical methods of cell disruption are often used in concert with the mechanical techniques. It must be noted that the mechanical processes can cause denaturation losses. A process that weakens the cell wall can result in disruption under much milder conditions.

3.4. Solid–Liquid Extraction

Disruption of the cell wall may not itself be adequate for complete release of the product molecule. Protein may be incorporated into cell walls, contained in organelles, or in complexes with other cell components, such as lipids and carbohydrates. Recombinant proteins expressed by bacteria frequently are present as inclusion bodies, compact solid masses that must be solubilized. If the clarified supernatant from cell disruption does not contain the product molecule, the solid mass must be concentrated and extracted.

Extraction is routinely employed in the isolation of proteins, such as enzymes, from plant or animal tissue. The starting material is chopped or ground to expose sites and reduce the diffusional path through the tissue mass. Extractants are most commonly aqueous, and various salts are used to optimize protein solubility parameters, as described in the chromatography chapter. The extraction formulation may also achieve some selectivity. The extraction formulation should have a pH well removed from product isoelectric point, since this is a region of minimum product solubility.

In the process of scale-up of extraction, one must carefully examine small-scale laboratory methods. In particular, note the bench-scale solvent:protein ratios. For the production process, you want to avoid handling large liquid volumes and low product concentrations, both of which will add to processing complexity and cost. The extraction system that provides a high concentration liquor is much more preferable. Proteolysis is always a concern with protein extraction. The source tissue probably contains proteolytic enzymes. Microbial contamination of the materials is another potential source of proteolytic agents. Extractions should be carried out at low temperatures and as expeditiously as possible, so as to minimize proteolysis.

Extraction from solids is a simple operation. Limitations include solubility of material to be extracted, and access of the product molecule through tissue structure and to the extractant liquid.

A low level of fluid shear provided by mixing of an extractant bath will greatly aid mass transfer kinetics. Avoid air entrainment to minimize protein denaturation.

4. Product Purification

Presuming that product has been isolated as a liquid fraction from the crude biomass, the next phase of processing is a selection of some set of operations that isolates the product molecule from a myriad of other components. Typically, the product is a small fraction of the present components at this stage, and it is most effective to use some method that will drop out the majority of undesired materials, thus reducing the mass to be final processed. The highly selective separation processes generally function best when the product molecules are a major feed constituent.

4.1. Precipitation and Crystallization

Selective separation based on precipitation of desired polypeptides can be a very effective preliminary operation. Precipitation is a technique that simultaneously provides concentration and purification. It relies on selective solubility parameters of the product molecule. When these solubility parameters are favorable, one can obtain a considerable degree of purification. Various recombinant products have been isolated at high purity with a simple selective precipitation step.

A classic example is the Cohn process (3) for fractionation and isolation of blood products. The various proteins in plasma are differentially affected by the use of water miscible organic solvent (ethanol). In the Cohn process, plasma is taken through a series of ethanol/water precipitations and resuspensions to segregate the plasma proteins into useful therapeutic fractions, i.e., albumin, γ globulins, and the coagulation factors. Segregation of the various plasma proteins is affected by both ethanol content and pH.

The solubility of a protein reaches a minimum when the protein net charge in solution is zero (i.e., the isoelectric point of the protein). Altering pH on a process scale can pose some problems not encountered in bench-scale purification. In process operations, it is desirable to keep fluid volumes low, so as to reduce capital equipment costs and to operate at high concentrations of product solute. To reduce liquid volumes, one must use high concentrations of acids/bases. Effective mixing thus plays an important role. With poor mixing, localized high concentrations of acids or bases can cause undesired precipitation and denaturation. Mixing techniques are empirical and imperfect. One can readily encounter local zones of pH extremes, with resultant denaturation (4). Although good mixing is favored by vigorous agitation, this can also generate aeration and foaming, which will result in product denaturation. Vigorous mixing processes may also generate hydraulic shear that can affect some labile proteins.

The kinetics of the precipitation process can have a significant effect on product purity and recovery. The process precipitates should be aggregates of identical molecules. Molecules in solution diffuse, collide, associate, and aggregate into large entities. To the extent that the process is driven by an abrupt shift in solubility conditions, one will favor the formation of a large number of small aggregates. A more gradual process will result in growth of larger aggregates (which will be easier to process). An expression for growth of aggregates by such a diffusive process is given by Eq. (5):

$$M_w(t) = M_w(t_0)\,(1 + 8Dd\pi m_o t)$$

where
$M_w(t)$ = aggregate molecular weight at time (t),
$M_w(t_0)$ = initial molecular weight,
D = particle diffusivity,
d = particle diameter,
t = time, and
m_o = molar conc. of aggregate species.

As aggregation proceeds, the process slows because of lower concentration of unaggregated species and the decreased diffusivity of larger entities. Fluid mixing can promote collisions of aggregates, and to a point, fluid mixing can accelerate the precipitate aggregation. At high fluid shears, however, the opposite can occur: Aggregates are torn apart and dispersed. Some level of mixing is therefore beneficial, but too vigorous a mixing process can reverse the desired effect. Development of an effective precipitation process remains an art.

Various other materials besides ethanol are commonly used in protein precipitation. Ammonium sulfate is regularly used to salt out proteins without extensive denaturation. Ammonium sulfate precipitation is a low-cost operation and even stabilizes certain proteins (6). Various other lyotropic series salts are also employed. Hydrophobic effects are important here, and the process will be a function of pH, salt concentration, and process temperature.

Water-soluble, high-mol-wt neutral polymers, such as polyethylene glycol (PEG) and dextrans, can also generate selective precipitation. These materials compete for water, causing the protein to precipitate out.

4.2. Process Chromatography

Chromatography constitutes the heart of a purification train in the processing of proteins and peptides. Compared to many of the traditional operations used in chemical processing (precipitation, crystallization, and so on), chromatography is a costly operation. For high-value products, such as biologic and biosynthetic products, the resolving ability of chromatographic separation readily justifies the process cost.

4.2.1. Chromatography Modes

Most of the analytical chromatography modes of analytical procedures have been employed at the process level. Normal-phase and ion-exchange chromatography are regularly used by the pharmaceutical industry for purification of a variety of products. Biologics production routinely uses ion exchange and GPC in purification trains. Ion exchange is frequently used in the early steps of biochemical purification. IEC packings are capable of processing high mass loads and are chemically rugged with respect to regeneration. Inasmuch as ion exchange can be viewed as an adsorption process controlled by the elution chemistry, it is not necessary to use small particle sizes to achieve the desired resolution. IEC is often used to adsorb undesired species from the crude feed, allowing product solutes to pass through with a reduced burden of these solutes.

GPC is not commonly used for mol-wt fractionation at the process level; loading and achievable resolution are not readily attained in scale-up. GPC is most commonly used as a desalting method or for a final-step segregation of low-mol-wt impurities in a product isolate.

Process scale reverse-phase HPLC and, to a lesser extent, normal-phase HPLC are used for purification of therapeutic peptides. Hydrophobic interaction chromatography is used in the purification of high-mol-wt proteins, such as biologics.

To date, affinity separations have found limited process acceptance. Although such systems can be highly specific, they give rise to some issues, particularly with respect to processing of therapeutic products. Issues include:

- Material and operating cost;
- Regulatory agency acceptance;
- Ligand leaching or bleeding; and
- Microbial contamination.

Notwithstanding the above, affinity chromatography is being employed with several biologic and biotechnology products, and we will doubtless see wider usage in the future. Examples of present use include the purification of Factor VIII, various monoclonal antibodies, and biosynthetic hormones. A number of people viewed affinity processing as a one-step process, but most present applications employ affinity in the middle of a process train. The front-end isolation methods are used to segregate the bulk of the crude feed stream from the product solute. Affinity is used to separate the desired product from other species difficult to separate by more conventional methods. The affinity separation is then followed by steps that can remove trace contaminants from the affinity eluant. The advent of affinity matrices that can scavenge small quantities of contaminants, such as endotoxins, from a product will doubtless find increased usage.

4.2.2. Media

The soft gel chromatography media (predominantly IEC and GPC) have been the most commonly used production processes for biologics and biochemicals. The pharmaceutical industry has more commonly used silicas (normal-phase) and ion-exchange resins in the production of organic synthetics and biosynthetic antibiotics. More recently, HPLC media has seen scale-up, notably in the purification of peptides. The introduction of rigid polymer, small-particle packings with their superior resolution and flow properties can be expected to displace some of the soft gels.

Microporous membrane affinity systems are a novel media introduction. These systems rely on the extremely rapid turnover of a ligand–substrate reaction owing to the direct convective contact of ligand and substrate during through-pore flow. This approach is highly conservative for high-cost ligand systems and can demonstrate very attractive process economics.

At present, there is a new class of packing materials being introduced that claim similarly rapid kinetics. These packings possess a bimodal pore distribution, and it is postulated that they can sustain convective flow directly through the large pore channels that traverse the particle. Limited data to date do indeed demonstrate very rapid reverse-phase separation performance of these materials.

4.2.3. Equipment

A range of columns, pumps, detectors, and other equipment is commercially available for large-scale chromatography, both with low-pressure soft gels and process HPLC. For HPLC, one can obtain a range of empty columns that can be packed with a medium of choice. Alternately, there are some HPLC systems available that utilize prepacked cartridges.

Proper column packing is a critical performance factor in HPLC. One of the main benefits of HPLC is its high attainable resolution, and any irregularity in bed packing or in flow distribution can vitiate this benefit. There are several systems available that address this issue by techniques to apply compression to the bed after packing, eliminating column voids and maintaining high efficiency.

Analytical columns typically have length:diameter ratios of ca. 50:1. As one scales up column size, there is no compelling reason to maintain such design. There are, in fact, severe flow and pressure penalties with large high-aspect ratio columns. As a result, large-scale columns have aspect ratios closer to 1:1. Some of the large gel columns, in fact, operate with diameters many times greater than the bed depth.

Process chromatography is a repetitive operation and generates a need for automation. The process system vendors have been able to adapt industrial PC computers and custom software to this requirement. The task is often complicated by the need to operate in an explosion-proof environment. Most commonly, this is addressed by designing with explosion-proof electrical equipment and sensors, and isolating the computer control in a separate room from the operating LC.

4.2.4. Scale-Up Techniques

The methods of analytical chromatography can often be applied to process scale-up, provided that one recognizes the fundamentally different objective of analytical and process separations. The analytical separation should achieve the maximum practical resolution of sample components to permit unequivocal identification and accurate quantitation of solute peaks. Column loads are generally kept as low as the detection system will allow to achieve these ends.

By contrast, the process chromatographic separation has the isolation of a solute to a selected purity level as its goal at the lowest possible cost. This modifies a chromatographic separation strategy over the analytical separation.

In a chromatographic separation, as the sample load (mass sample/mass packing) is increased, closely eluting peaks will broaden and tend to overlap. Figure 5 depicts a detector response for two chromatographic peaks that are incompletely resolved. This would be an unsatisfactory analytical separation. It would be difficult to extract accurate quantitation here or even to say with certainty that one is dealing only with two components.

Note, however, that if fraction cuts were made on the leading and trailing sections of these peaks, those cuts would be isolates of very high purity. Figure 6 provides some perspective of recovery from partially resolved peaks. Two Gaussian peaks are shown with various degrees of overlap. The unshaded extremes of the overlapped peaks represent cuts of >98% purity. It is apparent from this exercise that very poorly resolved peak pairs can still yield considerable fraction cuts of high purity. In a process situation, the mid-cut of such overlapped peaks could be rechromatographed to obtain more of the desired product fraction.

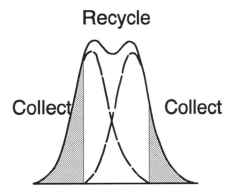

Fig. 5. Two partially resolved chromatographic peaks. Selective fraction cutting can yield high-purity isolates, even with incomplete resolution. Reproduced with the permission of MILLIPORE Corp.

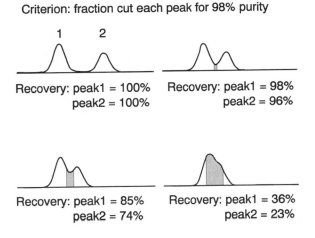

Fig. 6. Achievable yields from two (Gaussian) peaks showing various degrees of overlap.

In a process chromatographic separation, then, the column is considerably overloaded from the standpoint of analytical criteria. Fraction cuts are made in accordance with the above criteria. If more of the crude feed needs to be obtained, the reject fraction can be rechromatographed. A product cut can also be rechromatographed to increase purity. The reduced mass load on the column

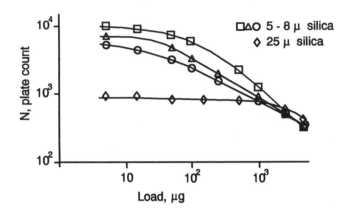

Fig. 7. Column plate count as a function of load for various packing diameters. Reproduced with the permission of MILLIPORE Corp.

will tend to increase separation efficiency, and in many situations,
rechromatography of a cut can yield better than 99.9% purity.

In parallel with this overload strategy of process chromatography is the selection of process packings. In an analytical separation, the resolving power of the system is a strong function of
the packing particle diameter. As a rule of thumb, a well-packed
column should yield a plate height of approx two particle diameters. As a column is highly loaded with sample, its resolution is
severely degraded.

Our process strategy calls for column overload. There is little
point then in using a high-resolution packing: The small-particle
diameter packings have much higher unit cost and impose high
operating pressure design. Figure 7 shows column plate count as a
function of load for various diameter silica packings. At high loads,
these packings all exhibit decreased plate counts, but it is the small-
diameter packings that show the most severe degradation. By contrast, the large-diameter packings show less resolution loss and, at
high loadings, have a performance equivalent to that of the small-
diameter materials.

For high-load separations then, do not use small-diameter
packings with their high unit costs and their imposition of high-
pressure, large-scale equipment. As a corollary to this, if one needs

to make a high-resolution process separation of a high-value product, one can do so with small-diameter packings. Loads must be reduced commensurately to achieve the higher degree of resolution.

Chromatographic separations are amenable to scale-up to process level. With data obtained on laboratory-scale chromatography equipment, it is quite feasible to predict process scale performance accurately.

The above material addresses the relationship of packing particle diameter to achievable resolution. The initial scale-up step then is to select the particular packing chemistry to be used in the separation and develop an analytical separation. The three most important parameters in this separation are the capacity factor (k'), the selectivity (α) for a product peak from its neighbors, and the saturation load (q_s, equivalent to a monolayer of solute or complete exchange in an IEC column) of the packing. The optimum separation for scale-up has a high α, high q_s, and low k'. Use conditions that produce a high selectivity; as column loads are increased, the peaks will broaden. The saturation load increases in proportion to the useful surface area of the packing. For separation of large molecules, there may be a trade-off of separation factor and pore size of the packing. The solvency of the mobile phase should be such that a k' of 2–4 will be achieved. Separations with a low k' leave little opportunity for resolution, whereas high k' means long cycle times and correspondingly low productivity.

Having selected a packing and a preliminary cycle, conduct a series of experiments at increasing loads (expressed as mass of crude feed per mass or vol of packing in the column). You will observe the following general phenomena:

1. The k' of all solutes will decrease (k' is not meaningful for GPC);
2. Peaks will broaden and tend to merge. Peaks will also become skewed, with steep leading profiles and trailing or tailing of the eluant; and
3. Solubility may become a problem at higher concentrations.

A given solute is less retained when in the presence of a high concentration of other compounds than when alone. In general, the amount of any solute found at equilibrium with the stationary phase increases less rapidly than its concentration in the mobile phase. This means that the higher the concentration of a solute in a band, the larger the remaining proportion in the mobile phase. This explains why the elution profile of single solute bands at high concentration is skewed, with a sharp front and a tail.

In unresolved bands, the proportion of each solute in the stationary phase is lower than if they were alone. Therefore, their average velocity is larger. The first component is less retained than it would be if alone. It is pushed or displaced by the second component.

What is the appropriate process load level? One will attain higher purity from solute bands that are just "touching" (on a detector trace). On the other hand, overloading to overlap can provide 3X–7X productivity increase. It is generally worthwhile to load to overlap and sacrifice some recovery by peak shaving (7).

Use a step gradient to purge residual solutes after the product molecule has eluted. During these high-load experiments, do mass balances of injected sample and recovered product solute to verify recovery. With unresolved bands, one should make several fraction cuts, assaying each for purity and recovery. This information will tell you where to make the process cuts.

Mobile-phase column velocity is another factor to be fixed during the scale-up experiments. Productivity is a direct function of flow rate; one can expect the plate count to drop at higher flow rates. Run the column at as high a flow rate as is consistent with acceptable product recovery and purity.

Complex feed stocks containing different compounds with a broad range of k' are costly and time-consuming to chromatograph. Column load is limited by the competing species. In this situation, more favorable economics may be achieved by doing a first-pass, low-resolution cut and then rechromatographing the reduced load.

The detector elution profile of the resulting separation will closely approximate that of the scaled-up separation. The analytical column that has been used for this process simulation will be considerably shorter than the process column. Although the plate count of a column is a function of length, it is a much stronger function of particle size and load. There may be band-broadening effects associated with plumbing and equipment in the process equipment. The separation resolution obtained in the simulation will probably apply to the process equipment. If there is a question about this:

1. Determine plate count *(N)* for the laboratory separation at process load conditions;
2. Similarly determine *N* for the process column, and compare;
3. If the plate count of the process column is significantly lower, add mixing to the laboratory system to match process column plate count; and
4. Rerun the separation on the lab system to find out if the separation is plate sensitive. If so, find out how much extra column length is needed (or reduced flow, and so on). Alternately, clean up the large system hydraulics.

Other factors that can be determined at lab scale include expected column life, regeneration and cleaning cycles, and mobile-phase recovery economics. Use of an automated LC system that can run unattended for 12–24 h periods allows one to simulate the multiple cycles of a process.

Using this approach, one can determine process operating parameters and gain a surprisingly accurate prediction of the full-scale performance. Such factors as solvent consumption, cycle times, packing life, and so on can be used to provide an economic model of the process. One can calculate the economic consequences of various loading and recycling strategies. Further, these data are obtained at laboratory scale and do not require massive quantities of mobile-phase constituents or (possibly difficult to obtain) product feed materials.

5. Ultrafiltration Techniques

Protein processing can be regarded as a cascade of operations that separate species and concentrate product. Often, the resolution operations serve to dilute the product molecule significantly in its suspending matrix, and there is a need to reconcentrate the product as it flows through a process. Membrane methods are primarily employed for (1) product concentration (water removal) and (2) removal of solutes of markedly differing molecular weight from the product molecule. In the processing of food proteins, water removal is commonly done by drying and evaporation methods, but this is seldom suitable for processing of biologically active polypeptides. The membrane separation methods are not high-resolution techniques, but are rather complementary to those techniques.

The use of microporous membranes for cell separation was described in Section 3. The microporous membrane is effective in separating solids from a liquid system. These membranes have no inherent capability to resolve various solutes. The ultrafiltration membrane can make certain solute separations.

Conceptually, an ultrafiltration membrane can be regarded as a filter that can allow small dissolved molecule flow, but that sieves and retains large dissolved molecules. Both microporous and ultrafiltration membranes are formed by processes that produce thin, highly porous structures. In the case of the microporous membrane, such structures are reasonably symmetric from "upstream" to "downstream" regions of the membrane. The ultrafiltration membrane by contrast has a highly asymmetric architecture. The greater part of the ultrafiltration membrane consists of a microporous structure. The upstream face of the membrane, however, consists of an apparent continuum of swollen polymer gel, a fraction of a micron in thickness, which is permeable to water and small solute molecules, but impermeable to large dissolved molecules. When a protein solution is introduced to the ultrafiltration membrane and pressure is applied, water and low-mol-wt solutes are driven through the membrane. The large protein molecules cannot permeate the dense polymer layer of the membrane face and are retained.

Ultrafiltration membranes are available for molecular-size cutoffs ranging from 1000–1,000,000 daltons. For a given mol-wt cutoff membrane, there is a broad spread between the size of molecules that are completely retained vs those that are completely passed. As a rule of thumb, this range is about one order of magnitude—a membrane that retains a 300,000 mol-wt solute at >95% will partially retain molecules as small as 30,000 mol-wt. The resolution for linear, nonbranched polymers is much poorer than for a branched molecule or globular protein. Completely linear high-molecular-weight solutes can apparently "snake" their way through the membrane. Solutes can also interact to influence retention. Blatt et al. (8) demonstrated that an ultrafiltration membrane capable of passing a significant fraction of albumin showed a marked increase in albumin retention when a γ globulin was present in the feed.

The ultrafiltration membrane is similar to a dialysis membrane in its principal of exclusion. Because of the very thin retention layer of the asymmetric ultrafiltration membrane, hydraulic pressure can generate the activity gradient that drives material through the membrane. By contrast, the symmetric dialysis membrane has a much higher flux resistance and relies on a solute concentration gradient to establish fluxes that are orders of magnitude lower. Water flux rates through ultrafiltration membranes are typically 1000 L/(m²-atm/h).

Such flux rates are not obtained in processing protein solutions, because of membrane polarization. Refer to Fig. 3 in the previous section covering tangential flow filtration. The same process occurs during ultrafiltration. Retained solutes form a continuum on the upstream face of the membrane, offering greatly increased hydraulic resistance to flow. Osmotic effects of this highly concentrated solute layer can also provide flux resistance. This is known as membrane polarization (a term derived from reverse osmosis membrane processing).

If a recirculating flow of solute (tangential flow, crossflow) is generated parallel to the membrane face, much of the retained solute will be swept away from the membrane face, thus increasing flux. Figure 8 shows the relationship between pressure drop across an ultrafiltration membrane (transmembrane pressure), tangential flow, and resultant flux for a given retained solute concentration. At low transmembrane pressures and fluxes, there is not a large degree of polarization. Retained solutes can diffuse away from the membrane face into the bulk solution. As pressure is increased, polarization becomes dominant, with the result that a limiting flux rate is achieved. Higher pressures will not significantly increase flux.

Increasing the tangential flow serves to remove solutes from the membrane face, thus increasing the achievable flux. The use of tangential flow is essential to the practical operation of ultrafiltration systems. Obviously, the flux is also dependent on concentration of retained solute, and in a batch process, flux will decline as solute concentration builds up.

Process ultrafiltration modules are designed for tangential flow operation. Membrane configurations include hollow fiber modules, spiral membrane modules, parallel flat plates, and tubular mod-

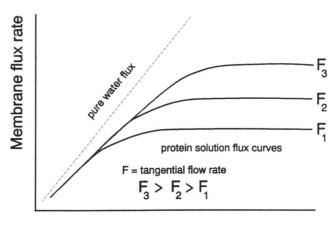

Fig. 8. Relationships of flux, differential pressure, and fluid shear in ultra-filtration.

ules. Each of these configurations provides a large surface area of membrane in a compact geometry and also permits good tangential flow development across the membrane face. Process configurations are recirculating loops, including a feed tank, pump, membrane modules, and receiver vessels. In process systems, heat exchangers are also included to remove heat generated by the mechanical energy input of the pumping system. Membrane systems are sanitized in place after use, and with proper care, membrane modules can be kept in service for many months. Ultrafiltration then is not a high-resolution tool, but can be quite effective for concentration of large-volume preparations of proteins.

The other significant application of ultrafiltration is in the removal of salts and organic solvents from a protein preparation. This process is called diafiltration. A volume of fluid is pressurized and circulated across the membrane. Water, salts, and low-mol-wt solutes flux through the membrane, while the product protein is retained. The fluxed liquid is continually replaced with water (or an appropriate strength buffer). The result is an exponential decrease in concentration of salts/solvents, while maintaining the retained protein at a constant concentration. Four replacement volumes are sufficient to obtain a 98% reduction in the salt/solvent. Diafiltration is a rapid process, quite amenable to scale-up. It has largely replaced the use of gel-filtration columns for desalting of protein preparations. Ultrafiltration thus simultaneously provides protein concentration while removing small molecules and salts.

Membrane systems are beginning to be employed for the desalting and concentration of low-mol-wt peptides. The membranes used in this application are modified reverse osmosis membranes. Although most ultrafiltration processes operate at modest transmembrane pressures (1–4 atm), the processing of peptides utilizes considerably higher pressures. The osmotic pressure of concentrated protein solutions is negligible. Because of the low molecular weight of the peptides, they can develop considerable osmotic pressures at mass concentrations exceeding 10%. This will create a driving force for water to backflux from the downstream side of the membrane. The net driving pressure for water flux is thus:

$$DP = P - \pi \tag{3}$$

where DP = driving pressure for water flux, P = hydraulic pressure, and π = osmotic pressure differential. The osmotic pressure to be overcome is several atmospheres, so that peptide processing takes place at much higher operating pressure than high-mol-wt protein processing.

Ultrafiltration methods are practical process complements to higher resolution isolation techniques, such as chromatography. Two other aspects of membrane processing should be mentioned. The first is the problem of membrane fouling. This is the same issue as was described in Section 3., covering crossflow filtration, as is addressed in the same manner. Very often ion-exchange chromatography upstream of a membrane system will serve to reduce fouling to a manageable level. If fouling is significant and intractable, a membrane process may not be an option.

A concern sometimes voiced about membrane processing is the effect of hydraulic shear on proteins. In well-designed membrane systems, the shear energy imparted is quite low (considerably less than encountered in feed and egress of centrifuges). Dunhill (9) and others have done considerable study of various shear effects on proteins. It appears that protein degradation attributed to shear in early membrane experiments was actually the result of stirred protein solutions denaturing at the gas–liquid interface, aeration and foaming, and adsorption of proteins on solid surfaces of the experimental apparatus. Most proteins can withstand process times of recirculation in membrane systems with little or no apparent degradation. The routine membrane processing of Factor VIII, a quite labile protein, tends to support this. In general, the cytoplasmic proteins are hardier than the cell-wall proteins with respect to shear effects.

6. Liquid–Liquid Extraction

The partitioning of solutes between two immiscible liquid phases is regularly applied to chemical purification. The insolubility of proteins in immiscible nonaqueous systems precludes this type of extraction as a process operation. The water-soluble polymers used for protein precipitation are also used in novel two-phase aqueous extraction separations (10). At the proper concentrations, dextran and PEG will simultaneously partition into two aqueous phases. The two phases have differential solubilizing power for various proteins and can be used for liquid–liquid extraction techniques. Other water-soluble polymer systems have been investigated (11).

Partitioning and selectivity are influenced by pH, temperature, solute molecular weight and hydrophobicity, and added salts. The partitioning of proteins in a PEG-Dextran system is markedly influenced by molecular weight of the PEG used. The polymers have a stabilizing influence on proteins, and extractions are commonly carried out at room temperature.

Extraction can proceed as a mixed-batch process, but this limits to completeness of transfer of a solute to the partition ratio of the solute in the two phases. The density difference of the two phases is not great, and the separation of extract phases can be abetted by continuous centrifugation.

To date, two-phase aqueous extraction has limited utilization at the process scale. Issues posed include separation efficiency, recovery of the polymers, efficiency of extraction from low-con-

centration liquors, added complexity to downstream purification, and removal of traces of polymers.

Reversed-phase micellular extraction is another novel process that has seen some application *(12)*. Reversed-phase micelles are surfactant aggregates suspended in an organic solvent. These micelles contain an aqueous core. Solubilization into the micelles occurs when the pH is lower than the protein's isolelectric point, and when the surfactant and protein are of opposite polarity. A protein-containing aqueous solution is contacted with a nonpolar solvent containing such micelles. Proteins can transfer to the micelles with some degree of selectivity and transfer back to a second aqueous phase. Partitioning is a function of solvent polarity, ionic strength and pH of the aqueous phase, and surfactant concentration and type. The micelles additionally serve to protect the proteins from inactivation by the organic milieu. Conventional liquid–liquid extraction process equipment can be employed.

Selectivity can be altered by adjustment of pH, solvent:aqueous phase ratio, ionic strength, and surfactant concentration. By attaching affinity ligands to long alkyl chains and adding them to the system, selectivity can be further enhanced *(13)*.

The features of this operation (selectivity, scalability) are attractive, but this process has seen limited scale-up to date. The micelles tend to grow to multishell, "onion-skin" structures with poor access to the core. Consistent large-scale formation of the micellular structure is balked by the difficult mass transport control required.

7. Process Development

Process purification of the polypeptides proceeds as a series of discrete unit operations. Although protein processing in the food industry is often continuous, in the production of high-value therapeutics, the process is most often batch-oriented.

Purification techniques developed at the bench scale may not be suitable for process. Electrophoresis is routinely used to isolate proteins at the milligram scale, but has yet to prove practical for kilogram production quantities, despite development of various large-scale electrophoretic equipment. One of the first steps in pro-

cess design is the development of a bench purification protocol satisfying the following requirements:

- All steps in the protocol will scale-up to process design capacity;
- All steps will meet environmental and safety requirements, and are consistent with the manufacturing environment. For pharmaceutical and diagnostic processes, selected steps will be consistent with good manufacturing practice (GMP) requirements of the regulatory agencies;
- The protocol, when scaled up, will meet your criteria of acceptable yield and product purity profile; and
- The process will meet benchmarks of operating and capital costs.

7.1. Cost Estimation: Individual Process Steps

Each unit operation (centrifugation, chromatography, extraction, and so on) has particular methods for estimating the physical parameters of scale-up and the associated operating costs from laboratory experiments. Many of these processes provide a direct linear scale-up. Where this is not the case, empirical factors based on past experience must be used. The process engineering literature generally can provide good guidelines for well-established unit operations. Section 4. of this chapter described the technique used in determining scaled-up performance of a chromatographic separation.

Table 2 provides an example of how one can estimate process performance and costs, based on laboratory-scale chromatographic data. Although the example is simply illustrative, it could be typical of a reverse-phase purification of a high-value peptide. The scale-up here is a linear scale-up of column volumes, with the same equivalent mass loadings.

In the cost calculation, one must know (or assume) the monetary value of the crude feed material and the value of the refined product. The product loss in the process step factors into the crude cost and finished product value. In the early stages of process development these costs may not be known, but it should be possible to use reasonable estimates.

In this example, the large-scale column will accept an injection of 71 g of crude material and yield 43 g of purified product (71* 65% *93%). Labor costs assume one operator at a burdened

Table 2
Scale-Up of a Chromatography Purification Process
Performance and Economics Prediction

Variable	Laboratory Preparative Column	Process System
Experimental		
Column diameter, cm	0.46	20.00
Column height, cm	20.00	50.00
Flow rate, cc/min	3.32	15,708.00
Separation time, min	20.00	20.00
Regeneration time, min	5.00	5.00
Crude load, g	0.015	70.89
Crude purity, %	65%	65%
Recovery, %	93%	93%
Recovery, g/cycle	0.00907	42.85
Column life, cycles	100	100
Cost factors		
Production cycles/yr		4000
Solvent cost, separation	$15.00	$1.00
Solvent cost, regeneration	$60.00	$6.50
Labor rate, $/h		$50.00
Packing cost $/cc		$1.25
Equipment cost		$570,000.00
Value crude, $/g		$15.00
Value purified product, $/g		$50.00
Predicted process economics		
Value product/cycle		$2,143.00
Costs		
Labor	$120.00	
Solvents	$149.00	
Equipment (Amortized 5 yr)	$29.00	
Value of crude feed	$1,063.00	
Cost/cycle		$1,557.00
Process contribution/cycle		$586.00
Process annual contribution		$2,342,246.00

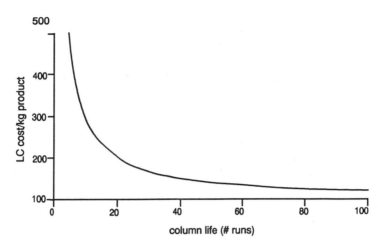

Fig. 9. Sensitivity analysis: process cost as a function of column life.

labor rate of $50/h. For this example, the capital equipment was simply amortized by straight-line depreciation over a 5-yr life. The annual contribution calculation provides the information required for process payback.

Note that the above model is amenable to sensitivity analysis. Figure 9 plots the column cost and the total process cost as a function of the number of cycles obtained on the column. This provides a good perspective on the number of cycles that one must obtain for reasonable process economics.

When product recovery losses contribute to a large portion of total costs (in the example, the recovery loss would amount to ca. $75), it may prove worthwhile to rechromatograph some of the reject cuts. In this example, solvent costs are a large component of the direct operating cost. Chromatography uses large volumes of solvents, and solvent reuse operations may prove worthwhile, particularly when disposal of spent materials proves a problem.

This type of analysis of data obtained at bench scale provides insights early in the product development stage and may suggest alternate purification strategies. The ability to simulate process economics with small quantities of materials is very advantageous.

Table 3
Exponential R Factors for Some Types and Sizes
of Bioprocess and Related Equipment

| | Size range | | |
Equipment	Minimum	Maximum	Exponent
Autoclaves	6500 in³	32,000 in³	0.37
Centrifuges, disk	4000 l/h	60,000 l/h	0.72
Centrifuges, tubular			0.68
Columns, anion-exchange	5 ft³	200 ft³	0.94
Columns, cation-exchange	5 ft³	200 ft³	0.73
Freeze-driers	15 l/cycle	450 l/cycle	0.41
Fermentors	200 l	900 l	0.18
	900 l	14,000 l	0.34
	14,000 l	200,000 l	0.70

7.2. Capital Equipment Costs

A process will be scaled up by several orders of magnitude from the bench-level development, and it is necessary to be able to anticipate installed process equipment costs. In the absence of actual vendor quotes for specific equipment, an approach to this is the use of exponential scaling factors (14), as follows:

$$(\text{cost}_2 \ / \ \text{cost}_1) = [(\text{size}_2 \ / \ \text{size}_1)]^R \qquad (4)$$

The original work in this area used a constant R factor of 0.6 for all types of equipment, but R factor data are available for a number of types of process equipment. Below, Table 3 illustrates R values for several equipment types (15).

R values for an equipment type are determined by obtaining prices or cost estimates for several different operating capacities, and doing a least-squares regression analysis of plotted data. It helps to obtain cost estimates from several vendors. Most bioprocess equipment have R factors <1. This means that, if capacity is doubled, the price will be less than double.

This is a simple technique for obtaining rough estimates of process equipment costs at an early stage. Care should be used in extrapolating past the range of available R factor data. Peripheral equipment (tanks, pumps) and facilities costs should also be considered.

8. Overall Process Design

Protein purification is virtually always a multistep process. The various process operations selected (and their order of implementation) are dependent on the physical/chemical characteristics of the product molecule and on the composition of the source. Listed below are some general guidelines for effective process design.

- Choose operations based on different physical separation parameters.
- Does the operation scale? What are scale-up economics?
- Choose processes that exploit the greatest differences in physical/chemical properties of product and impurities.
- Minimize the number of steps.
- Take the biggest biomass elimination step first.
- Do the most costly step last.

1. Choose operations based on different physical separation parameters. Multiple precipitations and resolubilizations are the basis of the Cohn blood fractionation process. The technique was developed at a time when there was not a range of protein separation operations available. Given the choice, one will generally find that higher purification factors can be achieved by a combination of isolation operations that are based on differing principles of separation. The combination of separations based on, say, ionic character and on hydrophobic character will most often prove more effective than trying to affect isolation by multiple ion exchanges or by a very high-resolution hydrophobic chromatography. Two-dimensional electrophoresis, although not a process operation, provides a graphic example (Chapter 9, Fig. 4) of this principle.

2. Does the operation scale? What are scale-up economics? A number of techniques used at the bench level prove impractical at production levels. Ultracentrifugation or electrophoresis

are examples of methods that do not scale up readily. An affinity purification that is perfectly satisfactory at bench scale may prove impractical if large quantities of the affinity ligand are not available or if the ligand cost is too great. Carefully examine each operation in a bench-scale protocol, and determine that these operations will scale-up. If not, substitute different operations, and prove out, at bench scale, that the modified process train will produce adequate purity and recovery.

3. Choose processes that exploit the greatest differences in physical/chemical properties of product and impurities. The closer in molecular structure and energy, the more difficult is the separation of species. When isolating a product from highly similar species, closely examine the physical data on the materials. Where there is a significant difference in molecular weight of product and impurities, an ultrafiltration or size-exclusion chromatography separation probably will work well. If two species showed similar hydrophobic chromatography elution times, but markedly different isoelectric points, this would suggest that an ion-exchange separation would prove more effective. Further, it would be useful to examine titration data on the materials and select a buffer pK in the region where there is the broadest divergence of the titration curves.

Each process step will result in <100% recovery of product mass and activity. In a multistep purification train, these losses accrue geometrically. Table 4 *(16)* below characterizes a hypothetical five-step purification train. In each step, it is assumed that there is 80% recovery of the product molecule. The table assumes a monetary value for a "unit" of starting material and an operating cost associated with each step.

Note some characteristics in this table. Because of the multiplicative effect of recovery losses, it requires more than threefold quantity of the product molecule in the starting biomass to end up with one unit of final product. Even though this example assumes equal cost/unit for each step, the total costs are loaded toward the front end of the process (again, because of the larger mass of material handled). This example serves to illustrate some general principles in the design of a purification process.

Table 4
Economics of a Model Multistep Purification Train

Step	Loss factor	Units product	Units loss	Process cost/unit
	20%	3.05		$1.00
1	20%	2.44	0.61	$0.50
2	20%	1.95	0.49	$0.50
3	20%	1.56	0.39	$0.50
4	20%	1.25	0.31	$0.50
5	20%	1.00	0.25	$0.50

	Process cost	Cumulative cost	Recovery increment value/unit
	$3.05	$3.05	$1.00
1	$1.22	$4.27	$1.75
2	$0.98	$5.25	$2.69
3	$0.78	$6.03	$3.87
4	$0.63	$6.65	$5.32
5	$0.50	$7.15	$7.15

4. Design the process with a minimum number of unit operations. Most process operations will result in some loss of product mass and/or activity, and provide another possible entry
   of contaminants. These losses are multiplicative. In batch-oriented processes, each operation is a separate campaign, with
   attendant record keeping and assays. It imposes additional
   time in which biological or chemical degradation of the product can ensue. In general, it is preferable to design a simple
   process based around a few operations that yield a high purification factor.
5. Take the biggest biomass elimination step first. The desired
   product is usually a small fraction of the starting biomass. By
   selecting early process steps of low resolution, but that can
   separate most of the waste fraction, the downstream operations can operate at a smaller and more economic scale. Ion-
   exchange chromatography is an example of this. Ion-exchange

media are tolerant of a range of materials. They have a high loading capacity, but moderate resolution. When the product is harvested on an ion-exchange resin, most of the contaminating species can be separated, and the eluted product fraction comes off as a concentrated mass. A high selectivity process (with high unit cost) can then be used much more economically.

6. Do the most costly step last. This is a corollary to the above principle. With a high unit cost operation, reducing the feed mass will provide operating economics.

The above is a set of guidelines, rather than hard and fast rules. The particulars of any process situation will influence the process strategy, but it can be seen that, given the many possible choices, a rational scheme to process design can result in significant reduction of operations costs.

Bibliography

1. Naveh D. (1990) Industrial-scale downstream processing of biotechnology products. *BioPharm*. May.
2. Wimpenny J. W. T. (1967) *Process Biochemistry*. Morgan-Grampian Ltd. London.
3. Cohn E. J., Strong W. L., Hughes W. L., Mulford D. J., Ashworth J. N., Melin M., and Taylor H. L. (1946) *J. Am. Chem. Soc.* **68**, 459–475.
4. Bell D. J., Hoare M., and Dunhill P. (1983) *Adv. Biochem. Eng. Biotechnol.* **26**, 1–18.
5. Ives K. J., ed. (1978) *The Scientific Basis of Flocculation*. Sijthoff and Noordhoff, Alphen aan den Rijn, the Netherlands.
6. Dixon M. and Webb E. C. (1979) *Enzymes*. Longmans, London.
7. Optimization in Preparative Liquid Chromatography, Golsha-Shirazi, Saroddin, Guiochon, Georges, ABL, June 90.
8. Blatt W. F., David A., Michaels A. S., and Nelson L. (1970) *Membrane Science and Technology*, (J. E. Flinn, ed.) Plenum, New York.
9. Dunhill P. (1983) Trends in downstream processing of proteins and enzymes. *Process Biochemistry*, **Oct.**, 9–13.
10. Mathiasson B. and Ling T. C/L. (1988) *Extraction in Aqueous Two-Phase Systems, Separation for Biotechnology*. Ellis Horwood, Chicester, UK, pp. 281,282.
11. Hughes P. and Lowe C. R. (1988) *Enz. Microb. Technol.* **10(2)**, pp. 115–122.
12. Burgess R. (1987) *Protein Separations Using Reversed Micelles, Protein Purification: Micro to Macro*. Alan R. Liss, New York, pp. 117–130.

13. Woll J. M. et al. (1987) *Protein Separations Using Reversed Micelles, Protein Purification: Micro to Macro*, (Burgess R., ed.), Alan R. Liss, New York, pp. 117–130.
14. Williams R. (1947) *Chem. Eng.* **54 (12),** pp. 124–125.
15. Remer D. S. and Idrovo J. H. (1990) Cost-estimating factors for biopharmaceutical process equipment. *Biopparm* **October,** 36–42.
16. Dwyer J. L. (1984) Scaling up bio-product separation with high performance liquid chromatography. *Biotechnology* **Nov.**

Index

573